Student Edition

Textbook Lite | Activities | Study Guide

BIOLOGY
FOR NGSS

BIOLOGY
FOR NGSS

Meet the Writing Team

Tracey
Senior Author

Tracey Greenwood
I have been writing resources for students since 1993. I have a Ph.D in biology, specialising in lake ecology and I have taught both graduate and undergraduate biology.

Lissa
Author

Lissa Bainbridge-Smith
I worked in industry in a research and development capacity for 8 years before joining BIOZONE in 2006. I have an M.Sc from Waikato University.

Kent
Author

Kent Pryor
I have a BSc from Massey University majoring in zoology and ecology and taught secondary school biology and chemistry for 9 years before joining BIOZONE as an author in 2009.

Richard
Founder & CEO

Richard Allan
I have had 11 years experience teaching senior secondary school biology. I have a Masters degree in biology and founded BIOZONE in the 1980s after developing resources for my own students.

Second edition 2016
Fourth printing

ISBN 978-1-927309-46-9

Copyright © 2016 Richard Allan
Published by BIOZONE International Ltd

Printed by REPLIKA PRESS PVT LTD using paper produced from renewable and waste materials

Next Generation Science Standards (NGSS) is a registered trademark of Achieve. Neither Achieve nor the lead states and partners that developed the Next Generation Science Standards were involved in the production of this product and do not endorse it.

Purchases of this workbook may be made direct from the publisher:

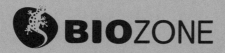

BIOZONE Corporation
USA and Canada

FREE phone: 1-855-246-4555
FREE fax: 1-855-935-3555
Email: sales@thebiozone.com
Web: www.thebiozone.com

Cover photograph

The ring-tailed lemur (*Lemur catta*) is one of the most recognized primates with a distinctive black and white ringed tail. It belongs to Lemuridae, one of five lemur families, and is the only member of the *Lemur* genus. Like all lemurs it is endemic to the island of Madagascar. It is a highly social species, living in groups of up to 30, with a female dominance hierarchy. Although it is listed as near threatened, it breeds readily in captivity and there are more than 2000 in captive breeding programs in zoos worldwide.

PHOTO: © irakite/www.istockphoto.com

Thanks to:

The staff at BIOZONE, including Holly Coon and Mike Campbell for design and graphics support, Paolo Curray and Malaki Toleafoa for IT support, Debbie Antoniadis and Arahi Hippolite for office handling and logistics, and the BIOZONE sales team.

Contents

NGSS: A Map of Core Ideas (DCIs) vi
Using the Student Edition viii
What are Crosscutting Concepts? x
Addressing Science and Engineering Practices xii
Nature of Science and Engineering Design xiv
Using the Tab System xv
Using BIOZONE'S Weblinks and Biolinks xvi

Science Practices

DCIs, CCCs, SEPs 1
1   How Do We Do Science? 2
2   Systems and Systems Models 4
3   Observations, Hypotheses, and Assumptions .. 3
4   Accuracy and Precision 6
5   Working with Numbers 7
6   Tallies, Percentages, and Rates 8
7   Fractions and Ratios 9
8   Dealing with Large Numbers........................... 10
9   Apparatus and Measurement 11
10 Practicing Data Transformation 15
11 A Case Study: Catalase Activity 14
12 Drawing Line Graphs 17
13 Recording Results 15
14 Practicing Data Manipulations........................ 16
15 Constructing Tables.. 17
16 Which Graph to Use?...................................... 18
17 Drawing Line Graphs....................................... 19
18 Interpreting Line Graphs 20
19 Drawing Scatter Graphs.................................. 21
20 Correlation or Causation? 22
21 Drawing Bar Graphs.. 23
22 Drawing Histograms.. 24
23 Mean, Median, and Mode 25
24 What is Standard Deviation? 27
25 Detecting Bias in Samples 28
26 Biological Drawings.. 29
27 Practicing Biological Drawings 31
28 Test Your Understanding 32

Concept Map: From Molecules to Organisms: Structures and Processes........................... 35

LS1.A
Cell Specialization and Organization

DCIs, CCCs, SEPs 36
29 The Hierarchy of Life 37
30 Introduction to Cells 38
31 Studying Cells 39
32 Plant Cells 41
33 Animal Cells 43
34 Identifying Organelles 45
35 The Structure of Membranes 46
36 Diffusion in Cells 47
37 Osmosis in Cells 48
38 Diffusion and Cell Size 49
39 Calculating Diffusion Rates 50
40 Factors Affecting Membrane Permeability........ 51

41 Active Transport 52
42 What is an Ion Pump? 53
43 Specialization in Plant Cells 54
44 Specialization in Animal Cells 55
45 What is DNA? 56
46 Nucleotides 57
47 DNA and RNA 58
48 Modeling the Structure of DNA 59
49 Genes Code for Proteins 63
50 Cracking the Genetic Code 64
51 Amino Acids Make Up Proteins....................... 65
52 Proteins Have Many Roles in Cells 66
53 Reactions in Cells 68
54 Enzymes Catalyze Reactions In Cells 69
55 Enzymes Have Optimal Conditions to Work 70
56 Investigating Catalase Activity 71
57 Organ Systems Work Together 73
58 Circulation and Gas Exchange Interactions 74
59 Circulation and Digestive Interactions 76
60 Plant Organ Systems 78
61 Interacting Systems in Plants 79
62 Chapter Review 80
63 KEY TERMS AND IDEAS: Did You Get It?....... 81
64 Summative Assessment................................... 82

LS1.A
Feedback Mechanisms

DCIs, CCCs, SEPs 84
65 Homeostasis 85
66 Keeping in Balance 86
67 Negative Feedback Mechanisms 88
68 Positive Feedback Mechanisms 89
69 Sources of Body Heat 90
70 Thermoregulation... 91
71 Thermoregulation in Humans.......................... 93
72 Body Shape and Heat Loss 95
73 Controlling Blood Glucose 96
74 Type 2 Diabetes 98
75 Homeostasis During Exercise 99
76 Effect of Exercise on Breathing and Heart Rate 100
77 Is the Effect of Exercise on Heart Rate Significant?... 101
78 Homeostasis in Plants 102
79 Measuring Transpiration in Plants 103
80 Chapter Review.. 106
81 KEY TERMS AND IDEAS: Did You Get It?..... 107
82 Summative Assessment................................... 108

LS1.B
Growth and Development

DCIs, CCCs, SEPs 110
83 Growth and Development of Organisms 111
84 DNA Replication 112
85 Details of DNA Replication 113
86 Modeling DNA Replication 114

CODES: **Activity** is marked: • to be done   ✓ when completed

Contents

87 The Functions of Mitosis 117
88 The Cell Cycle 118
89 Mitosis 119
90 Mitosis and Cytokinesis 120
91 Modeling Mitosis 122
92 Differentiation of Cells 123
93 Stem Cells Give Rise to Other Cells 124
94 Tissues Work Together 125
95 Chapter Review 127
96 KEY TERMS AND IDEAS: Did You Get It? 128
97 Summative Assessment 129

LS1.C & PS3.D
Energy in Living Systems

DCIs, CCCs, SEPs 130
98 Energy in Cells 131
99 ATP 132
100 Introduction to Photosynthesis 133
101 Investigating Photosynthetic Rate 135
102 Chloroplasts 136
103 Stages in Photosynthesis 137
104 The Fate of Glucose 138
105 Energy Transfer Between Systems 140
106 Energy From Glucose 141
107 Aerobic Cellular Respiration 143
108 Measuring Respiration 145
109 Modeling Photosynthesis
    and Cell Respiration 146
110 Chapter Review 149
111 KEY TERMS AND IDEAS: Did You Get It? 150
112 Summative Assessment 151

Concept Map: Ecosystems:
Interaction, Energy, and Dynamics 154

LS2.A
Interdependence in Ecosystems

DCIs, CCCs, SEPs 155
113 What is an Ecosystem? 156
114 Habitat and Tolerance Range 157
115 The Ecological Niche 158
116 Dingo Habitats 159
117 Population Density and Distribution 160
118 Species Interactions 162
119 Competition for Resources 164
120 Intraspecific Competition 165
121 Interspecific Competition 167
122 Reducing Competition Between Species 169
123 Predator-Prey Relationships 171
124 The Carrying Capacity of an Ecosystem 173
125 A Case Study in Carrying Capacity 174
126 Home Range Size in Dingoes 175
127 Resources and Distribution 176
128 Population Growth 177
129 Plotting Bacterial Growth 179
130 Investigating Bacterial Growth 180

131 A Case Study in Population Growth 181
132 Chapter Review 182
133 KEY TERMS AND IDEAS: Did You Get It? 183
134 Summative Assessment 184

LS2.B
Energy Flow and Nutrient Cycles

DCIs, CCCs, SEPs 185
135 Energy in Ecosystems 186
136 Comparing Aerobic and Anaerobic Systems 187
137 Producers 189
138 Consumers 190
139 Food Chains 191
140 Food Webs 192
141 Constructing Food Webs 193
142 Energy Inputs and Outputs 195
143 Energy Flow in an Ecosystem 196
144 Ecological Pyramids 198
145 Cycles of Matter 200
146 The Hydrologic Cycle 201
147 The Carbon Cycle 202
148 Modeling the Carbon Cycle 203
149 The Oxygen Cycle 204
150 Role of Photosynthesis in Carbon Cycling 205
151 The Nitrogen Cycle 207
152 Chapter Review 208
153 KEY TERMS AND IDEAS: Did You Get It? 209
154 Summative Assessment 210

LS2.C & ETS1.B
The Dynamic Ecosystem

DCIs, CCCs, SEPs 212
155 Ecosystem Dynamics 213
156 The Resilient Ecosystem 215
157 A Case Study in Ecosystem Resilience 216
158 Keystone Species 217
159 Ecosystem Changes 219
160 Global Warming and Ecosystem Change 221
161 Human Impact on Ecosystems 223
162 The Effects of Damming 224
163 The Impact of Alien Species 226
164 Human Impact on Fish Stocks 227
165 Evaluating a Solution to Overfishing 229
166 Deforestation and Species Survival 230
167 Modeling a Solution 232
168 Chapter Review 233
169 Summative Assessment 234

LS2.D
Social Behavior

DCIs, CCCs, SEPs 235
170 Social Groupings 236
171 Swarming, Flocking, and Herding 237
172 Migration 239
173 Social Organization 240

CODES: **Activity** is marked: ● to be done   ✓ when completed

Contents

174 How Social Behavior Improves Survival 241
175 Cooperative Behaviors 242
176 Cooperative Defense 243
177 Cooperative Attack 244
178 Cooperative Food Gathering 245
179 Chapter Review 246
180 KEY TERMS AND IDEAS: Did You Get It? 247
181 Summative Assessment 248

Concept Map: Heredity:
Inheritance and Variation of Traits 250

LS3.A
Inheritance of Traits

DCIs, CCCs, SEPs 251
182 Chromosomes 252
183 DNA Carries the Code 253
184 Not All DNA Codes for Protein 254
185 The Outcomes of Differing Gene Expression 255
186 DNA Packaging and Control of Transcription 256
187 Changes after Transcription and Translation . 257
188 Chapter Review 258
189 KEY TERMS AND IDEAS: Did You Get It? 259
190 Summative Assessment 260

LS3.B
Variation of Traits

DCIs, CCCs, SEPs 261
191 What is a Trait? 262
192 Different Alleles for Different Traits 263
193 Why is Variation Important? 264
194 Sources of Variation 265
195 Examples of Genetic Variation 266
196 Meiosis 267
197 Meiosis and Variation 268
198 Mutations 270
199 The Effects of Mutations 271
200 The Evolution of Antibiotic Resistance 272
201 Beneficial Mutations in Humans 273
202 Harmful Effects of Mutations in Humans 274
203 Influences on Phenotype 275
204 Environment and Variation 276
205 Genes and Environment Interact 278
206 Predicting Traits: The Monohybrid Cross 280
207 Predicting Traits: The Test Cross 281
208 Practicing Monohybrid Crosses 282
209 Predicting Traits: The Dihybrid Cross 283
210 Practicing Dihybrid Crosses 284
211 Testing the Outcome of Genetic Crosses 285
212 Pedigree Analysis 287
213 Chapter Review 289
214 KEY TERMS AND IDEAS: Did You Get It? 291
215 Summative Assessment 292

Concept Map: Biological Evolution:
Unity and Diversity 293

LS4.A
Evidence for Evolution

DCIs, CCCs, SEPs 294
216 Evidence For Evolution 295
217 The Common Ancestry of Life 296
218 The Fossil Record 298
219 Interpreting the Fossil Record 299
220 Transitional Fossils 300
221 Case Study: Whale Evolution 301
222 Anatomical Evidence for Evolution 302
223 DNA Evidence for Evolution 303
224 Protein Evidence for Evolution 304
225 Developmental Evidence for Evolution 305
226 Chapter Review 306
227 KEY TERMS AND IDEAS: Did You Get It? 307
228 Summative Assessment 308

LS4.B & LS4.C
Natural Selection and Adaptation

DCIs, CCCs, SEPs 309
229 How Evolution Occurs 310
230 Adaptation 312
231 Similar Environments, Similar Adaptations ... 314
232 Natural Selection in Finches 315
233 Natural Selection in Pocket Mice 316
234 Insecticide Resistance 318
235 Gene Pool Exercise 319
236 Modeling Natural Selection 321
237 What is a Species? 322
238 How Species Form 323
239 Patterns of Evolution 325
240 Evolution and Biodiversity 326
241 Extinction is a Natural Process 328
242 Humans and Extinction 329
243 Chapter Review 331
244 KEY TERMS AND IDEAS: Did You Get It? 332
245 Summative Assessment 333

Biodiversity LS4.D & ETS1.B

DCIs, CCCs, SEPs 336
246 Biodiversity 337
247 Humans Depend on Biodiversity 338
248 Biodiversity Hotspots 340
249 How Humans Affect Biodiversity 341
250 Ex-Situ Conservation 342
251 In-Situ Conservation 344
252 Conservation and Genetic Diversity 345
253 Maasai-Mara Case Study 346
254 Chapter Review 347
255 KEY TERMS AND IDEAS: Did You Get It? 348
256 Summative Assessment 349

Questioning Terms and Photo Credits 350
Index 351

CODES: **Activity** is marked: ● to be done   ✓ when completed

NGSS:
A Map of Core Ideas

This map shows the structure of the NGSS Life Science program as represented in this book. The dark blue boxes indicate the book sections, each of which has its own concept map. The blue ovals are the chapters in each section. We have placed some major connections between topics. You can make more of your own.

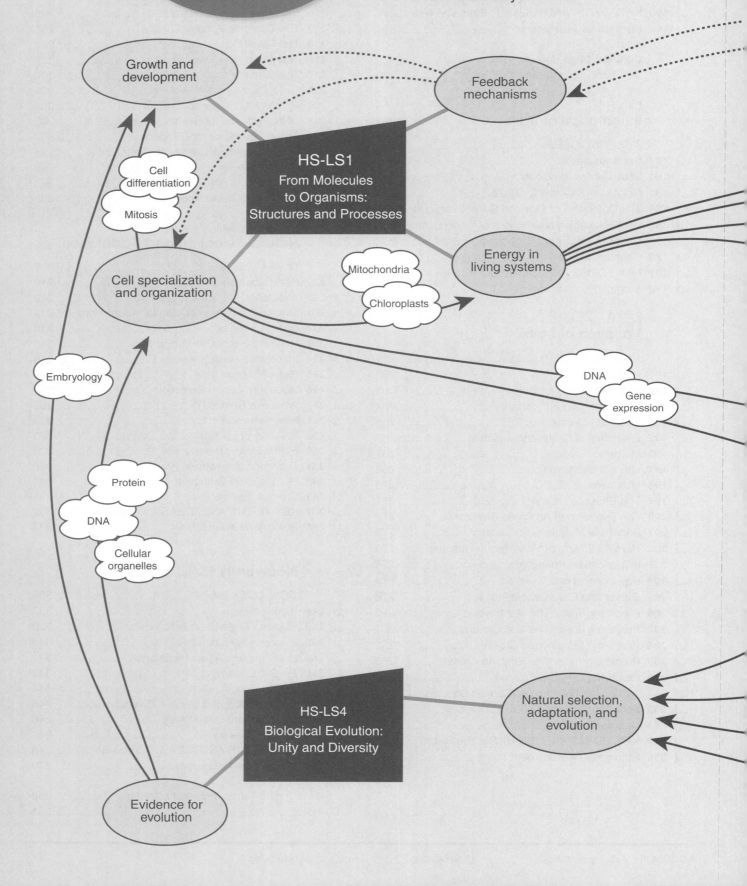

Growth and development

Feedback mechanisms

Cell differentiation

Mitosis

HS-LS1
From Molecules to Organisms: Structures and Processes

Mitochondria

Chloroplasts

Energy in living systems

Cell specialization and organization

Embryology

DNA

Gene expression

Protein

DNA

Cellular organelles

HS-LS4
Biological Evolution: Unity and Diversity

Natural selection, adaptation, and evolution

Evidence for evolution

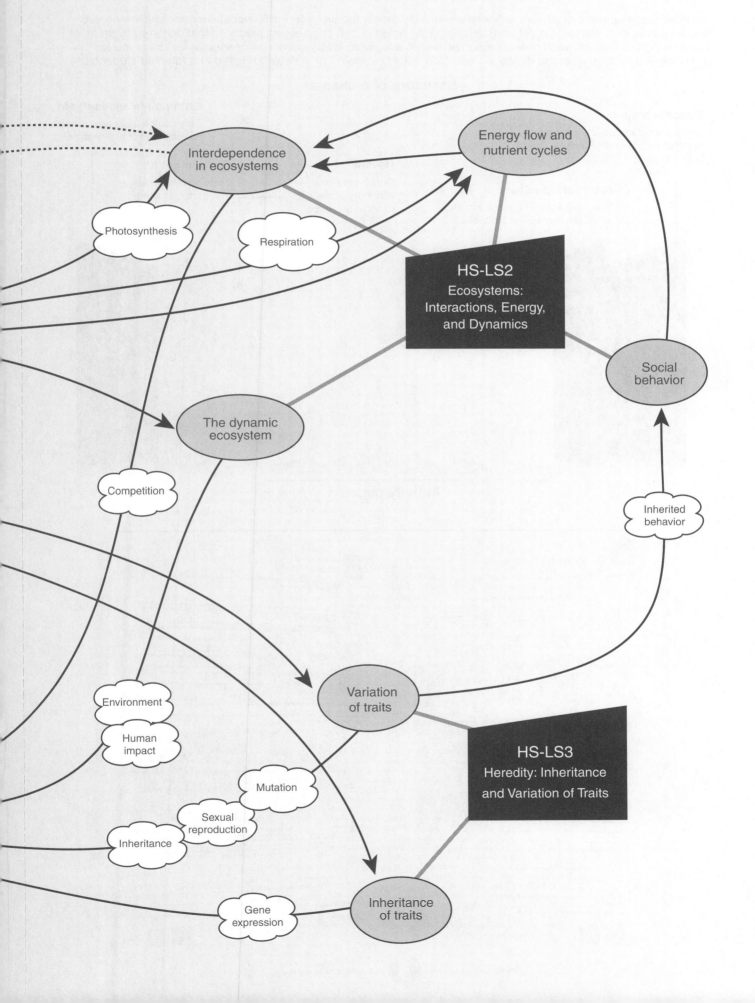

Interdependence in ecosystems

Energy flow and nutrient cycles

Photosynthesis

Respiration

HS-LS2
Ecosystems:
Interactions, Energy,
and Dynamics

Social behavior

The dynamic ecosystem

Competition

Inherited behavior

Environment

Human impact

Variation of traits

HS-LS3
Heredity: Inheritance
and Variation of Traits

Mutation

Sexual reproduction

Inheritance

Gene expression

Inheritance of traits

Using The Student Edition

Activities make up most of this book. These are usually presented as short instructional sequences allowing you to build a deeper understanding of core ideas and the science and engineering practices that accompany them as you progress through each chapter. Each activity is accompanied by questions or specific tasks for you to complete. A dark blue question denotes a question that is extension or suitable for gifted and talented students.

Structure of a chapter

Concept map
Use the word map of the content to make your own connections between parts of the course.

Chapter introduction
Identifies the activities relating to the DCIs, CCCs, and SEPs described.

Review
Create your own summary of material to help you revise.

Did you get it?
Tests your knowledge and understanding.

Summative assessment
This can be used as a formal assessment of the chapter.

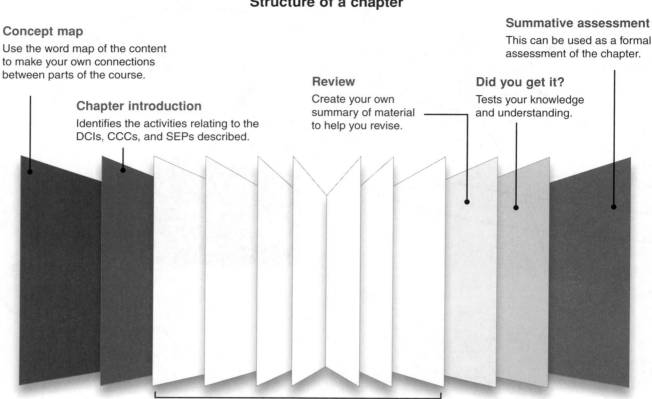

Activity pages

This identifies the major Disciplinary Core Idea to which this chapter applies.

The list of key terms can be used to create a glossary for revision. It will help you to use the appropriate terms when answering questions.

Mark the check boxes to indicate the outcomes you should complete. Check them off when you have finished.

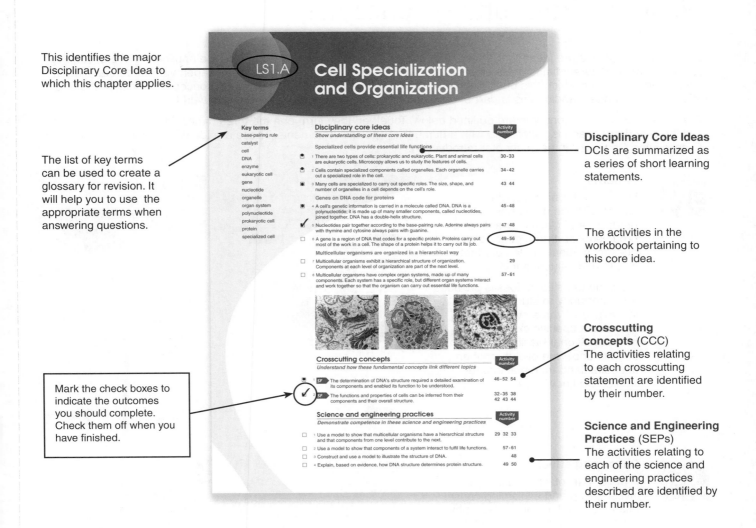

LS1.A

Cell Specialization and Organization

Key terms
- base-pairing rule
- catalyst
- cell
- DNA
- enzyme
- eukaryotic cell
- gene
- nucleotide
- organelle
- organ system
- polynucleotide
- prokaryotic cell
- protein
- specialized cell

Disciplinary core ideas
Show understanding of these core ideas

	Activity number

Specialized cells provide essential life functions

1. There are two types of cells: prokaryotic and eukaryotic. Plant and animal cells are eukaryotic cells. Microscopy allows us to study the features of cells. — 30-33

2. Cells contain specialized components called organelles. Each organelle carries out a specialized role in the cell. — 34-42

3. Many cells are specialized to carry out specific roles. The size, shape, and number of organelles in a cell depends on the cell's role. — 43 44

Genes on DNA code for proteins

4. A cell's genetic information is carried in a molecule called DNA. DNA is a polynucleotide; it is made up of many smaller components, called nucleotides, joined together. DNA has a double-helix structure. — 45-48

5. Nucleotides pair together according to the base-pairing rule. Adenine always pairs with thymine and cytosine always pairs with guanine. — 47 48

6. A gene is a region of DNA that codes for a specific protein. Proteins carry out most of the work in a cell. The shape of a protein helps it to carry out its job. — 49-56

Multicellular organisms are organized in a hierarchical way

7. Multicellular organisms exhibit a hierarchical structure of organization. Components at each level of organization are part of the next level. — 29

8. Multicellular organisms have complex organ systems, made up of many components. Each system has a specific role, but different organ systems interact and work together so that the organism can carry out essential life functions. — 57-61

Crosscutting concepts
Understand how these fundamental concepts link different topics

	Activity number

- **SF** The determination of DNA's structure required a detailed examination of its components and enabled its function to be understood. — 46-52 54

- **SF** The functions and properties of cells can be inferred from their components and their overall structure. — 32-35 38 42 43 44

Science and engineering practices
Demonstrate competence in these science and engineering practices

	Activity number

1. Use a model to show that multicellular organisms have a hierarchical structure and that components from one level contribute to the next. — 29 32 33

2. Use a model to show that components of a system interact to fulfil life functions. — 57-61

3. Construct and use a model to illustrate the structure of DNA. — 48

4. Explain, based on evidence, how DNA structure determines protein structure. — 49 50

Disciplinary Core Ideas
DCIs are summarized as a series of short learning statements.

The activities in the workbook pertaining to this core idea.

Crosscutting concepts (CCC)
The activities relating to each crosscutting statement are identified by their number.

Science and Engineering Practices (SEPs)
The activities relating to each of the science and engineering practices described are identified by their number.

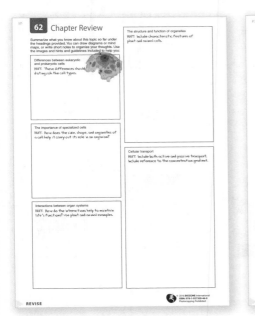

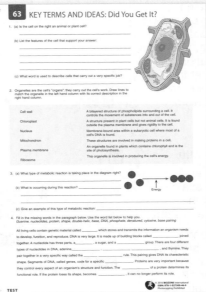

Chapter Review
Use the chapter reviews to summarize what you have learned. Use your notes to consolidate your understanding revise for tests and exams.

Key Terms and Ideas
Test your vocabulary and understanding of basic principles. These can be used as a self test or a class activity.

Summative Assessment
These activities allow you to demonstrate your understanding of a topic by combining knowledge and principles from the three dimensions. They are designed to be used as formal assessment tasks.

What Are Crosscutting Concepts?

Crosscutting concepts are ideas that are common to all fields of science. Recognizing the application of crosscutting concepts across different fields of life sciences, as well as other sciences, will help to deepen your understanding of the core ideas around which this book is structured. You will begin to see that common concepts link different areas of science and understanding this is part of developing a scientifically sound view of the world.

The seven crosscutting concepts are outlined below. Together, they form one the three dimensions of the Framework and the Standards. Each is associated with a number of specific statements. These are identified as they apply under 'Crosscutting Concepts' in each chapter introduction.

Patterns

We see patterns everywhere in science. These guide how we organize and classify events and organisms and prompt us to ask questions about the factors that create and influence them.

Studying transitional fossils provides an opportunity to study patterns in evolution. Whale evolution provides a specific example of how transitional fossils have been used to show the evolution of an ancestral land mammal into the whales we know today.

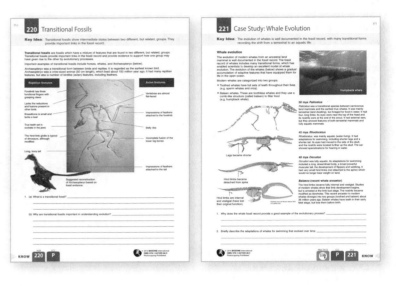

Cause and effect

A major part of science is investigating and explaining causal relationships. The mechanisms by which they occur can be tested in one context and used to explain and predict events in new contexts.

It is important to understand the relationship between variables. Does one variable cause an effect in the other? This relationship is explored in many areas including social behavior, ecosystem change, and natural selection and adaptation. Insecticide resistance looks at the cause of insecticide resistance and its beneficial effects to the organism.

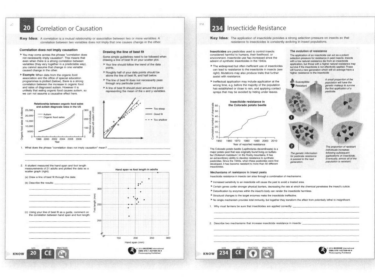

Scale, proportion, and quantity

Different things are relevant at different scales. Changes in scale, proportion, or quantity affect the structure or performance of a system.

Effects of scale, proportion and quantity are evident in many areas of biology, including factors controlling population size. The number of individuals relative to the resources available (e.g. food or space) limits population numbers.

You will notice that the last question in this activity is colored blue, denoting it is suitable as extension for gifted and talented students.

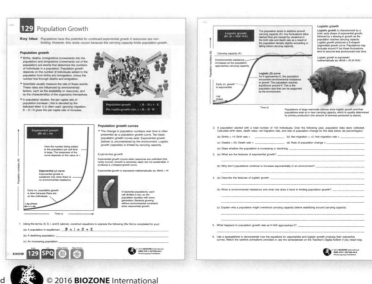

Stability and change

Science often deals with constructing explanations of how things change or how they remain stable.

Stability and change can be studied at several levels. An organism must respond to its environment to stay within normal operating limits required to carry out life's essential processes. Stability is also important at an ecosystem level. If an ecosystem can not resist or respond to environmental change, it may become unstable and vulnerable to degradation or lose its particular characteristics.

Energy and matter

Energy flows and matter cycles. Tracking these fluxes helps us understand how systems function.

At the cellular or organism level, the transformations of energy and matter can be examined in terms of their role in life's essential processes (e.g. generating ATP or glucose). These transformations can also be studied at an ecosystem level. We can investigate the transfer of energy and matter through the trophic levels of an ecosystem, including the sources of energy and the efficiency of transfers.

Systems and system models

Making a model of a system (e.g. physical, mathematical) provides a way to understand and test ideas.

Models can be used at different levels (molecular, cellular, organism, ecosystem) to understand how a system works. Specific models can be developed to understand the nature of these interactions at different scales.

Structure and function

The structure of an object or living thing determines many of its properties and functions.

Biology provides many opportunities to study how the structure of an object influences its function. The nucleotide sequence in DNA determines what type of protein is made. How the protein folds together to form its unique functional shape is determined by its amino acid sequence.

Addressing Science and Engineering Practices

Science and Engineering Practices for NGSS are supported throughout the workbook, beginning with an introductory chapter covering basic computational, analytical, and design skills, to the completion of activities focusing on the development of specific skills within the framework of the DCIs. The learning outcomes associated with the science and engineering practices for each chapter are identified in each chapter introduction, together with the activities that relate to those outcomes. These provide you with a concrete way in which to identify and gain competence in specific practices. We have supported some of the practices by providing an example for you to work through. This provides you with the necessary background to design and complete your own 'hands-on investigation.

PRACTICES

Asking questions (for science) and defining problems (for engineering)

Asking scientific questions about observations or content in texts helps to define problems and draw valid conclusions.

PRACTICES

Developing and using models

Models take on many forms. They can be used to represent a system or a part of a system. Using models can help to visualize a structure, process, or design and understand how it works. Models can also be used to improve a design.

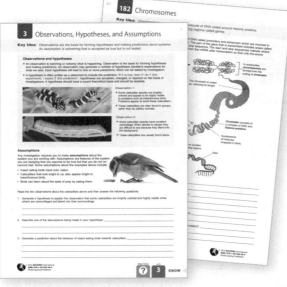

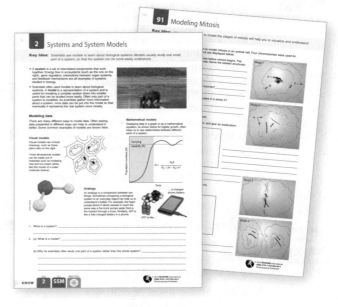

PRACTICES

Planning and carrying out investigations

Planning and carrying out investigations is an important part of independent research. Investigations allow ideas and models to be tested and refined.

PRACTICES

Analyzing and interpreting data

Once data is collected it must be analyzed to reveal any patterns or relationships. Tables and graphs are just two of the many ways to display and analyze data for trends.

PRACTICES

Using mathematics and computational thinking

Mathematics is a tool for understanding scientific data. Converting or transforming data helps to see relationships more easily while statistical analysis can help determine the significance of the results.

PRACTICES

Constructing explanations (for science) and designing solutions (for engineering)

Constructing explanations for observations and phenomena is an important part of science. It is a dynamic process and may involve drawing on existing knowledge as well as generating new ideas before an observation is explained or a problem is solved.

PRACTICES

Engaging in argument from evidence

Scientific argument based on evidence is an important part of gaining acceptance of new ideas in science. Logical reasoning based on empirical evidence is required when considering the merit of new claims or explanations of phenomena.

PRACTICES

Obtaining, evaluating, and communicating information

Evaluating information for scientific accuracy or bias is important in determining its validity and reliability. Communicating information in an effective way includes reports, graphics, oral presentation, and models. Visual models are a useful way of communicating complex scientific ideas.

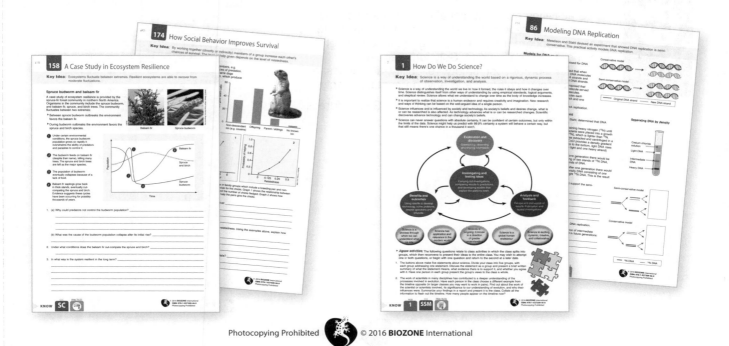

© 2016 **BIOZONE** International

Nature of Science and Engineering Design

Nature of Science in NGSS

Science provides a way of explaining the natural world. Scientific knowledge has accumulated over time and has involved many contributors. The nature of science combines established information with new knowledge to constantly refine what we know about the natural world. Eight nature of science understandings (below) are presented in the NGSS document. These understandings have been incorporated into most activities in your NGSS Biology Student Edition.

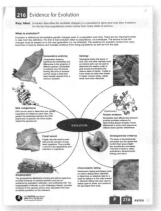

Scientific investigations use a variety of methods.

Scientific knowledge is based on empirical evidence.

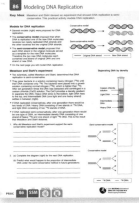

Scientific knowledge is open to revision in light of new evidence.

Science models, laws, mechanisms, and theories explain natural phenomena.

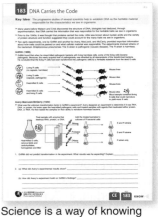

Science is a way of knowing

Scientific knowledge assumes an order and consistency in natural systems.

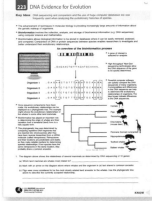

Science is a human endeavor.

Science addresses questions about the natural and material world.

ETS

Engineering Design (ETS-1)

ETS activities examine global challenges affecting society. They require you to design and/or evaluate solutions incorporating knowledge gained through science and engineering. Solutions require consideration of cost, safety, reliability, aesthetics, as well as social, cultural, and environmental impacts.

Several of the activities in Biology for NGSS address aspects of Engineering Design.

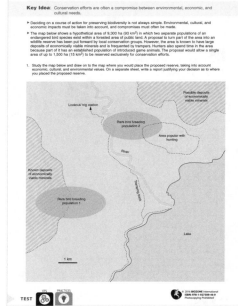

Photocopying Prohibited   © 2016 BIOZONE International

Using the Tab System

The tab system is a useful way to quickly identify crosscutting concepts and science and engineering practices pertaining to the disciplinary core idea of the activity. They also indicate whether or not the activity is supported online.

▶ The CCC tabs indicate activities (and core ideas) that share the same crosscutting concepts. Not all activities have a crosscutting code and some incorporate more than one. The PRACTICES picture codes identify which science and engineering practices (SEPs) are relevant to the activity (and core ideas). There may be more than one or none.

▶ The weblinks code is always the same as the activity number on which it is cited. On visiting the weblink page (below), find the number and it will correspond to one or more external websites providing a video or animation of some aspect of the activity's content. Occasionally, the weblink may provide a bank of photographs where images are provided in color, e.g. for plant and animal histology.

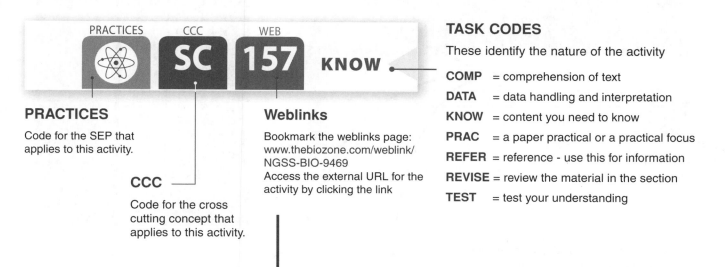

TASK CODES

These identify the nature of the activity

COMP = comprehension of text
DATA = data handling and interpretation
KNOW = content you need to know
PRAC = a paper practical or a practical focus
REFER = reference - use this for information
REVISE = review the material in the section
TEST = test your understanding

PRACTICES
Code for the SEP that applies to this activity.

CCC
Code for the cross cutting concept that applies to this activity.

Weblinks
Bookmark the weblinks page:
www.thebiozone.com/weblink/NGSS-BIO-9469
Access the external URL for the activity by clicking the link

www.thebiozone.com/weblink/NGSS-BIO-9469

This WEBLINKS page provides links to **external websites** with supporting information for the activities. These sites are distinct from those provided in the BIOLINKS area of BIOZONE's web site. For the most part, they are narrowly focussed animations and video clips directly relevant to some aspect of the activity on which they are cited. They provide great support to help your understanding of basic concepts.

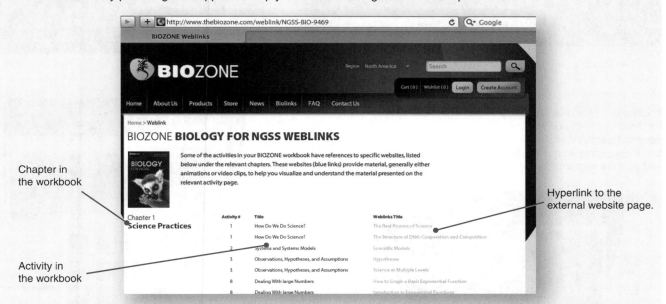

Chapter in the workbook

Activity in the workbook

Hyperlink to the external website page.

Bookmark weblinks by typing in the address: it is not accessible directly from BIOZONE's website
Corrections and clarifications to current editions are always posted on the weblinks page

Using BIOZONE's Weblinks and Biolinks

Weblinks

Weblinks is an online resource to support learning of specific core ideas and science and engineering practices, largely though explanatory animations and short videos. Weblinks also provide information relevant to research projects or as a starting point for group work. Weblinks is compilation of external URLS. The sites have been selected by BIOZONE for their suitability to the target audience and are regularly checked. Any errata for the student book, teacher's edition, or model answers are also posted on the Weblink page.

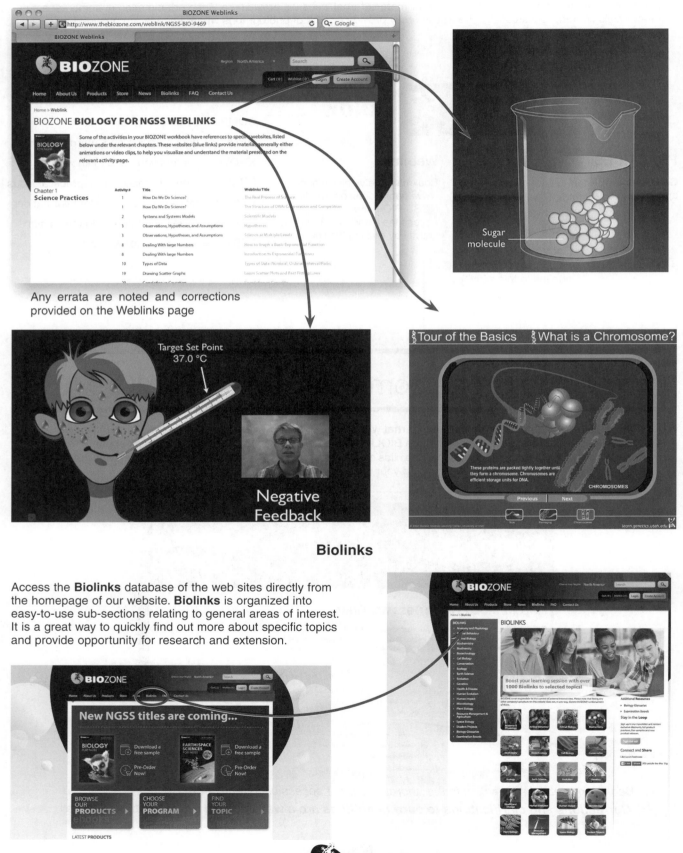

Any errata are noted and corrections provided on the Weblinks page

Biolinks

Access the **Biolinks** database of the web sites directly from the homepage of our website. **Biolinks** is organized into easy-to-use sub-sections relating to general areas of interest. It is a great way to quickly find out more about specific topics and provide opportunity for research and extension.

Science Practices

Key terms

accuracy

assumption

biological drawing

control

controlled variable

data

descriptive statistics

dependent variable

graph

hypothesis

independent variable

mean

median

mode

model

observation

precision

prediction

qualitative data

quantitative data

raw data

scientific method

table

variable

Science and engineering practices

Supported as noted and throughout subsequent chapters in context

Activity number

Asking questions and defining problems

☐ 1 Demonstrate an understanding of science as inquiry. Appreciate that unexpected results may lead to new hypotheses and to new discoveries. — 1

☐ 2 Formulate and evaluate questions that you can feasibly investigate. — 3

Develop and use models

☐ 3 Develop and use models based on evidence to describe systems or their components and how they work. — 2

☐ 4 Make accurate biological drawings to record important features of specimens and show how their components are related. — 26 27

Plan and carry out investigations

☐ 5 Plan and conduct investigations to provide data to test a hypothesis based on observations. Identify any assumptions in the design of your investigation. — 12 28

☐ 6 Consider and evaluate the accuracy and precision of the data that you collect. — 4 9

☐ 7 Use appropriate tools to collect and record data. Data may be quantitative (continuous or discontinuous), qualitative, or ranked. — 10 11 13

☐ 8 Variables are factors that can change or be changed in an experiment. Make and test hypotheses about the effect on a dependent variable when an independent variable is manipulated. Understand and use controls appropriately. — 11 28

Analyze and interpret data

☐ 9 Use graphs appropriate to the data to visualize data and identify trends. — 16 17 19 21 22

☐ 10 Summarize data and describe its features using descriptive statistics. — 23 24 25

☐ 11 Apply concepts of statistics and probability to answer questions and solve problems. — 77 211

Use mathematics and computational thinking

☐ 12 Demonstrate an ability to use mathematics and computational tools to analyze, represent, and model data. Recognize and use appropriate units in calculations. — 5 18 23

☐ 13 Demonstrate an ability to apply ratios, rates, percentages, and unit conversions. — 5 6 7 8

Construct explanations and design solutions

☐ 14 Explain results based on evidence and applying scientific ideas and principles. — 20 28

Engage in argument from evidence

☐ 15 Use evidence to defend and evaluate claims and explanations about science. — 28 56 76

Obtain, evaluate, and communicate information

☐ 16 Evaluate the validity and reliability of designs, methods, claims, and evidence. — 1 28

1 How Do We Do Science?

Key Idea: Science is a way of understanding the world based on a rigorous, dynamic process of observation, investigation, and analysis.

▶ Science is a way of understanding the world we live in: how it formed, the rules it obeys and how it changes over time. Science distinguishes itself from other ways of understanding by using empirical standards, logical arguments, and skeptical review. Science allows what we understand to change over time as the body of knowledge increases.

▶ It is important to realize that science is a human endeavor and requires creativity and imagination. New research and ways of thinking can be based on the well-argued idea of a single person.

▶ Science influences and is influenced by society and technology. As society's beliefs and desires change, what is or can be researched is also affected. As technology advances what is or can be researched changes. Scientific discoveries advance technology and can change society's beliefs.

▶ Science can never answer questions with absolute certainty. It can be confident of certain outcomes, but only within the limits of the data. Science might help us predict with 99.9% certainty a system will behave a certain way, but that still means there's one chance in a thousand it won't.

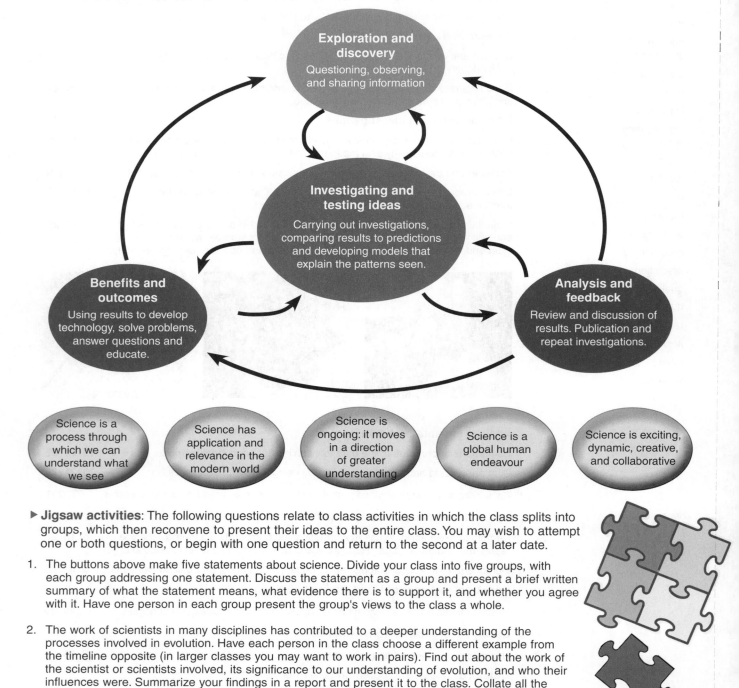

▶ **Jigsaw activities**: The following questions relate to class activities in which the class splits into groups, which then reconvene to present their ideas to the entire class. You may wish to attempt one or both questions, or begin with one question and return to the second at a later date.

1. The buttons above make five statements about science. Divide your class into five groups, with each group addressing one statement. Discuss the statement as a group and present a brief written summary of what the statement means, what evidence there is to support it, and whether you agree with it. Have one person in each group present the group's views to the class a whole.

2. The work of scientists in many disciplines has contributed to a deeper understanding of the processes involved in evolution. Have each person in the class choose a different example from the timeline opposite (in larger classes you may want to work in pairs). Find out about the work of the scientist or scientists involved, its significance to our understanding of evolution, and who their influences were. Summarize your findings in a report and present it to the class. Collate all the information to flesh out the timeline. How many people appear on the timeline now?

© 2016 **BIOZONE** International
ISBN: 978-1-927309-46-9
Photocopying Prohibited

- Although Charles Darwin is largely credited with the development of the theory of evolution by natural selection, his ideas did not develop in isolation, but within the context of the work of others before him. The **modern synthesis** (below) has a long history with contributors from all fields of science. Evolution by natural selection is one of the best substantiated theories in the history of science, supported by evidence from many disciplines, including paleontology, geology, genetics and developmental biology.

- The diagram below summarizes just some of the important players in the story of evolutionary biology. This is not to say they were collaborators or always agreed. Some of the work, such as Haeckel's work on embryology, was flawed and even untruthful.

- However, the work of many has contributed to a deeper understanding of evolutionary processes. This understanding continues to increase in the light of increasingly sophisticated molecular techniques and the collaboration of scientists internationally.

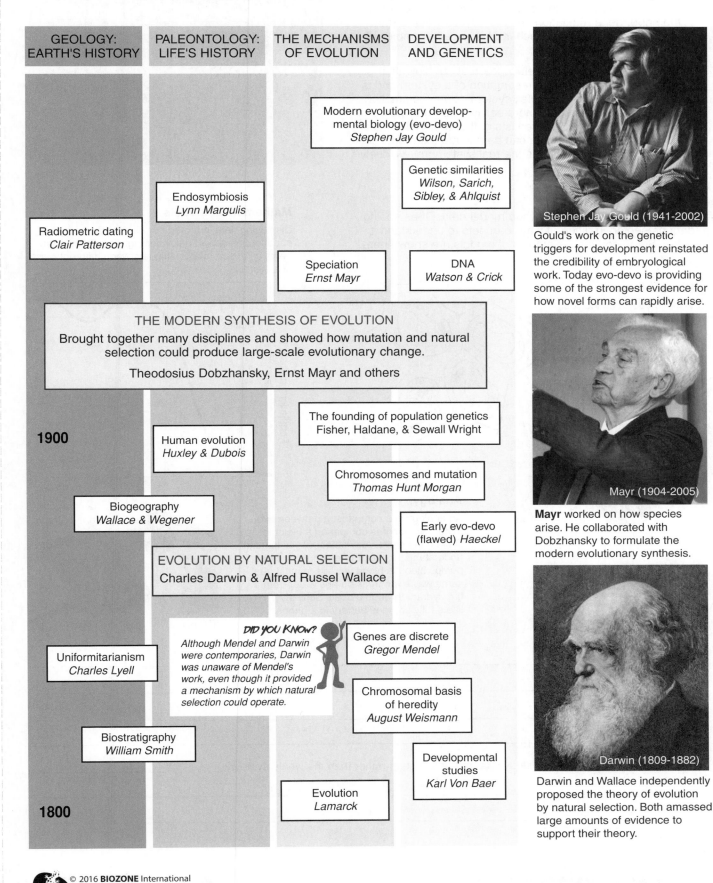

Stephen Jay Gould (1941-2002)

Gould's work on the genetic triggers for development reinstated the credibility of embryological work. Today evo-devo is providing some of the strongest evidence for how novel forms can rapidly arise.

Mayr (1904-2005)

Mayr worked on how species arise. He collaborated with Dobzhansky to formulate the modern evolutionary synthesis.

Darwin (1809-1882)

Darwin and Wallace independently proposed the theory of evolution by natural selection. Both amassed large amounts of evidence to support their theory.

© 2016 **BIOZONE** International
ISBN: 978-1-927309-46-9
Photocopying Prohibited

2 Systems and System Models

Key Idea: Scientists use models to learn about biological systems. Models usually study one small part of a system, so that the system can be more easily understood.

▶ A **system** is a set of interrelated components that work together. Energy flow in ecosystems (such as the one on the right), gene regulation, interactions between organ systems, and feedback mechanisms are all examples of systems studied in biology.

▶ Scientists often used models to learn about biological systems. A **model** is a representation of a system and is useful for breaking a complex system down into smaller parts that can be studied more easily. Often only part of a system is modelled. As scientists gather more information about a system, more data can be put into the model so that eventually it represents the real system more closely.

Modeling data

There are many different ways to model data. Often seeing data presented in different ways can help to understand it better. Some common examples of models are shown here.

Mathematical models

Displaying data in a graph or as a mathematical equation, as shown below for logistic growth, often helps us to see relationships between different parts of a system.

Visual models

Visual models can include drawings, such as these plant cells on the right.

Three dimensional models can be made out of materials such as modeling clay and ice-cream sticks, like this model of a water molecule (below).

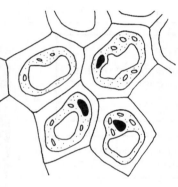

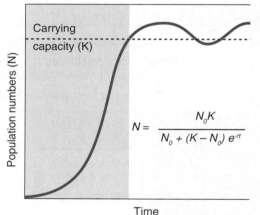

$$N = \frac{N_0 K}{N_0 + (K - N_0)\, e^{-rt}}$$

Population numbers (N)

Carrying capacity (K)

Time

James Hedberg

Analogy

An analogy is a comparison between two things. Sometimes comparing a biological system to an everyday object can help us to understand it better. For example, the heart pumps blood in blood vessels in much the same way a fire truck pumps water from a fire hydrant through a hose. Similarly, ATP is like a fully charged battery in a phone.

... a charged phone battery

ATP is like...

1. What is a system? _____

_____

2. (a) What is a model? _____

_____

(b) Why do scientists often study one part of a system rather than the whole system? _____

_____

_____

WEB   CCC   PRACTICES

SSM

© 2016 **BIOZONE** International
ISBN: 978-1-927309-46-9
Photocopying Prohibited

3 Observations, Hypotheses, and Assumptions

Key Idea: Observations are the basis for forming hypotheses and making predictions about systems. An assumption is something that is accepted as true but is not tested.

Observations and hypotheses

▶ An observation is watching or recording what is happening. Observation is the basis for forming hypotheses and making predictions. An observation may generate a number of hypotheses (tentative explanations for what we see). Each hypothesis will lead to one or more predictions, which can be tested by investigation.

▶ A hypothesis is often written as a statement to include the prediction: "If X is true, then if I do Y (the experiment), I expect Z (the prediction)". Hypotheses are accepted, changed, or rejected on the basis of investigations. A hypothesis should have a sound theoretical basis and should be testable.

Observation 1:

▶ Some caterpillar species are brightly colored and appear to be highly visible to predators such as insectivorous birds. Predators appear to avoid these caterpillars.

▶ These caterpillars are often found in groups, rather than as solitary animals.

Observation 2:

▶ Some caterpillar species have excellent camouflage. When alerted to danger they are difficult to see because they blend into the background.

▶ These caterpillars are usually found alone.

Assumptions

Any investigation requires you to make **assumptions** about the system you are working with. Assumptions are features of the system you are studying that you assume to be true but that you do not (or cannot) test. Some assumptions about the examples above include:

• Insect eating birds have color vision.
• Caterpillars that look bright to us, also appear bright to insectivorous birds.
• Birds can learn about the taste of prey by eating them.

Read the two observations about the caterpillars above and then answer the following questions:

1. Generate a hypothesis to explain the observation that some caterpillars are brightly colored and highly visible while others are camouflaged and blend into their surroundings:

   _____

   _____

2. Describe one of the assumptions being made in your hypothesis: _____

   _____

   _____

3. Generate a prediction about the behavior of insect eating birds towards caterpillars: _____

   _____

   _____

PRACTICES   WEB

 **KNOW**

4 Accuracy and Precision

Key Idea: Accuracy refers to the correctness of a measurement (how true it is to the real value). Precision refers to how close the measurements are to each other.

The terms accuracy and precision are often used when talking about measurements.

▶ **Accuracy** refers to how close a measured value is to its true value, i.e. the correctness of the measurement.

▶ **Precision** refers to the closeness of repeated measurements to each other, i.e. the ability to be exact. For example, a digital device, such as a pH meter (right) will give very precise measurements, but its accuracy depends on correct calibration.

Using the analogy of a target, repeated measurements are compared to arrows shot at a target. This analogy is useful when distinguishing between accuracy and precision.

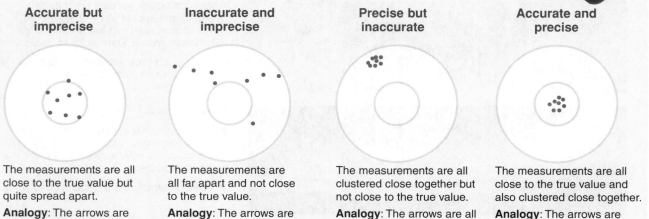

Accurate but imprecise	Inaccurate and imprecise	Precise but inaccurate	Accurate and precise
The measurements are all close to the true value but quite spread apart.	The measurements are all far apart and not close to the true value.	The measurements are all clustered close together but not close to the true value.	The measurements are all close to the true value and also clustered close together.
Analogy: The arrows are all close to the bullseye.	**Analogy**: The arrows are spread around the target.	**Analogy**: The arrows are all clustered close together but not near the bullseye.	**Analogy**: The arrows are clustered close together near the bullseye.

Significant figures

Significant figures (sf) are the digits of a number that carry meaning contributing to its precision. They communicate how well you could actually measure the data.

For example, you might measure the height of 100 people to the nearest cm. When you calculate their mean height, the answer is 175.0215 cm. If you reported this number, it implies that your measurement technique was accurate to 4 decimal places. You would have to round the result to the number of significant figures you had accurately measured. In this instance the answer is 175 cm.

Non-zero numbers (1-9) are always **significant**.

All zeros between non-zero numbers are always **significant**.

$$0.005704510$$

Zeros to the left of the first non-zero digit after a decimal point are **not significant**.

Zeros at the end of number where there is a decimal place are **significant** (e.g. 4600.0 has five sf).
BUT
Zeros at the end of a number where there is no decimal point are **not significant** (e.g. 4600 has two sf).

1. Why are precise but inaccurate measurements not helpful in a biological investigation? _____

_____

_____

_____

2. State the number of significant figures in the following examples:

(a) 3.15985 _____

(b) 0.0012 _____

(c) 1000 _____

(d) 1000.0 _____

(e) 42.3006 _____

(f) 120 _____

© 2016 **BIOZONE** International
ISBN: 978-1-927309-46-9
Photocopying Prohibited

5 | Working With Numbers

Key Idea: Using correct mathematical notation and being able to carry out simple calculations and conversions are fundamental skills in biology.

Commonly used mathematical symbols

In mathematics, universal symbols are used to represent mathematical concepts. They save time and space when writing. Some commonly used symbols are shown below.

=   Equal to

<   The value on the left is **less than** the value on the right

<<   The value on the left is **much less than** the value on the right

>   The value on the left is **greater than** the value on the right

>>   The value on the left is **much greater than** the value on the right

∝   Proportional to. A ∝ B means that A = a constant X B

~   Approximately equal to

Decimal and standard form

▶ **Decimal form** (also called ordinary form) is the longhand way of writing a number (e.g. 15,000,000). Very large or very small numbers can take up too much space if written in decimal form and are often expressed in a condensed **standard form**. For example, 15,000,000 is written as 1.5×10^7 in standard form.

▶ In standard form a number is always written as $A \times 10^n$, where A is a number between 1 and 10, and n (the exponent) indicates how many places to move the decimal point. n can be positive or negative.

▶ For the example above, A = 1.5 and n = 7 because the decimal point moved seven places (see below).

$$1\,5\,000\,000 = 1.5 \times 10^7$$

▶ Small numbers can also be written in standard form. The exponent (n) will be negative. For example, 0.00101 is written as 1.01×10^{-3}.

$$0.00101 = 1.01 \times 10^{-3}$$

▶ Converting can make calculations easier. Work through the following example to solve $4.5 \times 10^4 + 6.45 \times 10^5$.

1. Convert $4.5 \times 10^4 + 6.45 \times 10^5$ to decimal form:

_____

2. Add the two numbers together: _____

3. Convert to standard form: _____

Estimates

▶ When carrying out calculations, typing the wrong number into your calculator can put your answer out by several orders of magnitude. An **estimate** is a way of roughly calculating what answer you should get, and helps you decide if your final calculation is correct.

▶ Numbers are often rounded to help make estimation easier. The rounding rule is, if the next digit is 5 or more, round up. If the next digit is 4 or less, it stays as it is.

▶ For example, to estimate 6.8 x 704 you would round the numbers to 7 x 700 = 4900. The actual answer is 4787, so the estimate tells us the answer (4787) is probably right.

Use the following examples to practise estimating:

4. 43.2 x 1044: _____

5. 3.4 x 72 ÷ 15: _____

6. 658 ÷ 22: _____

Conversion factors and expressing units

▶ Measurements can be converted from one set of units to another by the use of a **conversion factor**. A conversion factor is a numerical factor that multiplies or divides one unit to convert it into another.

▶ Conversion factors are commonly used to convert non-SI units to SI units (e.g. converting pounds to kilograms). Note that mL and cm^3 are equivalent, as are L and dm^3.

In the space below, convert $5.6\ cm^3$ to mm^3 ($1\ cm^3 = 1000\ mm^3$):

7. _____

▶ The value of a variable must be written with its units where possible. SI units or their derivations should be used in recording measurements: volume in cm^3 (mL) or dm^3 (L), mass in kilograms (kg) or grams (g), length in meters (m), time in seconds (s). To denote 'per' use a negative exponent, e.g. per second is written as s^{-1} and per meter squared is written as m^{-2}.

▶ For example the rate of oxygen consumption should be expressed as:

Oxygen consumption ($cm^3\ g^{-1}\ s^{-1}$)

PRACTICES

DATA

6 Tallies, Percentages, and Rates

Key Idea: Unprocessed data is called raw data. A set of data is often manipulated or transformed to make it easier to understand and to identify important features.

The data collected by measuring or counting in the field or laboratory is called **raw data**. Raw data often needs to be processed into a form that makes it easier to identify its important features (e.g. trends) and make meaningful comparisons between samples or treatments. Basic calculations, such as totals (the sum of all data values for a variable), are commonly used to compare treatments. Some common methods of processing data include creating tally charts, and calculating percentages and rates. These are explained below.

Tally Chart

Records the number of times a value occurs in a data set

HEIGHT (cm)	TALLY	TOTAL
0-0.99	III	3
1-1.99	IIII I	6
2-2.99	IIII IIII	10
3-3.99	IIII IIII II	12
4-4.99	III	3
5-5.99	II	2

- A useful first step in analysis; a neatly constructed tally chart doubles as a simple histogram.

- Cross out each value on the list as you tally it to prevent double entries.

Percentages

Expressed as a fraction of 100

Men	Body mass (kg)	Lean body mass (kg)	% lean body mass
Athlete	70	60	85.7
Lean	68	56	82.3
Normal weight	83	65	78.3
Overweight	96	62	64.6
Obese	125	65	52.0

- Percentages express what proportion of data fall into any one category, e.g. for pie graphs.

- Allows meaningful comparison between different samples.

- Useful to monitor change (e.g. % increase from one year to the next).

Rates

Expressed as a measure per unit time

Time (minutes)	Cumulative sweat loss (mL)	Rate of sweat loss (mL min⁻¹)
0	0	0
10	50	5
20	130	8
30	220	9
60	560	11.3

- Rates show how a variable changes over a standard time period (e.g. one second, one minute, or one hour).

- Rates allow meaningful comparison of data that may have been recorded over different time periods.

Example: Height of 6 day old seedlings

Example: Percentage of lean body mass in men

Example: Rate of sweat loss during exercise in cyclists

1. What is raw data? _____

_____

2. Why is it useful to process raw data and express it differently, e.g. as a rate or a percentage? _____

_____

_____

3. Identify the best data transformation in each of the following examples:

(a) Comparing harvest (in kg) of different grain crops from a farm: _____

(b) Comparing amount of water loss from different plant species: _____

DATA

 © 2016 **BIOZONE** International
ISBN: 978-1-927309-46-9
Photocopying Prohibited

7 Fractions and Ratios

Key Idea: Fractions and ratios are widely used in biology and are often used to provide a meaningful comparison of sample data where the sample sizes are different.

Fractions

- ▶ Fractions express how many parts of a whole are present.
- ▶ Fractions are expressed as two numbers separated by a solidus (/). For example 1/2.
- ▶ The top number is the numerator. The bottom number is the denominator. The denominator can not be zero.

Simplifying fractions

- ▶ Fractions are often written in their simplest form (the top and bottom numbers cannot be any smaller, while still being whole numbers). Simplifying makes working with fractions easier.
- ▶ To simplify a fraction, the numerator and denominator are divided by the highest common number that divides into both numbers equally.
- ▶ For example, in a class of 20 students, five had blue eyes. This fraction is 5/20. To simplify this fraction 5 and 20 are divided by the highest common factor (5).

$$5 \div 5 = \textbf{1} \text{ and } 20 \div 5 = \textbf{4}$$

- ▶ The simplified fraction is 1/4.

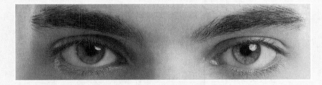

Adding fractions

- ▶ To add fractions the denominators must be the same. If the denominators are the same the numerators are simply added. E.g. 5/12 + 3/12 = 8/12
- ▶ When the denominators are different one (or both) fractions must be multiplied to give a common denominator, e.g. 4/10 + 1/2. By multiplying 1/2 by 5 the fraction becomes 5/10. The fractions can now be added together (4/10 + 5/10 = 9/10).

Ratios

- ▶ Ratios give the relative amount of two or more quantities (it shows how much of one thing there is relative to another).
- ▶ Ratios provide an easy way to identify patterns.
- ▶ Ratios do not require units.
- ▶ Ratios are usually expressed as *a : b*.
- ▶ In the example below, there are 3 blue squares and 1 gray square. The ratio would be written as 3:1.

Calculating ratios

- ▶ Ratios are calculated by dividing all the values by the smallest number.
- ▶ Ratios are often used in Mendelian genetics to calculate phenotype (appearance) ratios. Some examples for pea plants are given below.

882 inflated pod   299 constricted pod

To obtain the ratio divide both numbers by 299.
$299 \div 299 = 1$
$882 \div 299 = 2.95$
The ratio = 2.95 : 1

| 495 | 152 | 158 | 55 |
| round yellow | wrinkled yellow | round green | wrinkled green |

For the example above of pea seed shape and color, all of the values were divided by 55. The ratio obtained was: 9 : 2.8 : 2.9 : 1

---

1. (a) A student prepared a slide of the cells of an onion root tip and counted the cells at various stages in the cell cycle. The results are presented in the table (right). Calculate the ratio of cells in each stage (show your working):

_____

_____

Cell cycle stage	No. of cells counted	No. of cells calculated
Interphase	140	
Prophase	70	
Telophase	15	
Metaphase	10	
Anaphase	5	
Total	**240**	**4800**

(b) Assuming the same ratio applies in all the slides examined in the class, calculate the number of cells in each phase for a cell total count of 4800.

2. Simplify the following fractions:

   (a) 3/9 : _____ (b) 84/90: _____ (c) 11/121: _____

3. In the fraction example pictured above 5/20 students had blue eyes. In another class, 5/12 students had blue eyes. What fraction of students had blue eyes in both classes combined?

_____

 © 2016 **BIOZONE** International
ISBN: 978-1-927309-46-9
Photocopying Prohibited

PRACTICES

 DATA

8 Dealing with Large Numbers

Key Idea: Large scale changes in numerical data can be made more manageable by transforming the data using logarithms or plotting the data on log-log or log-linear (semi-log) paper.

▶ In biology, numerical data indicating scale can often decrease or increase exponentially. Examples include the exponential growth of populations, exponential decay of radioisotopes, and the pH scale.

▶ Exponential changes in numbers are defined by a function. A function is simply a rule that allows us to calculate an output for any given input. Exponential functions are common in biology and may involve very large numbers.

▶ Log transformations of exponential numbers can make them easier to handle.

Exponential function

▶ Exponential growth occurs at an increasingly rapid rate in proportion to the growing total number or size.

▶ In an exponential function, the base number is fixed (constant) and the exponent is variable.

▶ The equation for an exponential function is $y = c^x$.

▶ Exponential growth and decay (reduction) are possible.

▶ Exponential changes in numbers are easy to identify because the curve has a J-shape appearance due to its increasing steepness over time.

▶ An example of exponential growth is the growth of a microbial population in an unlimiting, optimal growth environment.

Log transformations

▶ A log transformation makes very large numbers easier to work with. The log of a number is the exponent to which a fixed value (the base) is raised to get that number. So $\log_{10}(1000) = 3$ because $10^3 = 1000$.

▶ Both $\log_{10}$ (common logs) and $\log_e$ (natural logs or *ln*) are commonly used.

▶ Log transformations are useful for data where there is an exponential increase or decrease in numbers. In this case, the transformation will produce a straight line plot.

▶ To find the $\log_{10}$ of a number, e.g. 32, using a calculator, key in log 32 = . The answer should be 1.51.

▶ Alternatively, the untransformed data can be plotted directly on a log-linear scale (as below). This is not difficult. You just need to remember that the log axis runs in exponential cycles. The paper makes the log for you.

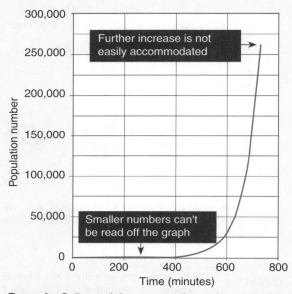

Example: Cell growth in a yeast culture where growth is not limited by lack of nutrients or build up of toxins.

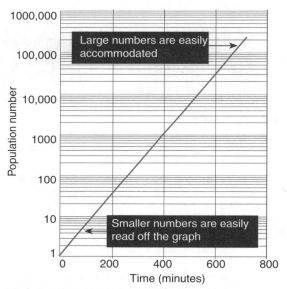

Example: The same yeast cell growth plotted on a log-linear scale. The y axis present 6 exponential cycles

1. Why is it useful to plot exponential growth using semi-log paper? _____

_____

_____

2. What would you do to show yeast exponential growth as a straight line plot on normal graph paper?

_____

3. Log transformations are often used when a value of interest ranges over several orders of magnitude. Can you think of another example of data from the natural world where the data collected might show this behavior?

_____

© 2016 **BIOZONE** International
ISBN: 978-1-927309-46-9
Photocopying Prohibited

9 Apparatus and Measurement

Key Idea: The apparatus used in experimental work must be appropriate for the experiment or analysis and it must be used correctly to eliminate experimental errors.

Selecting the correct equipment

It is important that you choose equipment that is appropriate for the type of measurement you want to take. For example, if you wanted to accurately weigh out 5.65 g of sucrose, you need a balance that accurately weighs to two decimal places. A balance that weighs to only one decimal place would not allow you to make an accurate enough measurement.

Study the glassware (right). Which would you use if you wanted to measure 225 mL? The graduated cylinder has graduations every 10 mL whereas the beaker has graduations every 50 mL. It would be more accurate to measure 225 mL in a graduated cylinder.

Percentage errors

Percentage error is a way of mathematically expressing how far out your result is from the ideal result. The equation for measuring percentage error is:

$$\frac{\text{experimental value - ideal value}}{\text{ideal value}} \times 100$$

For example, you want to know how accurate a 5 mL pipette is. You dispense 5 mL of water from a pipette and weigh the dispensed volume on a balance. The volume is 4.98 mL.

$$\frac{\text{experimental value (4.98) - ideal value (5.0)}}{\text{ideal value (5.0)}} \times 100$$

The percentage error = –0.4% (the negative sign tells you the pipette is dispensing **less** than it should).

Recognizing potential sources of error

It is important to know how to use equipment correctly to reduce errors. A spectrophotometer measures the amount of light absorbed by a solution at a certain wavelength. This information can be used to determine the concentration of the absorbing molecule (e.g. density of bacteria in a culture). The more concentrated the solution, the more light is absorbed. Incorrect use of the spectrophotometer can alter the results. Common mistakes include incorrect calibration, errors in sample preparation, and errors in sample measurement.

A cuvette (left) is a small clear tube designed to hold spectrophotometer samples. Inaccurate readings occur when:

- The cuvette is dirty or scratched (light is absorbed giving a falsely high reading).
- Some cuvettes have a frosted side to aid alignment. If the cuvette is aligned incorrectly, the frosted side absorbs light, giving a false reading.
- Not enough sample is in the cuvette and the beam passes over, rather than through the sample, giving a lower absorbance reading.

1. Assume that you have the following measuring devices available: 50 mL beaker, 50 mL graduated cylinder, 25 mL graduated cylinder, 10 mL pipette, 10 mL beaker. What would you use to accurately measure:

(a) 21 mL: _____ (b) 48 mL: _____ (c) 9 mL: _____

2. Calculate the percentage error for the following situations (show your working):

(a) A 1 mL pipette delivers a measured volume of 0.98 mL: _____

(b) A 10 mL pipette delivers a measured volume of 9.98 mL: _____

PRACTICES

DATA

10 Types of Data

Key Idea: Data is information collected during an investigation. Data may be quantitative, qualitative, or ranked.

Data is information collected during an investigation. Data may be quantitative, qualitative, or ranked. When planning a biological investigation, it is important to consider the type of data that will be collected. It is best to collect quantitative or numerical data, because it is easier to analyze it objectively (without bias).

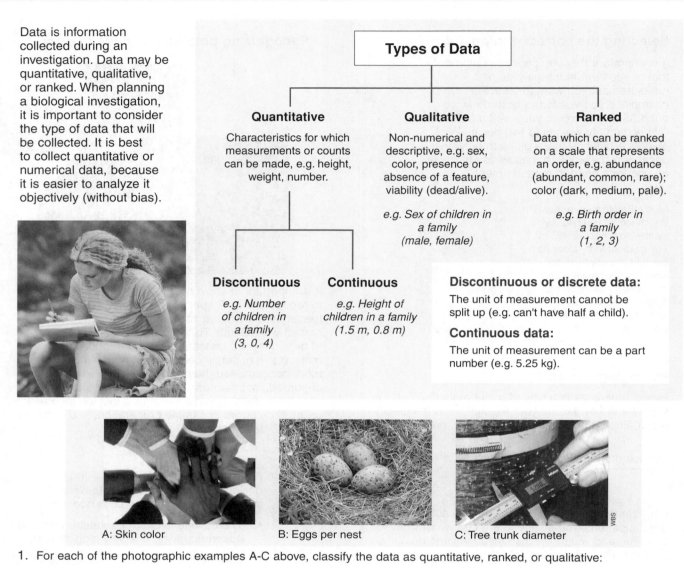

Types of Data

Quantitative
Characteristics for which measurements or counts can be made, e.g. height, weight, number.

Qualitative
Non-numerical and descriptive, e.g. sex, color, presence or absence of a feature, viability (dead/alive).

e.g. Sex of children in a family (male, female)

Ranked
Data which can be ranked on a scale that represents an order, e.g. abundance (abundant, common, rare); color (dark, medium, pale).

e.g. Birth order in a family (1, 2, 3)

Discontinuous
e.g. Number of children in a family (3, 0, 4)

Continuous
e.g. Height of children in a family (1.5 m, 0.8 m)

Discontinuous or discrete data:
The unit of measurement cannot be split up (e.g. can't have half a child).

Continuous data:
The unit of measurement can be a part number (e.g. 5.25 kg).

A: Skin color   B: Eggs per nest   C: Tree trunk diameter

1. For each of the photographic examples A-C above, classify the data as quantitative, ranked, or qualitative:

   (a) Skin color: _____

   (b) Number of eggs per nest: _____

   (c) Tree trunk diameter: _____

2. Why is it best to collect quantitative data where possible in biological studies? _____

   _____

   _____

   _____

3. Give an example of data that could not be collected quantitatively and explain your answer: _____

   _____

   _____

   _____

© 2016 **BIOZONE** International
ISBN: 978-1-927309-46-9
Photocopying Prohibited

11 Variables and Controls

Key Idea: Variables may be dependent, independent, or controlled. A control in an experiment allows you to determine the effect of the independent variable.

Types of variables

A **variable** is a factor that can be changed during an experiment (e.g. temperature). Investigations often look at how changing one variable affects another.

There are several types of variables:

- Independent
- Dependent
- Controlled

Only one variable should be changed at a time. Any changes seen are a result of the changed variable.

Remember! The dependent variable is 'dependent' on the independent variable.

Example: *When heating water, the temperature of the water depends on the time it is heated for. Temperature (dependent variable) depends on time (independent variable).*

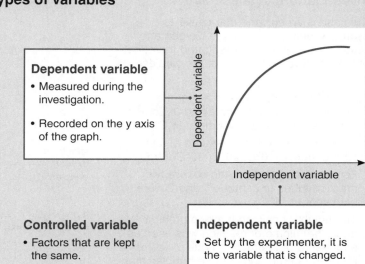

Dependent variable
- Measured during the investigation.
- Recorded on the y axis of the graph.

Controlled variable
- Factors that are kept the same.

Independent variable
- Set by the experimenter, it is the variable that is changed.
- Recorded on the graph's x axis.

Experimental controls

▶ A **control** is the standard or reference treatment in an experiment. Controls make sure that the results of an experiment are due to the variable being tested (e.g. nutrient level) and not due to another factor (e.g. equipment not working correctly).

▶ A control is identical to the original experiment except it lacks the altered variable. The control undergoes the same preparation, experimental conditions, observations, measurements, and analysis as the test group.

▶ If the control works as expected, it means the experiment has run correctly, and the results are due to the effect of the variable being tested.

Test plant (nutrient added)

Control plant (no nutrient added)

An experiment was designed to test the effect of a nutrient on plant growth. The control plant had no nutrient added to it. Its growth sets the baseline for the experiment. Any growth in the test plant above that seen in the control plant is due to the presence of the nutrient.

1. What is the difference between a dependent variable and an independent variable? _____

_____

_____

2. Why do we control the variables we are not investigating? _____

_____

_____

3. What is the purpose of the experimental control? _____

_____

_____

 © 2016 **BIOZONE** International
ISBN: 978-1-927309-46-9
Photocopying Prohibited

PRACTICES

KNOW

12 A Case Study: Catalase Activity

Key Idea: A simple experiment to test a hypothesis involves manipulating one variable (the independent variable) and recording the response.

Investigation: catalase activity

Catalase is an enzyme that converts hydrogen peroxide (H_2O_2) to oxygen and water. An experiment investigated the effect of temperature on the rate of the catalase reaction.

- 10 cm^3 test tubes were used for the reactions, each tube contained 0.5 cm^3 of catalase enzyme and 4 cm^3 of H_2O_2.
- Reaction rates were measured at four temperatures (10°C, 20°C, 30°C, 60°C).
- For each temperature, there were two reaction tubes (e.g. tubes 1 and 2 were both kept at 10°C).
- The height of oxygen bubbles present after one minute of reaction was used as a measure of the reaction rate. A faster reaction rate produced more bubbles than a slower reaction rate.
- The entire experiment, was repeated on two separate days.

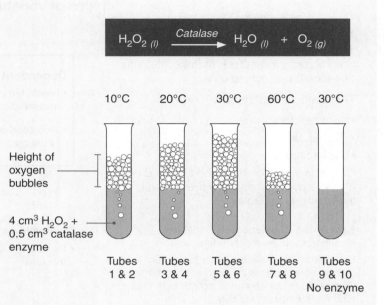

$$H_2O_2 {}_{(l)} \xrightarrow{\text{Catalase}} H_2O {}_{(l)} + O_2 {}_{(g)}$$

10°C   20°C   30°C   60°C   30°C

Height of oxygen bubbles

4 cm^3 H_2O_2 + 0.5 cm^3 catalase enzyme

Tubes 1 & 2   Tubes 3 & 4   Tubes 5 & 6   Tubes 7 & 8   Tubes 9 & 10 No enzyme

1. Write a suitable aim for this experiment: _____

_____

_____

2. Write an hypothesis for this experiment: _____

_____

_____

_____

3. (a) What is the independent variable in this experiment? _____

(b) What is the range of values for the independent variable? _____

(c) Name the unit for the independent variable: _____

(d) List the equipment needed to set the independent variable, and describe how it was used: _____

_____

_____

4. (a) What is the dependent variable in this experiment? _____

(b) Name the unit for the dependent variable: _____

(c) List the equipment needed to measure the dependent variable, and describe how it was used: _____

_____

_____

_____

5. Which tubes are the control for this experiment? _____

PRACTICES

TEST

 © 2016 **BIOZONE** International
ISBN: 978-1-927309-46-9
Photocopying Prohibited

13 Recording Results

Key Idea: Accurately recording results makes it easier to understand and analyze your data later. A table is a good way to record data but dataloggers will also record it automatically.

Ways to record data

Recording your results accurately is very important in any type of scientific investigation. If you have recorded your results accurately and in an organized way, it makes analyzing and understanding your data easier. Log books and dataloggers are two methods by which data can be recorded.

Log books

A log book records your ideas and results through your scientific investigation. It also provides proof that you have carried out the work.

- An A4 lined exercise book is a good choice for a log book. It gives enough space to write ideas and record results and provides space to paste in photos or extra material (such as printouts).

- Each entry must have the date recorded.

- Make sure that you can read what you write at a later date. A log book entry is meaningless if it is incomplete or cannot be read.

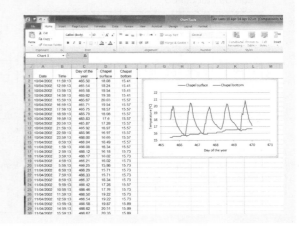

If you are not using a datalogger, a table (above) is often a good way to record and present your results as you collect them. Tables can also be useful for showing calculated values (such as rates and means). Recording data in a table as your experiment proceeds lets you identify any trends early on and change experimental conditions if necessary.

Dataloggers

A datalogger (also called a data recorder) is an electronic device that automatically records data over time.

- Dataloggers have a variety of sensors to measure different physical properties. Common sensors include light, temperature, pH, conductivity, and humidity.

- Dataloggers can be used in both field or laboratory experiments, and can be left to collect data without the experimenter being present.

- Information collected by the datalogger can be downloaded to a computer (below) so that the data can be accessed and analyzed.

1. Why is it important to accurately record your results? _____

2. Why must log book entries be well organized? _____

3. (a) What is a datalogger? _____

   (b) What are the advantages of using a datalogger over manually recording results? _____

PRACTICES

 KNOW

14 Practicing Data Manipulations

Key Idea: Percentages, rates, and frequencies are commonly used manipulations of raw data.

1. Complete the transformations for each of the tables on the right. The first value, and their working, is provided for each example.

   (a) TABLE: Incidence of red clover in different areas:

   Working: 124 ÷ 159 = 0.78 = 78%

   > This is the number of red clover out of the total.

Incidence of red and white clover in different areas

Clover plant type	Frost free area		Frost prone area		Totals
	Number	%	Number	%	
Red	124	78	26		
White	35		115		
Total	159				

   (b) TABLE: Plant water loss using a bubble potometer:

   Working: (9.0 – 8.0) ÷ 5 min = 0.2

   > This is the distance the bubble moved over the first 5 minutes. Note that there is no data entry possible for the first reading (0 min) because no difference can be calculated.

Plant water loss using a bubble potometer

Time (min)	Pipette arm reading (cm^3)	Plant water loss (cm^3 min^{-1})
0	9.0	–
5	8.0	0.20
10	7.2	
15	6.2	
20	4.9	

   (c) TABLE: Frequency of size classes in a sample of eels:

   Working: (7 ÷ 270) x 100 = 2.6 %

   > This is the number of individuals out of the total that appear in the size class 0-50 mm. The relative frequency is rounded to one decimal place.

Frequency of size classes in a sample of eels

Size class (mm)	Frequency	Relative frequency (%)
0-50	7	2.6
50-99	23	
100-149	59	
150-199	98	
200-249	50	
250-299	30	
300-349	3	
Total	270	

   (d) TABLE: Body composition in women:

   Working: (38 ÷ 50) x 100 = 76 %

   > This is lean body mass. The percentage lean body mass is calculated by dividing lean body mass by total body mass. It is multiplied by 100 to convert it into a percentage.

Body mass composition in women

Women	Body mass (kg)	Lean body mass (kg)	% lean body mass
Athlete	50	38	76
Lean	56	41	
Normal weight	65	46	
Overweight	80	48	
Obese	95	52	

© 2016 **BIOZONE** International
ISBN: 978-1-927309-46-9
Photocopying Prohibited

15 Constructing Tables

Key Idea: Tables are used to record and summarize data. Tables allow relationships and trends in data to be more easily recognized.

▶ Tables are used to record data during an investigation. Your log book should present neatly tabulated data (right).

▶ Tables allow a large amount of information to be condensed, and can provide a summary of the results.

▶ Presenting data in tables allows you to organize your data in a way that allows you to more easily see the relationships and trends.

▶ Columns can be provided to display the results of any data transformations such as rates. Basic descriptive statistics (such as mean or standard deviation) may also be included.

▶ Complex data sets tend to be graphed rather than tabulated.

Features of tables

Tables should have an accurate, descriptive title. Number tables consecutively through a report.

Heading and subheadings identify each set of data and show units of measurement.

Independent variable in the left column.

Control values should be placed at the beginning of the table.

Table 1: Length and growth of the third internode of bean plants receiving three different hormone treatments.

Treatment	Sample size	Mean rate of internode growth (mm day^{-1})	Mean internode length (mm)	Mean mass of tissue added (g day^{-1})
Control	50	0.60	32.3	0.36
Hormone 1	46	1.52	41.6	0.51
Hormone 2	98	0.82	38.4	0.56
Hormone 3	85	2.06	50.2	0.68

Each row should show a different experimental treatment, organism, sampling site etc.

Columns for comparison should be placed alongside each other. Show values only to the level of significance allowable by your measuring technique.

Organize the columns so that each category of like numbers or attributes is listed vertically.

1. What are two advantages of using a table format for data presentation?

   (a) _____

   _____

   (b) _____

   _____

2. Why might you tabulate data before you presented it in a graph? _____

   _____

   _____

© 2016 **BIOZONE** International
ISBN: 978-1-927309-46-9
Photocopying Prohibited

PRACTICES

KNOW

16 Which Graph to Use?

Key Idea: The type of graph you choose to display your data depends on the type of data you have collected.

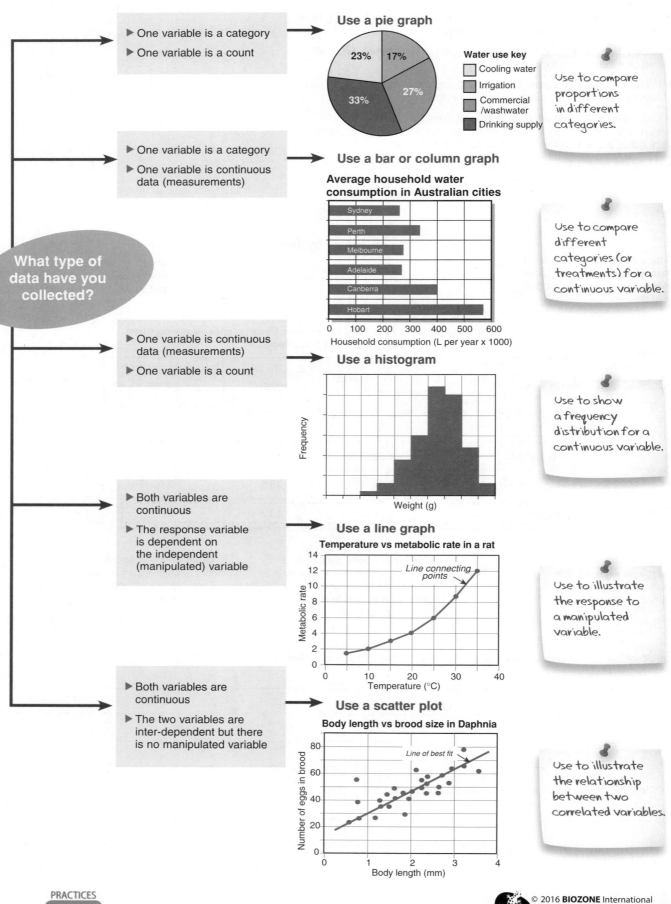

What type of data have you collected?

- ▶ One variable is a category
- ▶ One variable is a count

Use a pie graph

23% 17% 27% 33%

Water use key
- Cooling water
- Irrigation
- Commercial /washwater
- Drinking supply

Use to compare proportions in different categories.

- ▶ One variable is a category
- ▶ One variable is continuous data (measurements)

Use a bar or column graph

Average household water consumption in Australian cities

Sydney, Perth, Melbourne, Adelaide, Canberra, Hobart

0 100 200 300 400 500 600
Household consumption (L per year x 1000)

Use to compare different categories (or treatments) for a continuous variable.

- ▶ One variable is continuous data (measurements)
- ▶ One variable is a count

Use a histogram

Frequency
Weight (g)

Use to show a frequency distribution for a continuous variable.

- ▶ Both variables are continuous
- ▶ The response variable is dependent on the independent (manipulated) variable

Use a line graph

Temperature vs metabolic rate in a rat

Line connecting points

Metabolic rate
0 2 4 6 8 10 12 14
Temperature (°C)
0 10 20 30 40

Use to illustrate the response to a manipulated variable.

- ▶ Both variables are continuous
- ▶ The two variables are inter-dependent but there is no manipulated variable

Use a scatter plot

Body length vs brood size in Daphnia

Line of best fit

Number of eggs in brood
0 20 40 60 80
Body length (mm)
0 1 2 3 4

Use to illustrate the relationship between two correlated variables.

PRACTICES

KNOW

© 2016 **BIOZONE** International
ISBN: 978-1-927309-46-9
Photocopying Prohibited

17 Drawing Line Graphs

Key Idea: Line graphs are used to plot continuous data when one variable (the independent variable) affects another, the dependent variable.

Graphs provide a way to visually see data trends. **Line graphs** are used when one variable (the independent variable) affects another, the dependent variable. Important features of line graphs are:

▶ The data must be continuous for both variables.

▶ The dependent variable is usually a biological response.

▶ The independent variable is often time or the experimental treatment.

▶ The relationship between two variables can be represented as a continuum and the data points are plotted accurately and connected directly (point to point).

▶ Line graphs may be drawn with measure of error. The data are presented as points (the calculated means), with bars above and below, indicating a measure of variability or spread in the data (e.g. standard deviation).

▶ More than one curve can be plotted per set of axes. If the two data sets use the same measurement units and a similar range of values for the dependent variable, one scale on the y axis is used. If the two data sets use different units and/or have a very different range of values for the dependent variable, two scales for the y axis are used (see right). Distinguish between the two curves with a key.

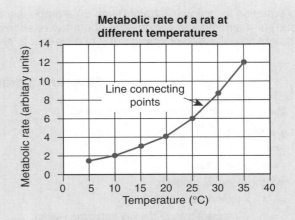

Metabolic rate of a rat at different temperatures

Line connecting points

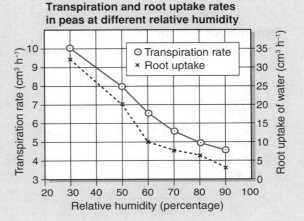

Transpiration and root uptake rates in peas at different relative humidity

○ Transpiration rate
× Root uptake

1. The results (shown right) were collected in a study investigating the effect of temperature on the activity of an enzyme.

   (a) Using the results provided, plot a line graph on the grid below:

   (b) Estimate the rate of reaction at 15°C: _____

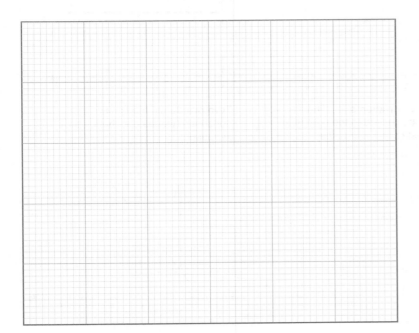

Lab Notebook

An enzyme's activity at different temperatures

Temperature (°C)	Rate of reaction (mg of product formed per minute)
10	1.0
20	2.1
30	3.2
35	3.7
40	4.1
45	3.7
50	2.7
60	0

PRACTICES

DATA

18 Interpreting Line Graphs

Key Idea: The equation for a straight line is y = mx + c. A line may have a positive, negative, or zero slope.

The equation for a linear (straight) line on a graph is y = mx + c. The equation can be used to calculate the gradient (slope) of a straight line and tells us about the relationship between x and y (how fast y is changing relative to x). For a straight line, the rate of change of y relative to x is always constant.

Measuring gradients and intercepts

The equation for a straight line is written as:

y = mx + c

Where :

y = the y-axis value

m = the slope (or gradient)

x = the x-axis value

c = the y intercept (where the line cross the y-axis).

Determining "m" and "c"

To find "c" just find where the line crosses the y-axis.

To find m:

1. Choose any two points on the line.
2. Draw a right-angled triangle between the two points on the line.
3. Use the scale on each axis to find the triangle's vertical length and horizontal length.
4. Calculate the gradient of the line using the following equation:

$$\frac{\text{change in y}}{\text{change in x}}$$

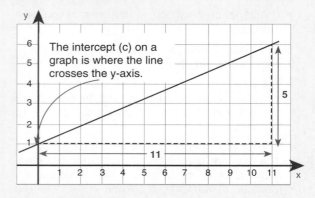

The intercept (c) on a graph is where the line crosses the y-axis.

For the example above:

c = 1

m = 0.45 (5 ÷11)

Once c and m have been determined you can choose any value for x and find the corresponding value for y.

For example, when x = 9, the equation would be:

y = 9 x 0.45 + 1

y = 5.05

A line may have a positive, negative, or zero slope

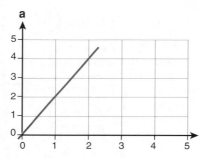
a

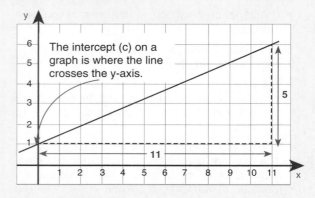
b

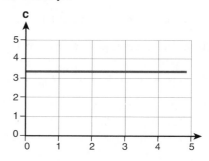
c

Positive gradients: the line slopes upward to the right (y is increasing as x increases).

Negative gradients: the line slopes downward to the right (y is decreasing as x increases).

Zero gradients: the line is horizontal (y does not change as x increases).

1. For the graph (right):

   (a) Identify the value of c: _____

   (b) Calculate the value of m: _____

   (c) Determine y if x = 2: _____

   _____

   (d) Describe the slope of the line: _____

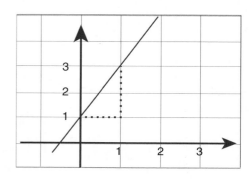

PRACTICES

DATA

© 2016 **BIOZONE** International
ISBN: 978-1-927309-46-9
Photocopying Prohibited

19 Drawing Scatter Graphs

Key Idea: Scatter graphs are used to plot continuous data where there is a relationship between two interdependent variables.

Scatter graphs are used to display continuous data where there is a relationship between two interdependent variables.

▶ The data must be continuous for both variables.

▶ There is no independent (manipulated) variable, but the variables are often correlated, i.e. they vary together in some predictable way.

▶ Scatter graphs are useful for determining the relationship between two variables.

▶ The points on the graph should not be connected, but a line of best fit is often drawn through the points to show the relationship between the variables.

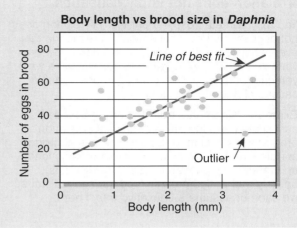

Body length vs brood size in *Daphnia*

1. In the example below, metabolic measurements were taken from seven Antarctic fish *Pagothenia borchgrevinski*. The fish are affected by a gill disease, which increases the thickness of the gas exchange surfaces and affects oxygen uptake. The results of oxygen consumption of fish with varying amounts of affected gill (at rest and swimming) are tabulated below.

(a) Plot the data on the grid (bottom right) to show the relationship between oxygen consumption and the amount of gill affected by disease. Use different symbols or colors for each set of data (at rest and swimming), and use only one scale for oxygen consumption.

(b) Draw a line of best fit through each set of points.

2. Describe the relationship between the amount of gill affected and oxygen consumption in the fish:

(a) For the at rest data set:

_____

_____

_____

_____

_____

_____

(b) For the swimming data set:

_____

_____

_____

_____

_____

_____

_____

_____

Oxygen consumption of fish with affected gills

Fish number	Percentage of gill affected	Oxygen consumption (cm³ g⁻¹ h⁻¹)	
		At rest	Swimming
1	0	0.05	0.29
2	95	0.04	0.11
3	60	0.04	0.14
4	30	0.05	0.22
5	90	0.05	0.08
6	65	0.04	0.18
7	45	0.04	0.20

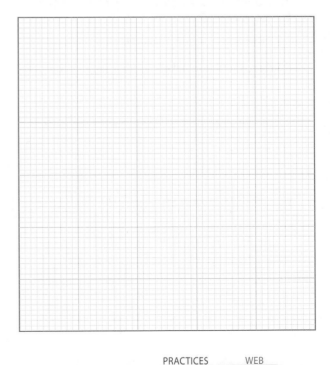

PRACTICES   WEB

19   **DATA**

20 Correlation or Causation

Key Idea: A correlation is a mutual relationship or association between two or more variables. A correlation between two variables does not imply that one causes change in the other.

Correlation does not imply causation

▶ You may come across the phrase "correlation does not necessarily imply causation". This means that even when there is a strong correlation between variables (they vary together in a predictable way), you cannot assume that change in one variable caused change in the other.

▶ **Example**: When data from the organic food association and the office of special education programmes is plotted (below), there is a strong correlation between the increase in organic food and rates of diagnosed autism. However it is unlikely that eating organic food causes autism, so we can not assume a causative effect here.

Drawing the line of best fit

Some simple guidelines need to be followed when drawing a line of best fit on your scatter plot.

▶ Your line should follow the trend of the data points.

▶ Roughly half of your data points should be above the line of best fit, and half below.

▶ The line of best fit does not necessarily pass through any particular point.

▶ A line of best fit should pivot around the point representing the mean of the x and y variables.

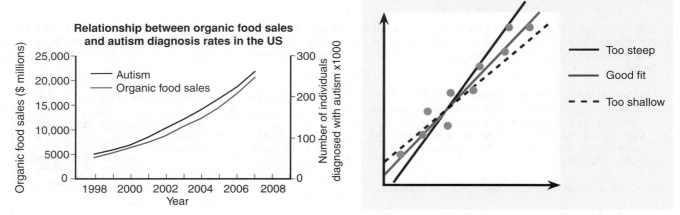

1. What does the phrase "correlation does not imply causation" mean? _____

_____

_____

2. A student measured the hand span and foot length measurements of 21 adults and plotted the data as a scatter graph (right).

    (a) Draw a line of best fit through the data:

    (b) Describe the results: _____

    _____

    _____

    _____

    _____

    (c) Using your line of best fit as a guide, comment on the correlation between hand span and foot length:

    _____

    _____

    _____

    _____

WEB   CCC   PRACTICES   PRACTICES

© 2016 **BIOZONE** International
ISBN: 978-1-927309-46-9
Photocopying Prohibited

21 Drawing Bar Graphs

Key Idea: Bar graphs are used to plot data that is non-numerical or discrete for at least one variable.

Bar graphs are appropriate for data that is non-numerical and discrete for at least one variable.

► There are no dependent or independent variables.

► Data is collected for discontinuous, non-numerical categories (e.g. place, color, and species), so the bars do not touch.

► Multiple sets of data can be displayed side by side for direct comparison.

► Axes may be reversed, i.e. the bars can be vertical or horizontal. When they are vertical, these graphs are called column graphs.

Size of woodlands in Britain

Woodland	Area
Cwm Clydach	20
Burnham Beeches	450
Scords Wood	350
Wyre Forest	500
Yarner Wood	400
Wistmans Wood	4

Area of woodland (Hectares): 0, 100, 200, 300, 400, 500, 600

1. Counts of eight mollusk species were made from a series of quadrat samples at two sites on a rocky shore. The summary data are presented on the right.

   (a) Tabulate the mean (average) numbers per square meter at each site in the table (below).

   (b) Plot a bar graph of the tabulated data on the grid below. For each species, plot the data from both sites side by side using different colors to distinguish the two sites.

Average abundance of 8 mollusk species from two sites along a rocky shore

Species	Mean (no. m^{-2})	
	Site 1	Site 2

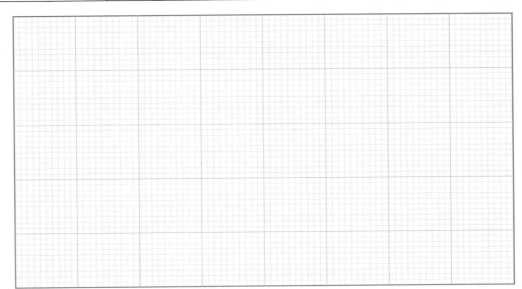

Field data notebook

Total counts at site 1 (11 quadrats) and site 2 (10 quadrats). Quadrats 1 sq m.

	Site 1		Site 2	
	No (m^{-2})		No (m^{-2})	
Species	Total	Mean	Total	Mean
Ornate limpet	232	21	299	30
Radiate limpet	68	6	344	34
Limpet sp. A	420	38	0	0
Cats-eye	68	6	16	2
Top shell	16	2	43	4
Limpet sp. B	628	57	389	39
Limpet sp. C	0	0	22	2
Chiton	12	1	30	3

© 2016 **BIOZONE** International
ISBN: 978-1-927309-46-9
Photocopying Prohibited

PRACTICES

 DATA

22 Drawing Histograms

Key Idea: Histograms graphically show the frequency distribution of continuous data.

Histograms are plots of continuous data and are often used to represent frequency distributions, where the y-axis shows the number of times a particular measurement or value was obtained. For this reason, they are often called frequency histograms. Important features of histograms include:

▶ The data are numerical and continuous (e.g. height or weight), so the bars touch.

▶ The x-axis usually records the class interval. The y-axis usually records the number of individuals in each class interval (frequency).

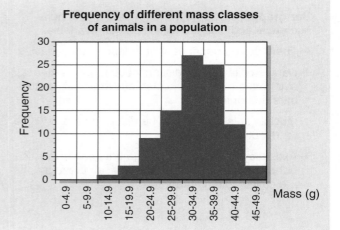

Frequency of different mass classes of animals in a population

1. The weight data provided below were recorded from 95 individuals (male and female), older than 17 years.
   (a) Create a tally chart (frequency table) in the table provided (right). An example of the tally for the weight grouping 55-59.9 kg has been completed for you. Note that the raw data values, once they are recorded as counts on the tally chart, are crossed off the data set in the notebook. It is important to do this in order to prevent data entry errors.

   (b) Plot a frequency histogram of the tallied data on the grid below.

Weight (kg)	Tally	Total
45 – 49.9		
50 – 54.9		
55 – 59.9	Ⅲⅼ Ⅱ	7
60 – 64.9		
65 – 69.9		
70 – 74.9		
75 – 79.9		
80 – 84.9		
85 – 89.9		
90 – 94.9		
95 – 99.9		
100 – 104.9		
105 – 109.9		

Lab notebook

Weight (in kg) of 95 individuals

63.4	81.2	65
56.5	83.3	75.6
84	95	76.8
81.5	105.5	67.8
73.4	82	68.3
56	73.5	63.5
60.4	75.2	58
83.5	63	58.5
82	70.4	50
61	82.2	92
55.2	87.8	91.5
48	86.5	88.3
53.5	85.5	81
63.8	87	72
69	98	66.5
82.8	71	61.5
68.5	76	66
67.2	72.5	65.5
82.5	61	67.4
83	60.5	73
78.4	67	67
76.5	86	71
83.4	85	70.5
77.5	93.5	65.5
77	62	68
87	62.5	90
89	63	83.5
93.4	60	73
83	71.5	66
80	73.8	57.5
76	77.5	76
56	74	

23 Mean, Median, and Mode

Key Idea: Descriptive statistics are used to summarize a data set and describe its basic features. The type of statistic calculated depends on the type of data and its distribution.

Descriptive statistics

▶ When we describe a set of data, it is usual to give a measure of **central tendency**. This is a single value identifying the central position within that set of data.

▶ **Descriptive statistics**, such as mean, median, and mode, are all valid measures of central tendency depending of the type of data and its distribution. They help to summarize features of the data, so are often called summary statistics.

▶ The appropriate statistic for different types of data variables and their distributions is described below.

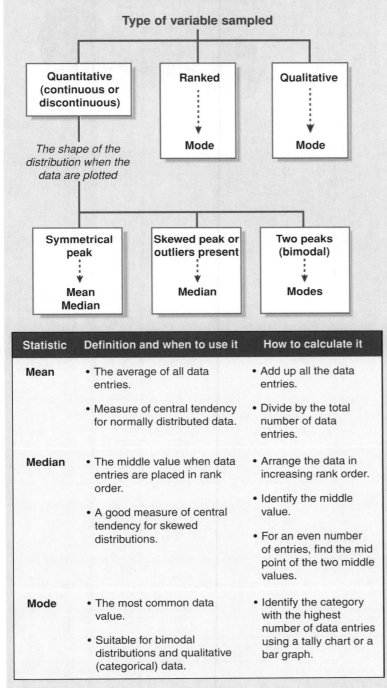

Type of variable sampled

- Quantitative (continuous or discontinuous)
- Ranked → Mode
- Qualitative → Mode

The shape of the distribution when the data are plotted

- Symmetrical peak → Mean / Median
- Skewed peak or outliers present → Median
- Two peaks (bimodal) → Modes

Statistic	Definition and when to use it	How to calculate it
Mean	• The average of all data entries. • Measure of central tendency for normally distributed data.	• Add up all the data entries. • Divide by the total number of data entries.
Median	• The middle value when data entries are placed in rank order. • A good measure of central tendency for skewed distributions.	• Arrange the data in increasing rank order. • Identify the middle value. • For an even number of entries, find the mid point of the two middle values.
Mode	• The most common data value. • Suitable for bimodal distributions and qualitative (categorical) data.	• Identify the category with the highest number of data entries using a tally chart or a bar graph.

Distribution of data

Variability in continuous data is often displayed as a frequency distribution. There are several types of distribution.

- Normal distribution (A): Data has a symmetrical spread about the mean. It has a classical bell shape when plotted.
- Skewed data (B): Data is not centered around the middle but has a "tail" to the left or right.
- Bimodal data (C): Data which has two peaks.

The shape of the distribution will determine which statistic (mean, median, or mode) should be used to describe the central tendency of the sample data.

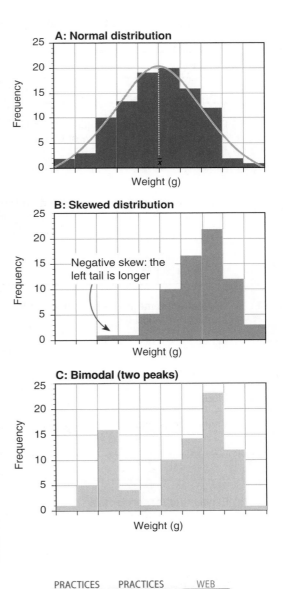

A: Normal distribution

B: Skewed distribution

Negative skew: the left tail is longer

C: Bimodal (two peaks)

© 2016 **BIOZONE** International
ISBN: 978-1-927309-46-9
Photocopying Prohibited

PRACTICES PRACTICES WEB

23 **DATA**

1. The birth weights of 60 newborn babies are provided right. Create a tally chart (frequency table) of the weights in the table provided below. Choose an appropriate grouping of weights.

Weight (kg)	Tally	Total

Birth weights (kg)

3.740	3.380	4.510	3.135	3.260
3.830	2.660	3.800	3.090	3.430
3.530	3.375	4.170	3.830	3.510
3.095	3.840	4.400	3.970	3.230
3.630	3.630	3.770	3.840	3.570
1.560	3.810	3.400	4.710	3.620
3.910	2.640	3.825	4.050	3.260
4.180	3.955	3.130	4.560	3.315
3.570	2.980	3.400	3.350	3.230
2.660	3.350	3.260	3.380	3.790
3.150	3.780	4.100	3.690	2.620
3.400	3.260	3.220	1.495	3.030

NEED HELP?
See Activity 22

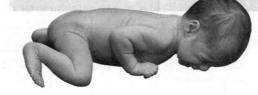

2. (a) On the graph paper (right) draw a frequency histogram for the birth weight data.

   (b) What type of distribution does the data have?

   _____

   _____

   _____

   (c) Predict whether mean, median, or mode would be the best measure of central tendency for the data:

   _____

   (d) Explain your reason for your answer in (c):

   _____

   _____

   _____

   (e) Calculate the mean, median, and mode for the birth weight data:

   Mean: _____

   Median: _____

   Mode: _____

   (f) What do you notice about the results in (e)? _____

   _____

   _____

   (g) Explain the reason for this: _____

   _____

   _____

24 What is Standard Deviation?

Key Idea: Standard deviation measures the variability (spread) in a set of data. It can be used to evaluate how reliably the mean represents the data.

▶ While it is important to know the mean of a data set, it is also important to know how well the mean represents the data set as a whole. This is evaluated using a simple measure of the spread in the data called **standard deviation**.

▶ In general, if the standard deviation is small, the mean will more accurately represent the data than if it is large.

Standard deviation

▶ Standard deviation is usually presented as $\bar{x} \pm s$. In normally distributed data, 68% of all data values will lie within one standard deviation (s) of the mean ($\bar{x}$) and 95% of all data values will lie within two standard deviations of the mean (right).

▶ Different sets of data can have the same mean and range, yet a different data distribution. In both the data sets below, 68% of the values lie within the range $\bar{x} \pm 1s$ and 95% of the values lie within $\bar{x} \pm 2s$. However, in B, the data values are more tightly clustered around the mean.

▶ Standard deviation is easily calculated using an spreadsheet. Data should be entered as columns. In a free cell, type the formula for standard deviation (this varies depending on the program) and select the cells containing the data values, enclosing them in parentheses.

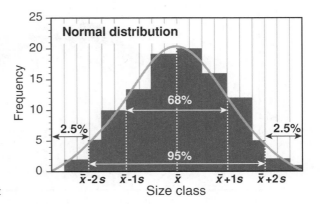

Normal distribution

Frequency / Size class / $\bar{x}$-2s $\bar{x}$-1s $\bar{x}$ $\bar{x}$+1s $\bar{x}$+2s / 68% / 95% / 2.5% / 2.5%

Histogram A has a larger standard deviation; the values are spread widely around the mean.

Both plots show a normal distribution with a symmetrical spread of values about the mean.

Histogram B has a smaller standard deviation; the values are clustered more tightly around the mean.

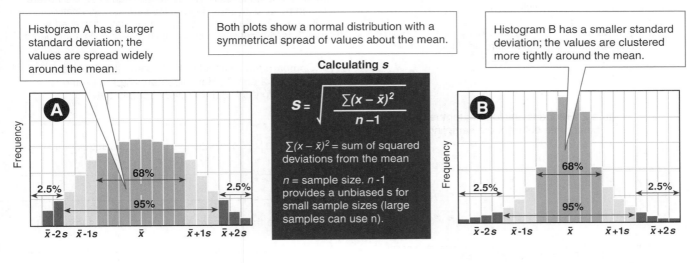

Calculating s

$$s = \sqrt{\frac{\sum(x - \bar{x})^2}{n-1}}$$

$\sum(x - \bar{x})^2$ = sum of squared deviations from the mean

n = sample size. $n-1$ provides a unbiased s for small sample sizes (large samples can use n).

1. Two data sets have the same mean. The first data set has a much larger standard deviation than the second data set. What does this tell you about the spread of data around the mean in each case? Which data set is most reliable?

_____

_____

_____

2. The data on the right shows the heights for 29 male swimmers.

(a) Calculate the mean for the data: _____

(b) Use manual calculation, a calculator, or a spreadsheet to calculate the standard deviation (s) for the data:

_____

(c) State the mean ± 1s: _____

(d) What percentage of values are within 1s of the mean? _____

(e) What does this tell you about the spread of the data? _____

_____

Raw data: Height (cm)					
178	177	188	176	186	175
180	181	178	178	176	175
180	185	185	175	189	174
178	186	176	185	177	176
176	188	180	186	177	

PRACTICES   PRACTICES   WEB

24 **KNOW**

25 Detecting Bias in Samples

Key Idea: Bias refers to the selection for or against one or more particular groups in such a way as to influence the findings of an investigation. Sampling method can affect the results of a study.

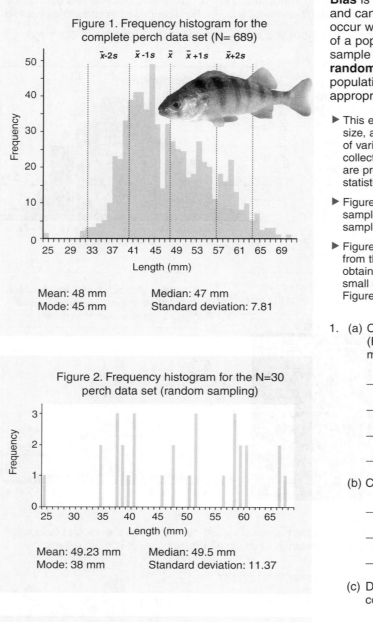

Figure 1. Frequency histogram for the complete perch data set (N= 689)

Mean: 48 mm   Median: 47 mm
Mode: 45 mm   Standard deviation: 7.81

Figure 2. Frequency histogram for the N=30 perch data set (random sampling)

Mean: 49.23 mm   Median: 49.5 mm
Mode: 38 mm   Standard deviation: 11.37

Figure 3. Frequency histogram for the N=50 perch data set (biased sampling)

Mean: 61.44 mm   Median: 63 mm
Mode: 64 mm   Standard deviation: 3.82

Bias is the selection for or against one particular group and can influence the findings of an investigation. Bias can occur when sampling is not random and certain members of a population are under- or overrepresented. Small sample sizes can also bias results. Bias can be reduced by **random sampling** (sampling in which all members of the population have an equal chance of being selected). Using appropriate collection methods will also reduce bias.

▶ This exercise illustrates how random sampling, large sample size, and sampling bias affect our statistical assessment of variation in a population. In this exercise, perch were collected and their body lengths (mm) were measured. Data are presented as a frequency histogram and with descriptive statistics (mean, median, mode and standard deviation).

▶ Figure 1 shows the results for the complete data set. The sample set was large (N= 689) and the perch were randomly sampled. The data are close to having a normal distribution.

▶ Figures 2 and 3 show results for two smaller sample sets drawn from the same population. The data collected in Figure 2 were obtained by random sampling but the sample was relatively small (N = 30). The person gathering the data displayed in Figure 3 used a net with a large mesh size to collect the perch.

1. (a) Compare the results for the two small data sets (Figures 2 and 3). How close are the mean and median to each other in each sample set?

_____

_____

_____

_____

(b) Compare the standard deviation for each sample set:

_____

_____

_____

(c) Describe how each of the smaller sample sets compares to the large sample set (Figure 1):

_____

_____

_____

_____

(d) Why do you think the two smaller sample sets look so different to each other?

_____

_____

_____

_____

 © 2016 **BIOZONE** International
ISBN: 978-1-927309-46-9
Photocopying Prohibited

26 Biological Drawings

Key Idea: Good biological drawings provide an accurate record of the specimen you are studying and enable you to make a record of its important features.

▶ Drawing is a very important skill to have in biology. Drawings record what a specimen looks like and give you an opportunity to record its important features. Often drawing something will help you remember its features at a later date (e.g. in a test).

▶ Biological drawings require you to pay attention to detail. It is very important that you draw what you actually see, and not what you think you should see.

▶ Biological drawings should include as much detail as you need to distinguish different structures and types of tissue, but avoid unnecessary detail which can make your drawing confusing.

▶ Attention should be given to the symmetry and proportions of your specimen. Accurate labeling, a statement of magnification or scale, the view (section type), and type of stain used (if applicable) should all be noted on your drawing.

▶ Some key points for making good biological drawing are described on the example below. The drawing of *Drosophila* (right) is well executed but lacks the information required to make it a good biological drawing.

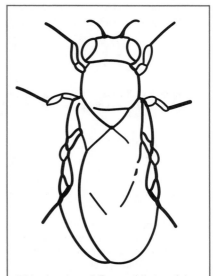

This drawing of *Drosophila* is a fair representation of the animal, but has no labels, title, or scale.

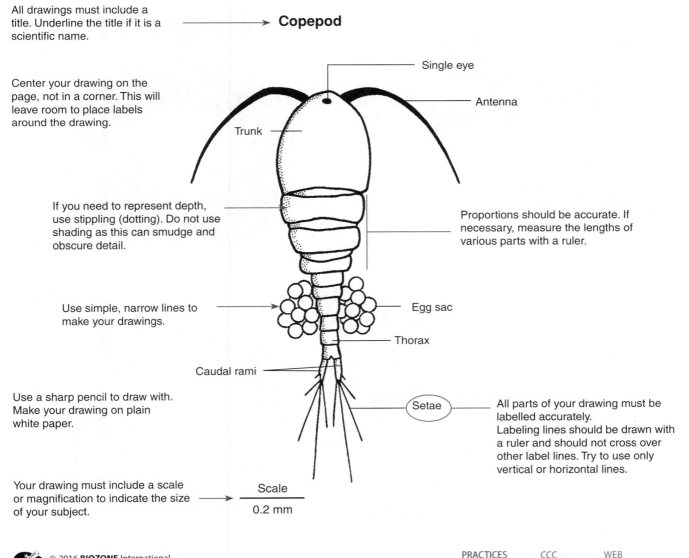

All drawings must include a title. Underline the title if it is a scientific name.

Copepod

Center your drawing on the page, not in a corner. This will leave room to place labels around the drawing.

Single eye

Antenna

Trunk

If you need to represent depth, use stippling (dotting). Do not use shading as this can smudge and obscure detail.

Proportions should be accurate. If necessary, measure the lengths of various parts with a ruler.

Use simple, narrow lines to make your drawings.

Egg sac

Thorax

Caudal rami

Use a sharp pencil to draw with. Make your drawing on plain white paper.

Setae

All parts of your drawing must be labelled accurately.
Labeling lines should be drawn with a ruler and should not cross over other label lines. Try to use only vertical or horizontal lines.

Your drawing must include a scale or magnification to indicate the size of your subject.

Scale

0.2 mm

PRACTICES   CCC   WEB

SSM   26   **REFER**

Annotated diagrams

An annotated diagram is a diagram that includes a series of explanatory notes. These provide important or useful information about your subject.

Transverse section through collenchyma of _Helianthus_ stem. Magnification x 450

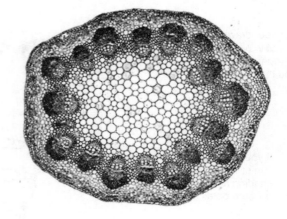

Primary wall with secondary thickening.

Cytoplasm
A watery solution containing dissolved substances, enzymes, and the cell organelles.

Nucleus
A large, visible organelle. It contains most of the cell's DNA.

Chloroplast
These are specialized plastids containing the green pigment chlorophyll. Photosynthesis occurs here.

Vacuole containing cell sap.

Plan diagrams

Plan diagrams are drawings made of samples viewed under a microscope at low or medium power. They are used to show the distribution of the different tissue types in a sample without any cellular detail. The tissues are identified, but no detail about the cells within them is included.

The example here shows a plan diagram produced after viewing a light micrograph of a transverse section through a dicot stem.

Light micrograph of a transverse section through a dicot stem.

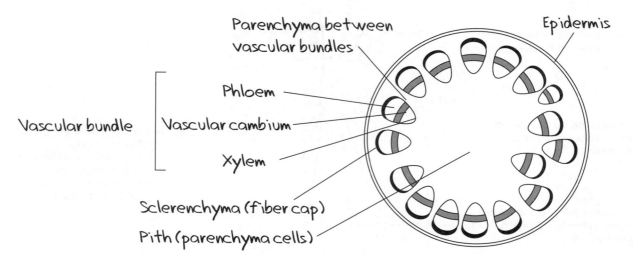

Parenchyma between vascular bundles

Epidermis

Phloem

Vascular bundle

Vascular cambium

Xylem

Sclerenchyma (fiber cap)

Pith (parenchyma cells)

27 Practicing Biological Drawings

Key Idea: Attention to detail is vital when making accurate and useful biological drawings.

Above: Use relaxed viewing when drawing at the microscope. Use one eye (the left for right handers) to view and the right eye to look at your drawing.

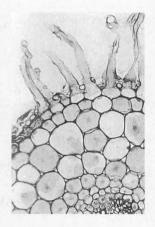

Above: Light micrograph Transverse section (TS) through a *Ranunculus* root. Right: A biological drawing of the same section.

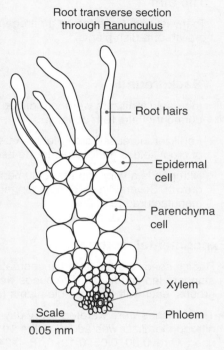

Root transverse section through <u>Ranunculus</u>

- Root hairs
- Epidermal cell
- Parenchyma cell
- Xylem
- Phloem

Scale
0.05 mm

1. The image below is a labelled photomicrograph (x50) showing a partial section through a dicot root. Use this image to construct a plan diagram:

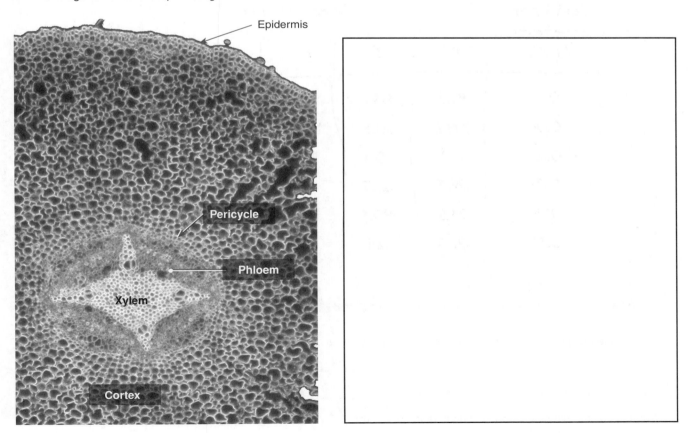

- Epidermis
- Pericycle
- Phloem
- Xylem
- Cortex

28 Test Your Understanding

Key Idea: Analyzing experimental data relating to the effect of fertilizer on the growth of radishes provides a way to test your understanding of data collection, handling, and analysis.

The aim

To investigate the effect of a nitrogen fertilizer on the growth of radish plants.

Radishes

Background

Inorganic fertilizers were introduced to crop farming during the late 19th century.

Fertilizer increased crop yields. An estimated 50% of crop yield is attributable to the use of fertilizer.

Nitrogen is a very important element for plant growth. Several types of nitrogen fertilizers are manufactured (e.g. urea).

Hypothesis

If plants need nitrogen to grow, radish growth will increase with increasing nitrogen concentration.

Experimental method

▶ Radish seeds were planted in separate identical pots (5 cm x 5 cm wide x 10 cm deep) and kept together in standard lab conditions. The seeds were planted into a commercial soil mixture and divided randomly into six groups, each with five sample plants (a total of 30 plants in six treatments).

▶ The radishes were watered every day at 10 am and 3 pm with 500 mL per treatment per watering. Water soluble nitrogen fertilizer was added to the 10 am watering on the 1st, 11th, and 21st days. The fertilizer concentrations used were: 0.00, 0.06, 0.12, 0.18, 0.24, and 0.30 g L-1 and each treatment received a different concentration.

▶ The plants were grown for 30 days before being removed from the pots, washed, and the radish root weighed. The results are presented below.

Fertilizer concentration (g L^{-1})	Sample number				
	1	2	3	4	5
0	80.1	83.2	82.0	79.1	84.1
0.06	109.2	110.3	108.2	107.9	110.7
0.12	117.9	118.9	118.3	119.1	117.2
0.18	128.3	127.3	127.7	126.8	DNG*
0.24	23.6	140.3	139.6	137.9	141.1
0.30	122.3	121.1	122.6	121.3	123.1

*DNG = did not germinate

† Based on data from M S Jilani, *et al* Journal Agricultural Research

1. Identify the independent variable for the experiment and its range: _____

   _____

2. Identify the dependent variable for the experiment: _____

3. What is the sample size for each concentration of fertilizer?

4. (a) One of the radishes recorded in the table on the previous page did not grow as expected and produced an extreme value. Record the outlying value here:

_____

(b) Why should this value not be included in future calculations? _____

_____

_____

5. Use Table 1 below to record the raw data from the experiment. You will need to include column and row headings and a title, and complete some simple calculations. Some headings have been entered for you.

Table 1: -

- -

	Mass of radish root (g)					Total mass	Mean mass

6. The students decided to collect more data by counting the number of leaves on each radish plant at day 30. The data are presented in Table 2.
   Use the space below to calculate the mean, median and mode for the leaf data. Add these data to table 2.

Table 2: Number of leaves on radish plants under six different fertilizer concentrations.

Fertilizer concentration (g L⁻¹)	Number of leaves at day 30							
	Sample (n)					Mean	Median	Mode
	1	2	3	4	5			
0	9	9	10	8	7			
0.06	15	16	15	16	16			
0.12	16	17	17	17	16			
0.18	18	18	19	18	DNG*			
0.24	6	19	19	18	18			
0.30	18	17	18	19	19			

* DNG: Did not germinate

7. Use the grid below to draw a line graph of the experimental results. Plot your calculated mean mass data from Table 1, and remember to include a title and correctly labelled axes.

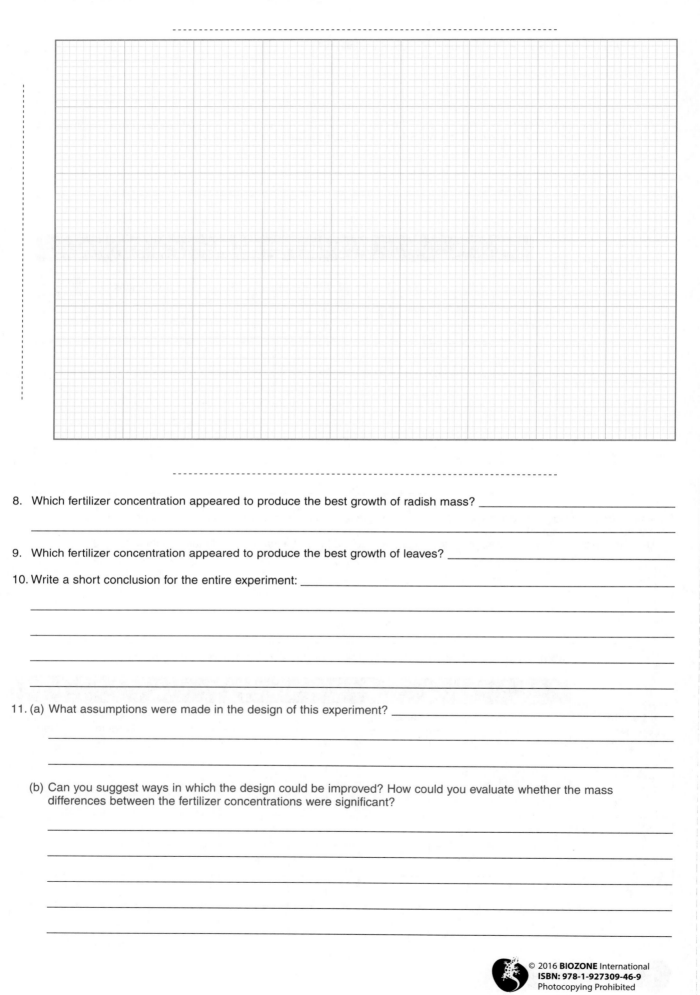

8. Which fertilizer concentration appeared to produce the best growth of radish mass? _____

_____

9. Which fertilizer concentration appeared to produce the best growth of leaves? _____

10. Write a short conclusion for the entire experiment: _____

_____

_____

_____

_____

11. (a) What assumptions were made in the design of this experiment? _____

_____

_____

(b) Can you suggest ways in which the design could be improved? How could you evaluate whether the mass differences between the fertilizer concentrations were significant?

_____

_____

_____

_____

_____

© 2016 **BIOZONE** International
ISBN: 978-1-927309-46-9
Photocopying Prohibited

From Molecules to Organisms: Structures and Processes

Concepts and connections
Use arrows to make your own connections between related concepts in this section of this book.

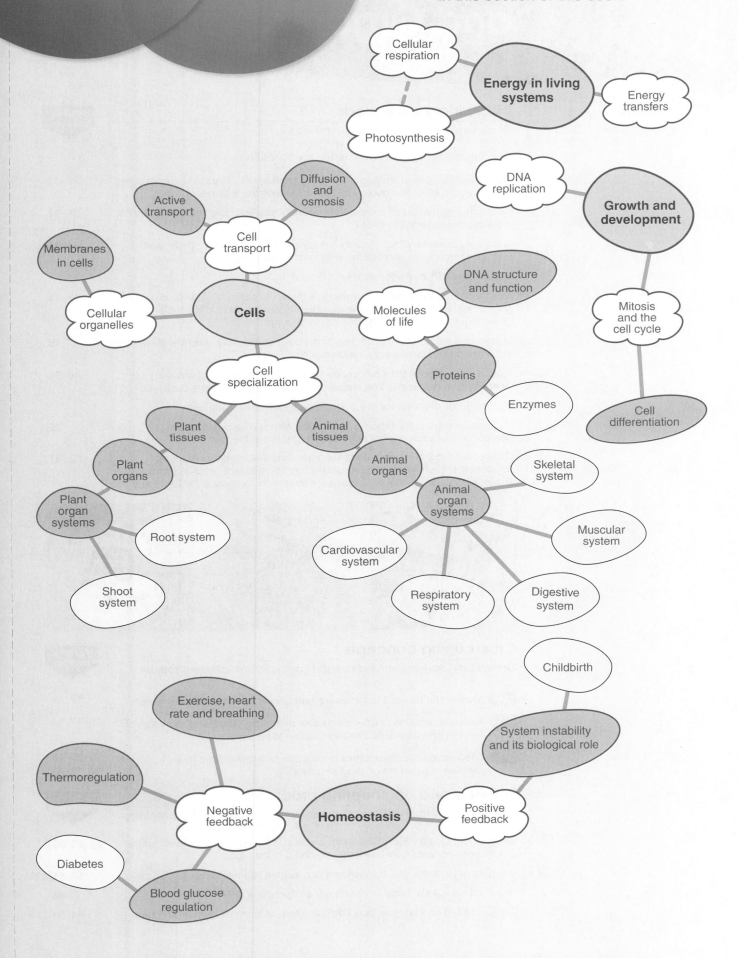

Cell Specialization and Organization

Key terms

base-pairing rule

catalyst

cell

DNA

enzyme

eukaryotic cell

gene

nucleotide

organelle

organ system

polynucleotide

prokaryotic cell

protein

specialized cell

Disciplinary core ideas

Show understanding of these core ideas

Activity number

Specialized cells provide essential life functions

☐ 1 There are two types of cells: prokaryotic and eukaryotic. Plant and animal cells are eukaryotic cells. Microscopy allows us to study the features of cells.
30-33

☐ 2 Cells contain specialized components called organelles. Each organelle carries out a specialized role in the cell.
34-42

☐ 3 Many cells are specialized to carry out specific roles. The size, shape, and number of organelles in a cell depends on the cell's role.
43 44

Genes on DNA code for proteins

☐ 4 A cell's genetic information is carried in a molecule called DNA. DNA is a polynucleotide; it is made up of many smaller components, called nucleotides, joined together. DNA has a double-helix structure.
45-48

☐ 5 Nucleotides pair together according to the base-pairing rule. Adenine always pairs with thymine and cytosine always pairs with guanine.
47 48

☐ 6 A gene is a region of DNA that codes for a specific protein. Proteins carry out most of the work in a cell. The shape of a protein helps it to carry out its job.
49-56

Multicellular organisms are organized in a hierarchical way

☐ 7 Multicellular organisms exhibit a hierarchical structure of organization. Components at each level of organization are part of the next level.
29

☐ 8 Multicellular organisms have complex organ systems, made up of many components. Each system has a specific role, but different organ systems interact and work together so that the organism can carry out essential life functions.
57-61

Crosscutting concepts

Understand how these fundamental concepts link different topics

Activity number

☐ 1 **SSM** Models can be used to represent interactions in a physiological system.
57-61

☐ 2 **SF** The determination of DNA's structure required a detailed examination of its components and enabled its function to be understood.
46-52 54

☐ 3 **SF** The functions and properties of cells can be inferred from their components and their overall structure.
32-35 38 42 43 44

Science and engineering practices

Demonstrate competence in these science and engineering practices

Activity number

☐ 1 Use a model to show that multicellular organisms have a hierarchical structure and that components from one level contribute to the next.
29 32 33

☐ 2 Use a model to show that components of a system interact to fulfil life functions.
57-61

☐ 3 Construct and use a model to illustrate the structure of DNA.
48

☐ 4 Explain, based on evidence, how DNA structure determines protein structure.
49 50

29 The Hierarchy of Life

Key Idea: The structural organization of multicellular organisms is hierarchical. Components at each level of organization are part of the next level.

All multicellular organisms are organized in a hierarchy of structural levels, where each level builds on the one below it. It is traditional to start with the simplest components (parts) and build from there. Higher levels of organization are more complex than lower levels.

Hierarchical organization enables **specialization** so that individual components perform a specific function or set of related functions. Specialization enables organisms to function more efficiently.

The diagram below explains this hierarchical organization for a human.

The cellular level

Cells are the basic structural and functional units of an organism. Cells are specialized to carry out specific functions, e.g. cardiac (heart) muscle cells (below).

DNA

1

Atoms and molecules

The chemical level

All the chemicals essential for maintaining life, e.g. water, ions, fats, carbohydrates, amino acids, proteins, and nucleic acids.

2

The organelle level

Molecules associate together to form the organelles and structural components of cells, e.g. the nucleus (above).

3

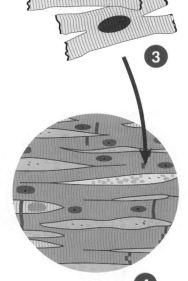

7 The organism

The cooperating organ systems make up the organism, e.g. a human.

The tissue level   **4**

Groups of cells with related functions form tissues, e.g. cardiac (heart) muscle (above). The cells of tissue often have a similar origin.

6 The system level

Groups of organs with a common function form an organ system, e.g. cardiovascular system (right).

5 The organ level

An organ is made up of two or more types of tissues to carry out a particular function. Organs have a definite form and structure, e.g. heart (left).

© 2016 **BIOZONE** International
ISBN: 978-1-927309-46-9
Photocopying Prohibited

PRACTICES   WEB

29   **REFER**

30 Introduction to Cells

Key Idea: Cells are classified as either prokaryotic or eukaryotic. They are distinguished on the basis of their size and internal organization and complexity.

▶ The **cell** is the smallest unit of life. Cells are often called the building blocks of life.

▶ Cells are either prokaryotic cells or eukaryotic cells. Within each of these groups, cells may vary greatly in their size, shape, and functional role.

Prokaryotic cells

▶ Prokaryotic cells are bacterial cells.

▶ Prokaryotic cells lack a membrane-bound nucleus or any membrane-bound organelles.

▶ They are small (generally 0.5-10 μm) single cells (unicellular).

▶ They are relatively basic cells and have very little cellular organization (their DNA, ribosomes, and enzymes are free floating within the cell cytoplasm).

▶ Single, circular chromosome of naked DNA.

▶ Prokaryotes have a cell wall, but it is different to the cell walls that some eukaryotes have.

Eukaryotic cells

▶ Eukaryotic cells have a membrane-bound nucleus, and other membrane-bound organelles.

▶ Plant cells, animals cells, fungal cells, and protists are all eukaryotic cells.

▶ Eukaryotic cells are large (30-150 μm). They may exist as single cells or as part of a multicellular organism.

▶ Multiple linear chromosomes consisting of DNA and associated proteins.

▶ They are more complex than prokaryotic cells. They have more structure and internal organization.

Nuclear membrane absent. Single, naked chromosome is free in cytoplasm within a nucleoid region.

Simple cell structure (limited organization)

Membrane-bound organelles are absent

Nucleoid region (pale)

Peptidoglycan cell wall

A prokaryotic cell: *E.coli*

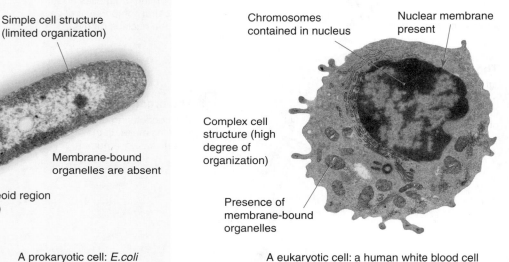

Chromosomes contained in nucleus

Nuclear membrane present

Complex cell structure (high degree of organization)

Presence of membrane-bound organelles

A eukaryotic cell: a human white blood cell

1. What are the main features of a prokaryotic cell? _____

_____

_____

2. (a) What are the main features of a eukaryotic cell? _____

    _____

    _____

  (b) Name examples of eukaryotic cells: _____

    _____

© 2016 **BIOZONE** International
ISBN: 978-1-927309-46-9
Photocopying Prohibited

31 Studying Cells

Key Idea: The light microscope is a valuable tool for examining cells and tissues in detail. Preparing and viewing slides correctly is an essential skill for practical studies of cells and tissues.

▶ The light (or optical) microscope (LM) is an important tool in biology. Light microscopy involves illuminating a sample and passing the light that is transmitted or reflected through lenses to give a magnified view of the sample.

▶ High power compound light microscopes use visible light and a combination of lenses to magnify objects up to several 100 times. Bright field microscopy (below) is the simplest and involves illuminating the specimen from below and viewing from above. Specimens must be thin and mostly transparent so that light can pass through. No detail will be seen in specimens that are thick or opaque.

▶ The wavelength of light limits the resolution of light microscopes to around 0.2 μm. Objects closer than this will not be distinguished as separate. Electron microscopes provide higher resolutions as they use a shorter wavelength electron beam, rather than light.

Structure of a typical compound light microscope

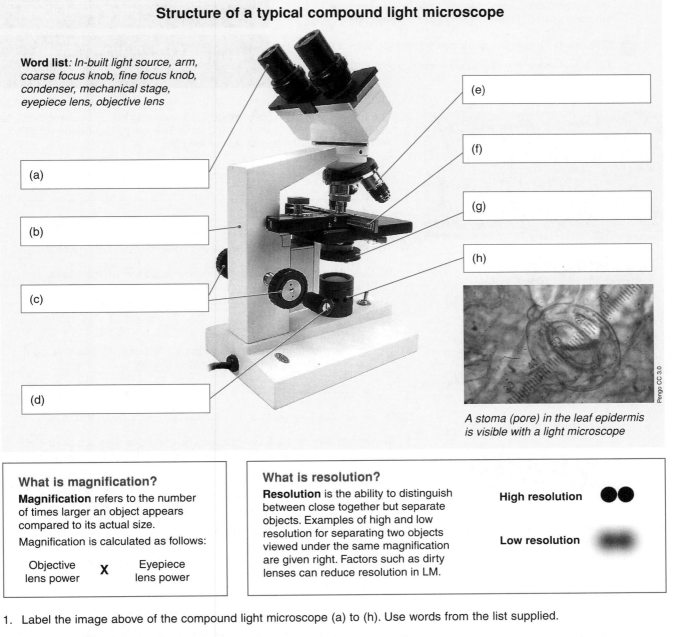

Word list: In-built light source, arm, coarse focus knob, fine focus knob, condenser, mechanical stage, eyepiece lens, objective lens

(a)

(b)

(c)

(d)

(e)

(f)

(g)

(h)

A stoma (pore) in the leaf epidermis is visible with a light microscope

Pengo CC 3.0

What is magnification?

Magnification refers to the number of times larger an object appears compared to its actual size.

Magnification is calculated as follows:

| Objective lens power | **X** | Eyepiece lens power |

What is resolution?

Resolution is the ability to distinguish between close together but separate objects. Examples of high and low resolution for separating two objects viewed under the same magnification are given right. Factors such as dirty lenses can reduce resolution in LM.

High resolution ●●

Low resolution

1. Label the image above of the compound light microscope (a) to (h). Use words from the list supplied.

2. Determine the magnification of a microscope using:

   (a) 15 X eyepiece and 40 X objective lens: _____

   (b) 10 X eyepiece and 60 X objective lens: _____

PRACTICES   WEB

31   **KNOW**

How do we calculate linear magnification?

▶ Magnification is how much larger an object appears compared to its actual size. It can be calculated from the ratio of image height to object height.

▶ If the ratio is greater than one, the image is enlarged. If it is less than one, it is reduced. In calculating magnification, all measurements should be in the same units.

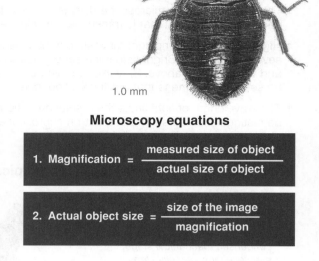

1.0 mm

Worked example

 1 Measure the body length of the bed bug image (right). Your measurement should be 40 mm (not including the body hairs and antennae).

2 Measure the length of the scale line marked 1.0 mm. You will find it is 10 mm long. The magnification of the scale line can be calculated using equation 1.

The magnification of the scale line is **10** (10 mm ÷ 1 mm)

The magnification of the image will also be X10 because the scale and image are magnified to the same degree.

3 Calculate the actual size of the bed bug using equation 2.

The actual size of the bed bug is **4 mm** (40 mm ÷ 10)

Microscopy equations

$$1. \ \text{Magnification} = \frac{\text{measured size of object}}{\text{actual size of object}}$$

$$2. \ \text{Actual object size} = \frac{\text{size of the image}}{\text{magnification}}$$

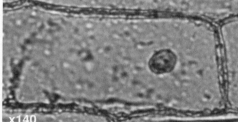
x140

3. The bright field microscopy image (left) shows an onion epidermal cell. Its measured length is 52,000 µm (52 mm). The image has been magnified 140 X. Calculate the actual size of the cell:

_____

_____

_____

4. The image of the flea (left) has been captured using light microscopy.

(a) Calculate the magnification using the scale line on the image:

_____

_____

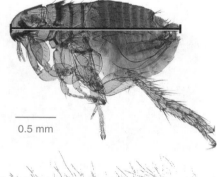
0.5 mm

(b) The body length of the flea is indicated by a line. Measure along the line and calculate the actual length of the flea:

_____

_____

5. The image size of the *E.coli* cell (left) is 43 mm, and its actual size is 2 µm. Using this information, calculate the magnification of the image:

_____

_____

6. Explain why a higher magnification is not particularly useful if the resolution is poor: _____

_____

_____

_____

7. When focusing a specimen it necessary to focus on the lowest magnification first, before switching to higher magnifications. Why do you think this is important?

_____

_____

32 Plant Cells

Key Idea: Plant cells are eukaryotic cells. They have many types of organelles, each of which has a specific role within the cell.

What is an organelle?

▶ The word organelle means "small organ". Therefore, organelles are the cell's "organs" and carry out the cell's work.

▶ Organelles represent one level of organization in a multicellular organism. One component (the cell) is made up of many smaller parts (organelles).

▶ Eukaryotic cells contain many different types of organelles. Each type of organelle has a specific role in the cell to help it function.

▶ Plant cells have several types of membrane-bound organelles called plastids. These make and store food and pigments. Some of the organelles found in a plant cell are shown below.

Features of a plant cell

▶ Plant cells are eukaryotic cells. Features that identify plant cells as eukaryotic cells include:

▶ A membrane-bound nucleus.

▶ Membrane-bound organelles (e.g. nucleus, mitochondria, endoplasmic reticulum).

▶ Features that can be used to identify a plant cell include the presence of:

• Cellulose cell wall

• Chloroplasts and other plastids

• Large vacuole (often centrally located)

A generalized plant cell

Chloroplast

A specialized plastid containing the green pigment chlorophyll. Chloroplasts are the site for photosynthesis. Photosynthesis uses light energy to convert carbon dioxide to glucose.

Cellulose cell wall

A semi-rigid structure that lies outside the plasma membrane. It has several roles including protecting the cell and providing shape. Many materials pass freely through the cell wall.

Plasma membrane

Located inside the cell wall in plants. It controls the movement of materials into and out of the cell.

Large central vacuole:

Plant vacuoles contain cell sap. Sap is a watery solution containing dissolved food material, ions, waste products, and pigments. Functions include storage, waste disposal, and growth.

Mitochondrion

Mitochondria are the cell's energy producers. They use the chemical energy in glucose to make ATP (the cell's usable energy currency).

Endoplasmic reticulum (ER)

A network of tubes and flattened sacs continuous with the nuclear membrane. There are two types of ER. Rough ER has ribosomes attached. Smooth ER has no ribosomes (so it appears smooth).

Nuclear pore

Nuclear membrane

Nucleus

Most of a plant cell's DNA is here.

Ribosomes

These small structures make proteins by joining amino acids.

Cytoplasm

A watery solution containing dissolved materials, enzymes, and the cell organelles.

Amyloplast

Specialized plastid that makes and stores starch (a glucose polymer).

Golgi apparatus

A structure made up of membranous sacs. It stores, modifies, and packages proteins.

PRACTICES · CCC · WEB

SF · 32 · **KNOW**

42

1. Use the diagram of a plant cell on the previous page to become familiar with the features of a plant cell. Use your knowledge to label the ten structures in the transmission electron micrograph (TEM) of the cell below.

   Use the following list of terms to help you: *nuclear membrane, cytoplasm, endoplasmic reticulum, mitochondrion, starch granule, nucleus, vacuole, plasma membrane, cell wall, chloroplast.*

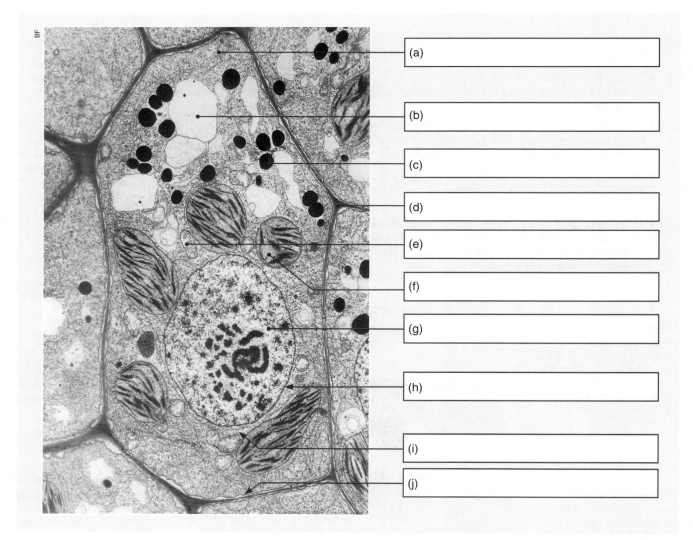

(a)

(b)

(c)

(d)

(e)

(f)

(g)

(h)

(i)

(j)

2. (a) Which features identify this plant cell as eukaryotic? _____

   _____

   _____

   (b) Use a colored marker to show the cell wall surrounding this cell.

3. (a) What is an organelle? _____

   _____

   _____

   (b) Why are there so many different types of organelles in eukaryotic cells (e.g. plant and animal cells)? _____

   _____

   _____

4. (a) Name the organelle where photosynthesis occurs: _____

   (b) How many of these organelles are present in the labeled cell above? _____

   _____

© 2016 **BIOZONE** International
ISBN: 978-1-927309-46-9
Photocopying Prohibited

33 Animal Cells

Key Idea: Animal cells are eukaryotic cells. They lack several of the structures and organelles found in plant cells.

Animal cells are eukaryotic cells. Features that identify them as eukaryotic cells include:

▶ A membrane-bound nucleus.

▶ Membrane-bound organelles.

> **DID YOU KNOW?**
> Animal cells lack the rigid cell wall found in plant cells, so their shape is more irregular and they can sometimes move about or change shape.

Features of an animal cell

Animal cells have many of the same structures and organelles that plant cells have, but several features help to identify them, including:

- No cell wall
- Often have an irregular shape
- No chloroplasts or other plastids
- No large vacuoles (if any)
- They have centrioles (not found in the cells of most plants)

A generalized animal cell

Mitochondrion
Organelles involved in the production of ATP (usable energy).

Smooth endoplasmic reticulum (smooth ER)
Its main role is to make lipids and phospholipids.

Plasma membrane
The cell boundary. The membrane is a semi-fluid phospholipid bilayer with embedded proteins. It separates the cell from its external environment and controls the movement of substances into and out of the cell.

Small vacuole
Not always present.

Ribosomes
These make proteins. They can be found floating free in the cytoplasm or attached to the surface of rough ER.

Golgi apparatus
The flattened, disc-shaped sacs of the Golgi are stacked one on top of each other, very near, and sometimes connected to, the ER. Vesicles bud off from the Golgi and transport protein products away.

Microvilli
Small finger-like extensions which increase the cell's surface area (not all animal cells have these).

Lysosome
Sac-like organelles containing enzymes that break down foreign material, cell debris, and worn-out organelles.

Rough endoplasmic reticulum (rough ER)
These have ribosomes attached to the surface. Proteins are made here.

Nuclear pore
This is a hole in the nuclear membrane. It allows molecules to pass between the nucleus and the rest of the cell.

Nucleus
A large organelle containing most of the cell's DNA. Within the nucleus, is a denser structure called the **nucleolus** (*n*).

Centrioles
Paired cylindrical structures contained within the centrosome (an organelle that organizes the cell's microtubules). The centrioles form the spindle fibers involved in nuclear division. They are made of protein microtubules and are always at 90° to each other.

Cytoplasm

PRACTICES   CCC   WEB

SF 33 **KNOW**

1. Study the diagram of an animal cell on the previous page to become familiar with the features of an animal cell. Use your knowledge to identify and label the structures in the transmission electron micrograph (TEM) of the cell below.

   Use the following list of terms to help you: *cytoplasm, plasma membrane, rough endoplasmic reticulum, mitochondrion, nucleus, centriole, Golgi apparatus, lysosome.*

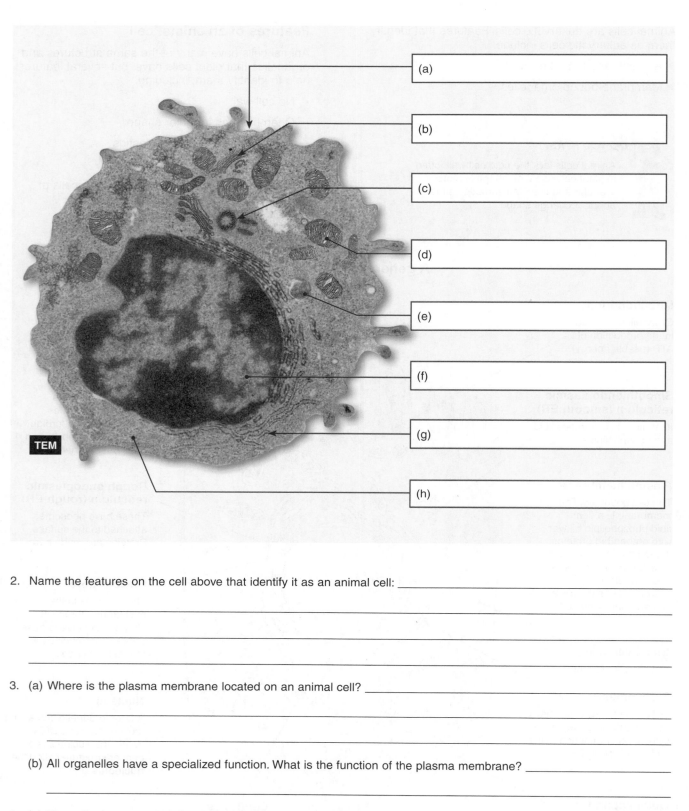

(a)

(b)

(c)

(d)

(e)

(f)

(g)

(h)

TEM

2. Name the features on the cell above that identify it as an animal cell: _____

   _____

   _____

   _____

3. (a) Where is the plasma membrane located on an animal cell? _____

   _____

   _____

   (b) All organelles have a specialized function. What is the function of the plasma membrane? _____

   _____

4. (a) Name the largest organelle visible on the animal cell above: _____

   (b) What important material does this organelle contain? _____

   _____

© 2016 **BIOZONE** International
ISBN: 978-1-927309-46-9
Photocopying Prohibited

34 Identifying Organelles

Key Idea: Cellular organelles can be identified in electron micrographs by their specific features.

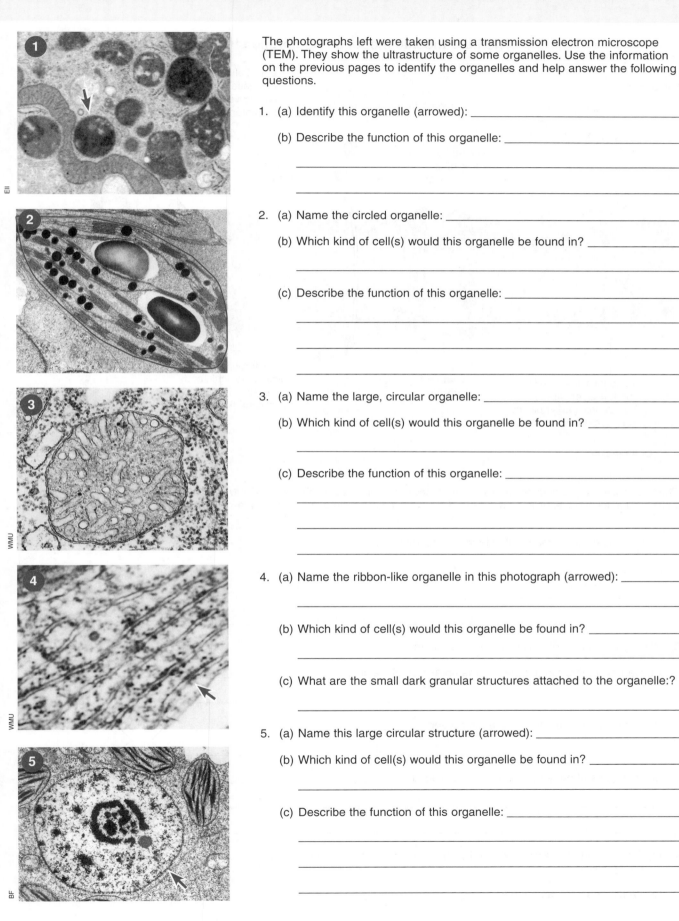

The photographs left were taken using a transmission electron microscope (TEM). They show the ultrastructure of some organelles. Use the information on the previous pages to identify the organelles and help answer the following questions.

1. (a) Identify this organelle (arrowed): _____

   (b) Describe the function of this organelle: _____

   _____

   _____

2. (a) Name the circled organelle: _____

   (b) Which kind of cell(s) would this organelle be found in? _____

   _____

   (c) Describe the function of this organelle: _____

   _____

   _____

   _____

3. (a) Name the large, circular organelle: _____

   (b) Which kind of cell(s) would this organelle be found in? _____

   _____

   (c) Describe the function of this organelle: _____

   _____

   _____

   _____

4. (a) Name the ribbon-like organelle in this photograph (arrowed): _____

   _____

   (b) Which kind of cell(s) would this organelle be found in? _____

   _____

   (c) What are the small dark granular structures attached to the organelle:?

   _____

5. (a) Name this large circular structure (arrowed): _____

   (b) Which kind of cell(s) would this organelle be found in? _____

   _____

   (c) Describe the function of this organelle: _____

   _____

   _____

   _____

CCC   WEB

SF **34** **TEST**

35 The Structure of Membranes

Key Idea: Plasma membranes are composed of a lipid bilayer in which proteins move freely.

The fluid mosaic model of membrane structure

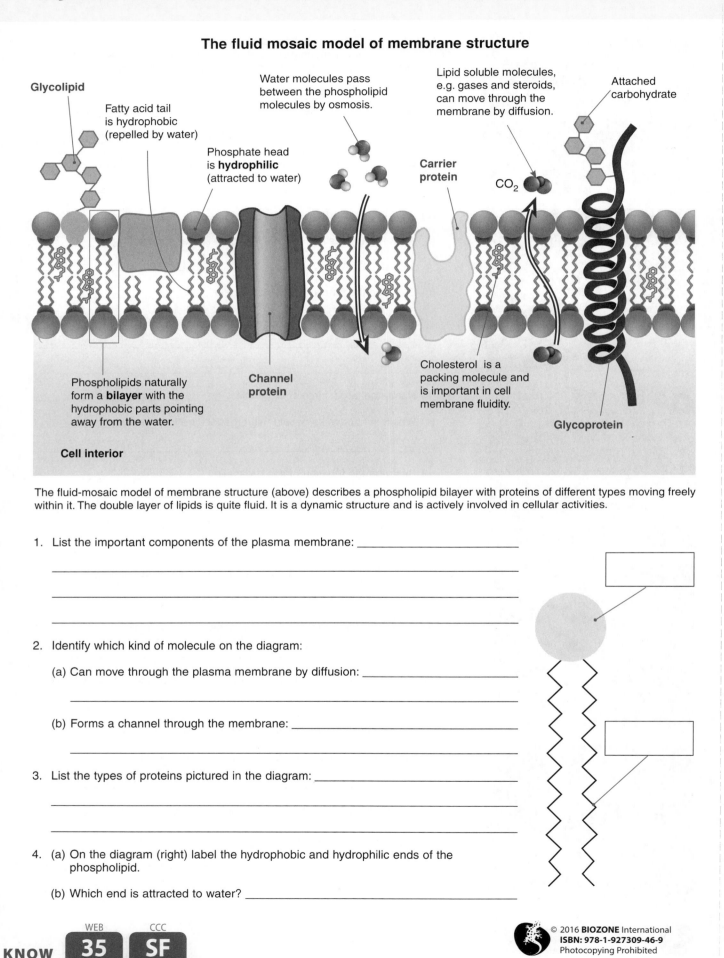

Glycolipid

Fatty acid tail is hydrophobic (repelled by water)

Phosphate head is **hydrophilic** (attracted to water)

Water molecules pass between the phospholipid molecules by osmosis.

Lipid soluble molecules, e.g. gases and steroids, can move through the membrane by diffusion.

Attached carbohydrate

Carrier protein

CO_2

Phospholipids naturally form a **bilayer** with the hydrophobic parts pointing away from the water.

Channel protein

Cholesterol is a packing molecule and is important in cell membrane fluidity.

Glycoprotein

Cell interior

The fluid-mosaic model of membrane structure (above) describes a phospholipid bilayer with proteins of different types moving freely within it. The double layer of lipids is quite fluid. It is a dynamic structure and is actively involved in cellular activities.

1. List the important components of the plasma membrane: _____

   _____

   _____

   _____

2. Identify which kind of molecule on the diagram:

   (a) Can move through the plasma membrane by diffusion: _____

   _____

   (b) Forms a channel through the membrane: _____

   _____

3. List the types of proteins pictured in the diagram: _____

   _____

   _____

4. (a) On the diagram (right) label the hydrophobic and hydrophilic ends of the phospholipid.

   (b) Which end is attracted to water? _____

WEB   CCC

KNOW **35** **SF**

© 2016 **BIOZONE** International
ISBN: 978-1-927309-46-9
Photocopying Prohibited

36 Diffusion in Cells

Key Idea: Diffusion is the movement of molecules from high concentration to a low concentration (i.e. down a concentration gradient).

What is diffusion?

▶ **Diffusion** is the movement of particles from regions of high concentration to regions of low concentration. Diffusion is a passive process, meaning it needs no input of energy to occur. During diffusion, molecules move randomly about, becoming evenly dispersed.

▶ Most diffusion in biological systems occurs across membranes. Simple diffusion occurs directly across a membrane, whereas facilitated diffusion involves helper proteins. Neither requires the cell to expend energy.

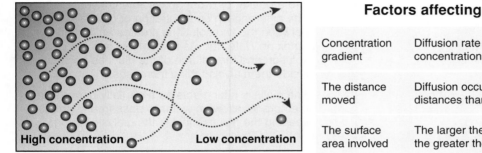

High concentration ⟶ **Low concentration**

Concentration gradient

If molecules can move freely, they move from high to low concentration (down a concentration gradient) until evenly dispersed. Net movement then stops.

Factors affecting the rate of diffusion

Concentration gradient	Diffusion rate is higher when there is a greater concentration difference between two regions.
The distance moved	Diffusion occurs at a greater rate over shorter distances than over a larger distances.
The surface area involved	The larger the area across which diffusion occurs, the greater the rate of diffusion.
Barriers to diffusion	Rate of diffusion is slower across thick barriers than across thin barriers.
Temperature	Rate of diffusion increases with temperature.

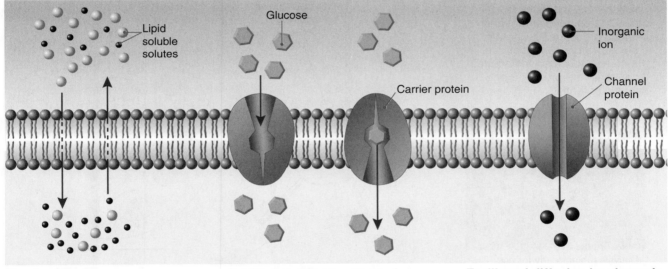

Simple diffusion

Molecules move directly through the membrane without assistance and without any energy expenditure. <u>Example</u>: O_2 diffuses into the blood and CO_2 diffuses out.

Facilitated diffusion by carriers

Carrier proteins allow large lipid-insoluble molecules that cannot cross the membrane by simple diffusion to be transported into the cell. <u>Example</u>: the transport of glucose into red blood cells.

Facilitated diffusion by channels

Channel proteins (hydrophilic pores) in the membrane allow inorganic ions to pass through the membrane. <u>Example</u>: K^+ ions leaving nerve cells to restore membrane resting potential.

1. What is diffusion? _____

_____

2. (a) How is facilitated diffusion different from simple diffusion? _____

_____

(b) How is it the same? _____

37 Osmosis in Cells

Key Idea: Osmosis is the diffusion of water molecules from a lower solute concentration to a higher solute concentration across a partially permeable membrane.

Osmosis

▶ Osmosis is the diffusion of water molecules from regions of lower solute concentration (higher free water concentration) to regions of higher solute concentration (lower free water concentration) across a partially permeable membrane.

▶ A partially permeable membrane lets some, but not all, molecules pass through. The plasma membrane of a cell is an example of a partially permeable membrane.

▶ Osmosis is a passive process (it requires no energy to occur).

Osmotic potential

The presence of solutes (dissolved substances) in a solution increases the tendency of water to move into that solution. This tendency is called the osmotic potential or osmotic pressure. The greater a solution's concentration (i.e. the more total dissolved solutes it contains) the greater the osmotic potential.

Demonstrating osmosis

Osmosis can be demonstrated using the simple experiment described below.

A glucose solution (high solute concentration) is placed into dialysis tubing, and the tubing is placed into a beaker of water (low solute concentration). The difference in concentration of glucose (solute) between the two solutions creates an osmotic gradient. Water moves by osmosis into the glucose solution and the volume of the glucose solution inside the dialysis tubing increases.

The dialysis tubing acts as a partially permeable membrane, allowing water to pass freely, while keeping the glucose inside the dialysis tubing.

Dialysis tubing ready for use

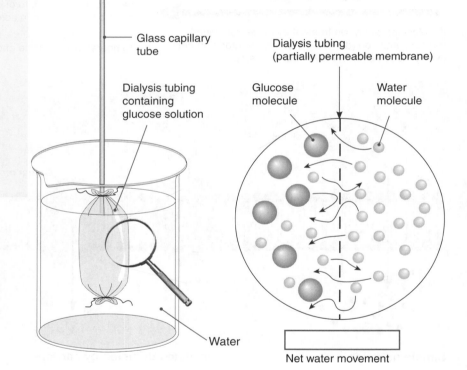

Glass capillary tube

Dialysis tubing containing glucose solution

Dialysis tubing (partially permeable membrane)

Glucose molecule

Water molecule

Water

Net water movement

1. What is osmosis? _____

_____

_____

2. (a) In the blue box on the diagram above, draw an arrow to show the direction of net water movement.

   (b) Why did water move in this direction? _____

_____

_____

_____

WEB

KNOW 37

 © 2016 **BIOZONE** International
ISBN: 978-1-927309-46-9
Photocopying Prohibited

38 Diffusion and Cell Size

Key Idea: Diffusion is more efficient at delivering materials to the interior of cells when cells have a large surface area relative to their volume.

Single-celled organisms

Single-celled organisms (e.g. *Amoeba*), are small and have a large surface area relative to the cell's volume. The cell's requirements can be met by the diffusion or active transport of materials directly into and out of the cell (below).

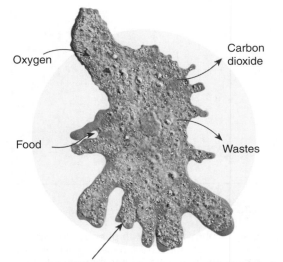

Oxygen

Carbon dioxide

Food

Wastes

The **plasma membrane**, which surrounds every cell, regulates the movement of substances into and out of the cell. For each square micrometer of membrane, only so much of a particular substance can cross per second.

Multicellular organisms

Multicellular organisms (e.g. plants and animals) are often large and large organisms have a relatively small surface area compared to their volume. Diffusion alone is not sufficient to supply their cells with everything they need, so multicellular organisms need specialized systems to transport materials to and from their cells.

In a multicellular organism, such as an elephant, the body's need for respiratory gases cannot be met by diffusion through the skin.

A specialized gas exchange surface (lungs) and circulatory (blood) system are required to supply the body's cells with oxygen and remove carbon dioxide.

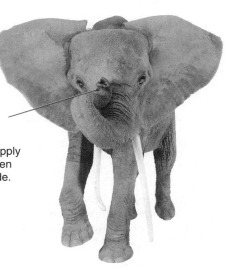

1. Calculate the volume, surface area, and the ratio of surface area to volume for each of the four cubes below (the first has been done for you). Show your calculations as you complete the table below.

2 cm cube

3 cm cube

4 cm cube

5 cm cube

Cube size	Surface area (cm²)	Volume (cm³)	Surface area to volume ratio
2 cm cube	2 x 2 x 6 = 24 cm² (2 cm x 2 cm x 6 sides)	2 x 2 x 2 = 8 cm³ (height x width x depth)	24 to 8 = 3:1
3 cm cube			
4 cm cube			
5 cm cube			

CCC WEB

SF **38** **DATA**

39 Calculating Diffusion Rates

Key Idea: The surface area to volume ratio decreases as cell volume increases.

NEED HELP?
See Activity 17

1. Use your calculations from the activity Diffusion and Cell Size (Q1) to create a graph of the surface area against the volume of each cube, on the grid on the right. Draw a line connecting the points and label axes and units.

2. Which increases the fastest with increasing size: the volume or the surface area?

   _____

   _____

3. Explain what happens to the ratio of surface area to volume with increasing size.

   _____

   _____

   _____

   _____

   _____

4. The diffusion of molecules into cells of varying sizes can be modelled using agar cubes infused with phenolphthalein indicator and soaked in sodium hydroxide (NaOH). Phenolphthalein turns pink in the presence of a base (NaOH). As the NaOH diffuses into the agar, the phenolphthalein changes to pink and indicates how far the NaOH has diffused into the agar. Agar blocks are cut into cubes of varying size, so the effect of cell size on diffusion can be studied.

   (a) Use the information below to fill in the table on the right:

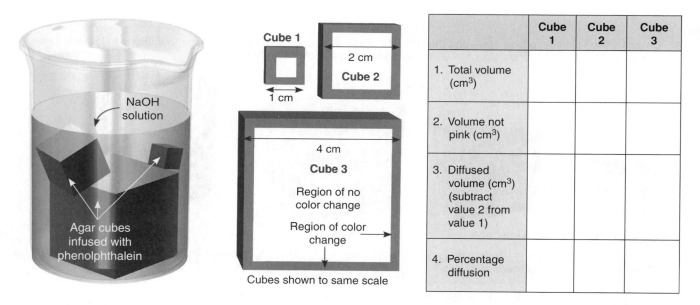

Cube 1 — 1 cm
Cube 2 — 2 cm
Cube 3 — 4 cm

NaOH solution

Agar cubes infused with phenolphthalein

Region of no color change

Region of color change

Cubes shown to same scale

	Cube 1	Cube 2	Cube 3
1. Total volume (cm³)			
2. Volume not pink (cm³)			
3. Diffused volume (cm³) (subtract value 2 from value 1)			
4. Percentage diffusion			

   (b) Diffusion of substances into and out of a cell occurs across the plasma membrane. For a cuboid cell, explain how increasing cell size affects the ability of diffusion to provide the materials required by the cell:

   _____

   _____

   _____

   _____

© 2016 **BIOZONE** International
ISBN: 978-1-927309-46-9
Photocopying Prohibited

PRACTICES   PRACTICES

DATA

40 Factors Affecting Membrane Permeability

Key Idea: High temperatures can disrupt the structure of cellular membranes and alter their permeability, making them leaky.

Membrane permeability can be disrupted if membranes are subjected to high temperatures. At temperatures above the optimum, the membrane proteins become denatured (they lose their structure). The denatured proteins no longer function properly and the membrane loses its selective permeability and becomes leaky.

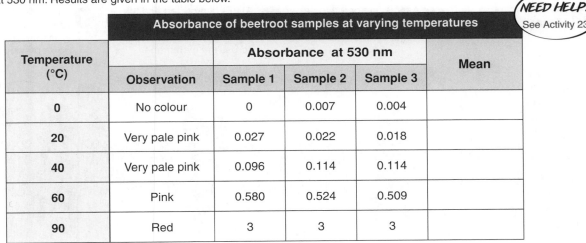

Beetroot cubes

Background

Plant cells often contain a large central vacuole surrounded by a membrane called a **tonoplast**. In beetroot plants, the vacuole contains a water-soluble red pigment called betacyanin, which gives beetroot its color. If the tonoplast is damaged, the red pigment leaks out into the surrounding environment. The amount of leaked pigment relates to the amount of damage to the tonoplast.

Experimental method

Raw beetroot was cut into uniform cubes using a cork borer with a 4 mm internal diameter. The cubes were trimmed to 20 mm lengths and placed in a beaker of distilled water for 30 minutes.

5 cm^3 of distilled water was added to 15 clean test tubes. Three were placed into a beaker containing ice. These were the 0°C samples. Three test tubes were placed into water baths at 20, 40, 60, or 90°C and equilibrated for 30 minutes. Once the tubes were at temperature, the beetroot cubes were removed from the distilled water and blotted dry on a paper towel. One beetroot cube was added to each of the test tubes. After 30 minutes, they were removed. The color of the solution in each test tube was observed by eye and then the absorbance of each sample was measured at 530 nm. Results are given in the table below.

The aim and hypothesis

The aim was to investigate the effect of temperature on membrane permeability. The students hypothesized that the amount of pigment leaking from the beetroot cubes would increase with increasing temperature.

NEED HELP?
See Activity 23

Absorbance of beetroot samples at varying temperatures

Temperature (°C)	Observation	Absorbance at 530 nm			Mean
		Sample 1	Sample 2	Sample 3	
0	No colour	0	0.007	0.004	
20	Very pale pink	0.027	0.022	0.018	
40	Very pale pink	0.096	0.114	0.114	
60	Pink	0.580	0.524	0.509	
90	Red	3	3	3	

1. Why is it important to wash the beetroot cubes in distilled water prior to carrying out the experiment? ____________

____

____

2. (a) Complete the table above by calculating the mean absorbance for each temperature:

(b) Based on the results in the table above, describe the effect of temperature on membrane permeability: ____________

____

____

(c) Explain why this effect occurs: ____________

____

ISBN: 978-1-927309-46-9

41 Active Transport

Key Idea: Active transport uses energy to transport molecules against their concentration gradient across a plasma membrane.

▶ Active transport is the movement of molecules (or ions) from regions of low concentration to regions of high concentration across a plasma membrane.

▶ Active transport needs energy to proceed because molecules are being moved against their concentration gradient.

▶ The energy for active transport comes from ATP (adenosine triphosphate). Energy is released when ATP is hydrolyzed (water is added) forming ADP (adenosine diphosphate) and inorganic phosphate (P_i).

▶ Transport (carrier) proteins in the plasma membrane use energy to transport molecules across a membrane (below).

▶ Active transport can be used to move molecules into and out of a cell.

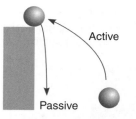

A ball falling is a passive process (it requires no energy input). Replacing the ball requires active energy input.

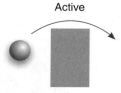

It requires energy to actively move an object across a physical barrier.

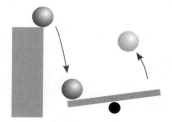

Sometimes the energy of a passively moving object can be used to actively move another. For example, a falling ball can be used to catapult another (left).

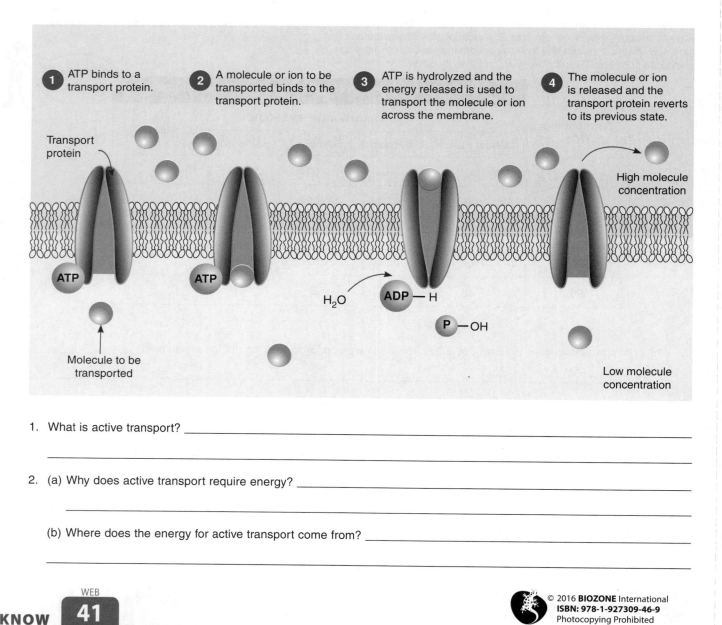

1 ATP binds to a transport protein.

2 A molecule or ion to be transported binds to the transport protein.

3 ATP is hydrolyzed and the energy released is used to transport the molecule or ion across the membrane.

4 The molecule or ion is released and the transport protein reverts to its previous state.

Transport protein

High molecule concentration

ATP

ATP

H_2O

ADP — H

P — OH

Molecule to be transported

Low molecule concentration

1. What is active transport? _____

_____

2. (a) Why does active transport require energy? _____

_____

(b) Where does the energy for active transport come from? _____

_____

© 2016 **BIOZONE** International
ISBN: 978-1-927309-46-9
Photocopying Prohibited

42 What is an Ion Pump?

Key Idea: Ion pumps are transmembrane proteins that use energy to move ions and other molecules across a membrane against their concentration gradient.

▶ Proteins play an important role in the movement of molecules into and out of cells. When molecules are moved against their concentration gradient, energy is needed.

▶ Some lipid-soluble molecules can readily cross the cell membrane. Water soluble or charged molecules cross the membrane by facilitated diffusion or by active transport involving special membrane proteins called **ion pumps**.

▶ Ion pumps directly or indirectly use energy to transport ions across the membrane against a concentration gradient.

▶ The sodium-potassium pump (below, left) is found in almost all animal cells and is also common in plant cells. The concentration gradient created by ion pumps is often coupled to the transport of other molecules, such as glucose or sucrose, across the membrane (below right).

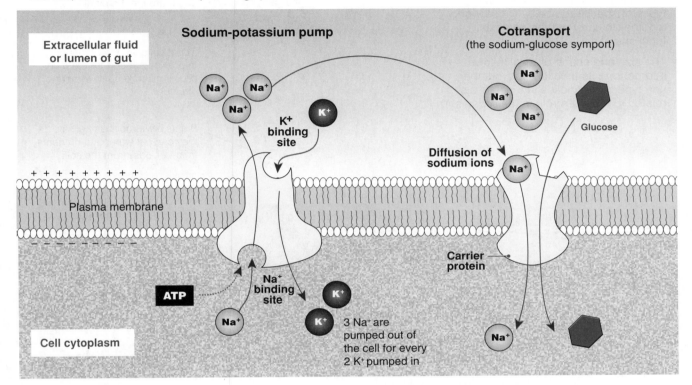

Sodium-potassium (Na⁺/K⁺) Pump

The Na⁺/K⁺ pump is a protein in the membrane that uses energy in the form of ATP to exchange sodium ions (Na⁺) for potassium ions (K⁺) across the membrane. The unequal balance of Na⁺ and K⁺ across the membrane creates large concentration gradients that can be used to drive transport of other substances (e.g. cotransport of glucose). The Na⁺/K⁺ pump also helps to maintain the right balance of ions and so helps regulate the cell's water balance.

Cotransport (coupled transport)

A specific carrier protein controls the entry of glucose into the intestinal epithelial cells from the gut where digestion is taking place. The energy for this is provided indirectly by a gradient in sodium ions. The carrier 'couples' the return of Na⁺ down its concentration gradient to the transport of glucose into the cell. The process is therefore called cotransport. A low intracellular concentration of Na⁺ (and therefore the concentration gradient for transport) is maintained by a sodium-potassium pump.

1. What is an ion pump? _____

2. (a) Explain what is meant by cotransport: _____

   (b) How is cotransport used to move glucose into the intestinal epithelial cells? _____

© 2016 **BIOZONE** International
ISBN: 978-1-927309-46-9
Photocopying Prohibited

CCC · WEB · SF · 42 · **KNOW**

43 Specialization in Plant Cells

Key Idea: The specialized cells in a plant have specific features associated with their particular roles in the plant.

Cell specialization

▶ A **specialized cell** is a cell with the specific features needed to perform a particular function in the organism.

▶ Cell specialization occurs during development when specific genes are switched on or off.

▶ Multicellular organisms have many types of specialized cells. These work together to carry out the essential functions of life.

▶ The size and shape of a cell allows it to perform its function. The number and type of organelles in a cell is also related to the cell's role in the organism.

Cells in the leaves of plants are often green because they contain the pigment chlorophyll which is needed for photosynthesis.

Specialized cells in vascular tissue are needed to transport water and sugar around the plant.

Some cells are strengthened to provide support for the plant, allowing it to keep its form and structure.

Plants have root-hair cells so they can get water and nutrients (mineral ions) from the soil.

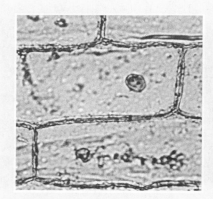

Many plant cells have a regular shape because of their semi-rigid cell wall. Specialized cells form different types of tissues. Simple tissues, like this onion epidermis, have only one cell type.

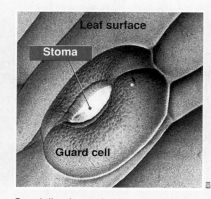

Leaf surface

Stoma

Guard cell

Specialized guard cells surround the stomata (pores) on plant leaves. The guard cells control the opening and closing of stomata and prevent too much water being lost from the plant.

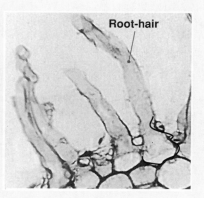

Root-hair

A plant root-hair is a tube-like outgrowth of a plant root cell. Their long, thin shape greatly increases their surface area. This allows the plant to absorb water and minerals efficiently.

1. What is a specialized cell? _____

_____

2. (a) Name the specialized cell that helps to prevent water loss in plants: _____

   (b) How does this cell prevent water being lost? _____

   _____

   _____

3. How do specialized root hairs help plants to absorb more water and minerals from the soil? _____

   _____

   _____

© 2016 **BIOZONE** International
ISBN: 978-1-927309-46-9
Photocopying Prohibited

44 Specialization in Animal Cells

Key Idea: There are many different types of animal cells, each with a specific role in the body. Animal cells are often highly modified for their specific role.

Specialization in animal cells

▶ There are over 200 different types of cells in the human body.

▶ Animal cells lack a cell wall, so they can take on many different shapes. Therefore, there many more types of animal cells than there are plant cells.

▶ Specialized cells often have modifications or exaggerations to a normal cell feature to help them do their job. For example, nerve cells have long, thin extensions to carry nerve impulses over long distances in the body.

▶ Specialization improves efficiency because each cell type is highly specialized to perform a particular task.

Fat cell

Thin, flat epithelial cells line the walls of blood vessels (arrow). Large fat cells store lipid.

Some nerve cells are over 1 m long.

Louisa Howard, Katherine Connolly Dartmouth College

SEM: White blood cell

TEM: Cellular projections of intestinal cell

RBC

SEM: Egg and sperm

Some animal cells can move or change shape. A sperm cell must be able to swim so that it can fertilize an egg. A white blood cell changes its shape to engulf and destroy foreign materials (e.g. bacteria).

Cells that line the intestine have extended cell membranes. This increases their surface area so that more food (nutrients) can be absorbed. Red blood cells (RBCs) have no nucleus so they have more room inside to carry oxygen around the body.

The egg (ovum) is the largest human cell. It is about 0.1 mm in diameter and can be seen with the naked eye. The smallest human cells are sperm cells and red blood cells.

1. What is the advantage of cell specialization in a multicellular organism? _____

_____

_____

_____

2. For each of the following specialized animal cells, name a feature that helps it carry out its function:

(a) White blood cell: _____

(b) Sperm cell: _____

(c) Nerve cell: _____

(d) Red blood cell: _____

© 2016 **BIOZONE** International
ISBN: 978-1-927309-46-9
Photocopying Prohibited

CCC   WEB

SF **44**   **KNOW**

45 What is DNA?

Key Idea: A cell's genetic information is called DNA. In eukaryotic cells, DNA is located in the cell nucleus.

About DNA

▶ **DNA** stands for **d**eoxyribo**n**ucleic **a**cid.

▶ DNA is called the blueprint for life because it contains all of the information an organism needs to develop, function, and reproduce.

▶ DNA stores and transmits genetic information.

▶ DNA is found in every cell of all living organisms.

▶ DNA has a double-helix structure (left). If the DNA in a single human cell was unwound, it would be more than two meters long! The long DNA molecules are tightly packed in an organized way so that they can fit into the nucleus.

DNA is found in every cell of all living organisms: animals, plants, fungi, protists, and bacteria.

DNA contains the instructions an organism needs to develop, survive, and reproduce. Small differences in DNA cause differences in appearance.

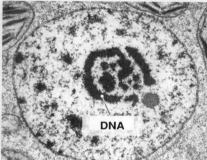

In eukaryotes, most of the cell's DNA is located in the nucleus (above). A very small amount is located in mitochondria and in the chloroplasts of plants.

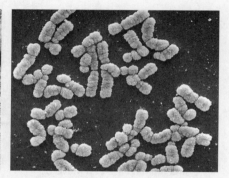

The DNA in eukaryotes is packaged into chromosomes (above). Each chromosome is made up of a DNA molecule and associated proteins. The proteins help to package the DNA into the nucleus.

1. (a) What does DNA stand for? _____

    (b) Where is most of the DNA found in eukaryotes? _____

    (c) What does DNA do? _____

    _____

    _____

2. (a) How is DNA packaged up in eukaryotes? _____

    _____

    (b) Why does DNA have to be tightly packaged up? _____

    _____

    _____

KNOW

 © 2016 **BIOZONE** International
ISBN: 978-1-927309-46-9
Photocopying Prohibited

46 Nucleotides

Key Idea: Nucleotides are the building blocks of DNA and RNA. A nucleotide has three components: a base, a sugar, and a phosphate group.

The structure of a nucleotide

Nucleotides are the building blocks of nucleic acids (DNA and RNA). Nucleotides have three parts to their structure (see diagrams below):

▶ A nitrogen containing base

▶ A five carbon sugar

▶ A phosphate group

Symbolic form of a nucleotide

(showing positions of the 5 C atoms on the sugar)

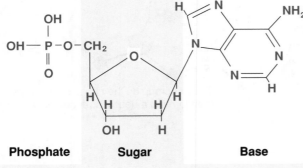

Phosphate group

Base

Sugar

Chemical structure of a nucleotide

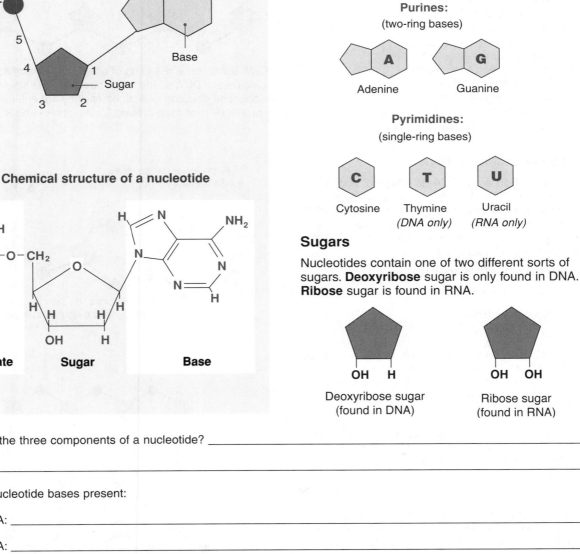

Phosphate   Sugar   Base

Nucleotide bases

Five different kinds of nitrogen bases are found in nucleotides. These are:

Adenine (A)

Guanine (G)

Cytosine (C)

Thymine (T)

Uracil (U)

DNA contains adenine, guanine, cytosine, and thymine.

RNA also contains adenine, guanine, and cytosine, but uracil (U) is present instead of thymine.

Purines:
(two-ring bases)

A   G

Adenine   Guanine

Pyrimidines:
(single-ring bases)

C   T   U

Cytosine   Thymine   Uracil
              (DNA only)   (RNA only)

Sugars

Nucleotides contain one of two different sorts of sugars. **Deoxyribose** sugar is only found in DNA. **Ribose** sugar is found in RNA.

OH   H   OH   OH

Deoxyribose sugar   Ribose sugar
(found in DNA)   (found in RNA)

1. What are the three components of a nucleotide? _____

_____

2. List the nucleotide bases present:

(a) In DNA: _____

(b) In RNA: _____

3. Name the sugar present:

(a) In DNA: _____ (b) In RNA: _____

CCC   WEB

SF   46   **KNOW**

47 DNA and RNA

Key Idea: DNA and RNA are nucleic acids made up of long chains of nucleotides, which store and transmit genetic information. DNA is double-stranded. RNA is single-stranded.

The structure of DNA

► Nucleotides join together to form **nucleic acids.**

► **Deoxyribonucleic acid** (DNA) is a nucleic acid.

► DNA consists of a two strands of nucleotides linked together to form a **double helix.** A double helix is like a ladder twisted into a corkscrew shape. The rungs of the ladder are the two nitrogen bases joined by hydrogen bonds. The double helix is 'unwound' to show its structure in the diagram (right).

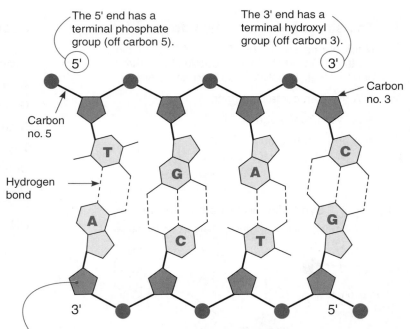

The 5' end has a terminal phosphate group (off carbon 5).

The 3' end has a terminal hydroxyl group (off carbon 3).

Carbon no. 5

Carbon no. 3

Hydrogen bond

The DNA backbone is made up of alternating phosphate and sugar molecules. Each DNA strand has a direction. The single strands run in the opposite direction to each other (they are anti-parallel). The ends of a DNA strand are labeled 5' (five prime) and 3' (three prime).

Who discovered the DNA double helix?

Two scientists, James Watson and Francis Crick, are credited with discovering the structure of DNA. However, they used X-ray pictures of DNA from another scientist, Rosalind Franklin, to confirm their hypothesis.

The structure of RNA

Ribonucleic acid (RNA) is a type of nucleic acid. Like DNA, the nucleotides are linked together through a condensation reaction. RNA is a single stranded, and has many functions including protein synthesis, and cell regulation. There are 3 types of RNA:

► Messenger RNA (mRNA)

► Transfer RNA (tRNA)

► Ribosomal RNA (rRNA)

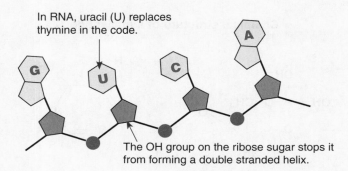

In RNA, uracil (U) replaces thymine in the code.

The OH group on the ribose sugar stops it from forming a double stranded helix.

1. The diagram on the right shows a double-stranded DNA molecule. Label the following:
   (a) Sugar group
   (b) Phosphate group
   (c) Hydrogen bonds
   (d) Purine bases
   (e) Pyrimidine bases

2. If you wanted to use a radioactive or fluorescent tag to label only the RNA in a cell and not the DNA, what molecule(s) would you label?

   _____

3. If you wanted to use a radioactive or fluorescent tag to label only the DNA in a cell and not the RNA, what molecule(s) would you label?

   _____

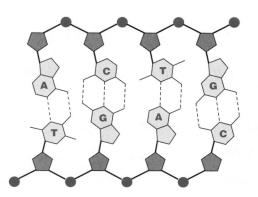

KNOW 47 SF

© 2016 **BIOZONE** International
ISBN: 978-1-927309-46-9
Photocopying Prohibited

48 Modeling the Structure of DNA

Key Idea: Nucleotides pair together in a specific way called the base pairing rule. In DNA, adenine always pairs with thymine, and cytosine always pairs with guanine.

The exercise on the following pages is designed to help you understand the structure of DNA, and learn the base pairing rule for DNA.

The way the nucleotide bases pair up between strands is very specific. The chemistry and shape of each base means they can only bond with one other DNA nucleotide. Use the information in the table below if you need help remembering the base pairing rule while you are constructing your DNA molecules.

DID YOU KNOW?

Chargaff's rules

Before Watson and Crick described the structure of DNA, an Austrian chemist called Chargaff analyzed the base composition of DNA from a number of organisms. He found that the base composition varies between species but that within a species the percentage of A and T bases are equal and the percentage of G and C bases are equal. Validation of Chargaff's rules was the basis of Watson and Crick's base pairs in the DNA double helix model.

DNA base pairing rule

Adenine	always pairs with	**Thymine**	A ⟷ T
Thymine	always pairs with	**Adenine**	T ⟷ A
Cytosine	always pairs with	**Guanine**	C ⟷ G
Guanine	always pairs with	**Cytosine**	G ⟷ C

1. Cut out each of the nucleotides on page 61 by cutting along the columns and rows (see arrows indicating two such cutting points). Although drawn as geometric shapes, these symbols represent chemical structures.

2. Place one of each of the four kinds of nucleotide on their correct spaces below:

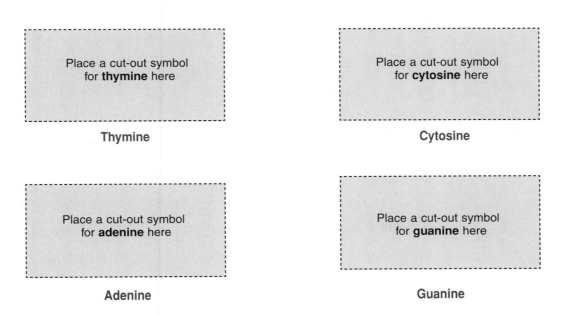

Place a cut-out symbol for **thymine** here

Thymine

Place a cut-out symbol for **cytosine** here

Cytosine

Place a cut-out symbol for **adenine** here

Adenine

Place a cut-out symbol for **guanine** here

Guanine

3. Identify and label each of the following features on the adenine nucleotide immediately above: **phosphate, sugar, base, hydrogen bonds**.

4. Create one strand of the DNA molecule by placing the 9 correct 'cut out' nucleotides in the labelled spaces on the following page (DNA molecule). Make sure these are the right way up (with the P on the left) and are aligned with the left hand edge of each box. Begin with thymine and end with guanine.

5. Create the complementary strand of DNA by using the base pairing rule above. Note that the nucleotides have to be arranged upside down.

6. Once you have checked that the arrangement is correct, glue, paste, or tape these nucleotides in place.

 © 2016 **BIOZONE** International
ISBN: 978-1-927309-46-9
Photocopying Prohibited

PRACTICES   CCC   CCC   WEB

SF   SSM   48   **PRAC**

DNA molecule

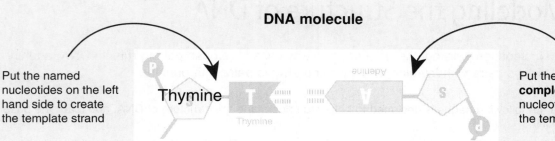

Put the named nucleotides on the left hand side to create the template strand

Thymine

Put the matching **complementary** nucleotides opposite the template strand

Cytosine

Adenine

Adenine

Guanine

Thymine

Thymine

Cytosine

Guanine

© 2016 **BIOZONE** International
ISBN: 978-1-927309-46-9

Nucleotides

Tear out this page and separate each of the 24 nucleotides
by cutting along the columns and rows (see arrows indicating the cutting points).

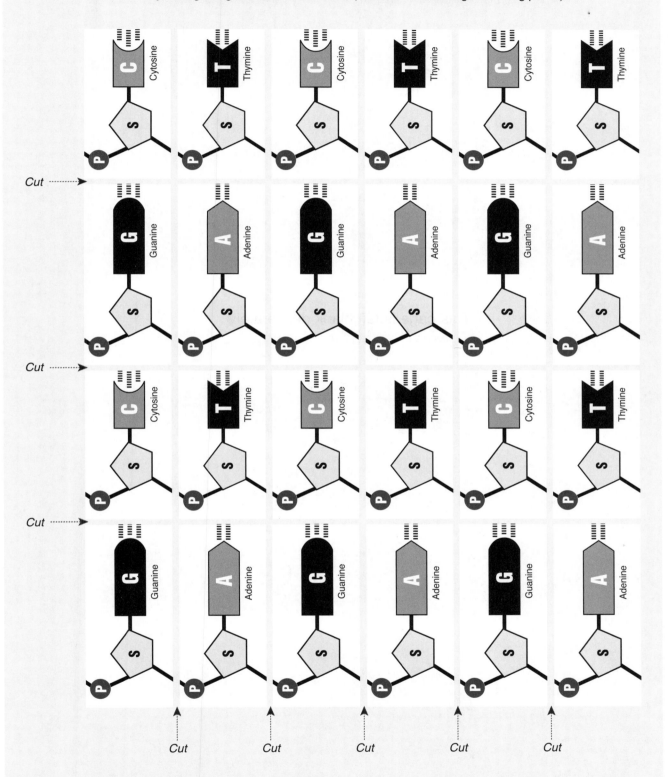

This page is left blank deliberately

49 Genes Code for Proteins

Key Idea: Genes are sections of DNA that code for proteins. Genes are expressed when they are transcribed into mRNA and then translated into a protein.

▶ A **gene** is a section of DNA that codes for a protein. **Gene expression** is the process of rewriting a gene into a protein. It involves two stages: **transcription** of the DNA and **translation** of the mRNA into protein.

▶ A gene is bounded by a start (promoter) region, upstream of the gene, and a terminator region, downstream of the gene. These regions control transcription by telling RNA polymerase where to start and stop.

▶ RNA polymerase binds to the promoter region to begin transcription of the gene.

▶ The one gene-one protein model is helpful for visualizing the processes involved in gene expression, although it is overly simplistic for eukaryotes. The information flow for gene to protein is shown below.

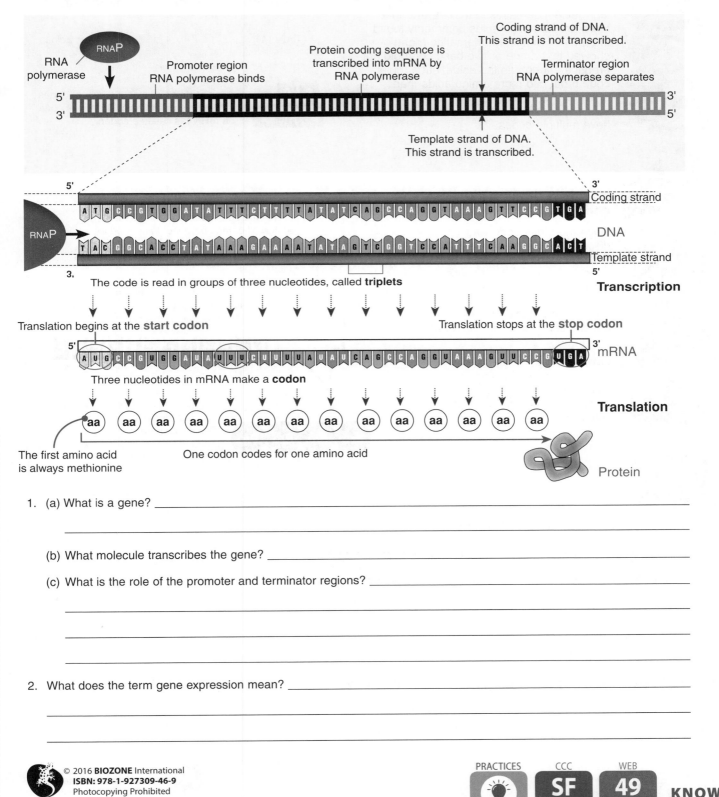

1. (a) What is a gene? _____

   _____

   (b) What molecule transcribes the gene? _____

   (c) What is the role of the promoter and terminator regions? _____

   _____

   _____

   _____

2. What does the term gene expression mean? _____

   _____

   _____

© 2016 **BIOZONE** International
ISBN: 978-1-927309-46-9
Photocopying Prohibited

PRACTICES   CCC   WEB

SF   49   **KNOW**

50 Cracking the Genetic Code

Key Idea: Scientists used mathematics and scientific experiments to unlock the genetic code. A series of three nucleotides, called a triplet, codes for a single amino acid.

The genetic code

Once it was discovered that DNA carries the genetic code needed to produce proteins, the race was on to "crack the code" and find out how it worked.

The first step was to find out how many nucleotide bases code for an amino acid. Scientists knew that there were four nucleotide bases in mRNA, and that there are 20 amino acids commonly found in proteins. Simple mathematics (right) showed that a one or two base code did not produce enough amino acids, but a triplet code produced more amino acids than existed. The triplet code was accepted once scientists confirmed that some amino acids have multiple codes.

Number of bases in the code	Working	Number of amino acids produced
Single (4^1)	4	4 amino acids
Double (4^2)	4 x 4	16 amino acids
Triple (4^3)	4 x 4 x 4	64 amino acids

A triplet (three nucleotide bases) codes for a single amino acid. The triplet code on mRNA is called a codon.

How was the genetic code cracked?

Once the triplet code was discovered, the next step was to find out which amino acid each codon produced. Two scientists, Marshall Nirenberg and Heinrich Matthaei, developed an experiment to crack the code. Their experiment is shown on the right.

Over the next few years, similar experiments were carried out using different combinations of nucleotides until all of the codes were known.

In a test tube, Nirenberg and Matthaei added all of the components needed to make proteins (except mRNA).

They then made an mRNA strand containing only one repeated nucleotide. The strand below shows cytosine (C).

Once the components were added together an amino acid was produced. The codon CCC produced the amino acid proline (Pro).

1. (a) How many types of nucleotide bases are there in mRNA? _____

(b) How many different amino acids are commonly found in proteins? _____

(c) Why did scientists reject a one or two base code when trying to work out the genetic code? _____

_____

_____

_____

2. A triplet code could potentially produce 64 amino acids. Why are only 20 amino acids produced? _____

_____

_____

© 2016 **BIOZONE** International
ISBN: 978-1-927309-46-9
Photocopying Prohibited

51 Amino Acids Make Up Proteins

Key Idea: The structure of a protein allows it to carry out its role in an organism. The sequence of amino acids and the chemical interactions between them determine a protein's shape.

Proteins are made up of amino acids

▶ Proteins are large molecules made up of many smaller units called amino acids joined together. The amino acids are joined together by peptide bonds. The sequence of amino acids in a protein is determined by the order of nucleotides in DNA.

▶ All amino acids have a common structure (right) consisting of an amine group, a carboxyl group, a hydrogen atom, and an 'R' group. Each type of amino acid has a different 'R' group (side chain). Each "R" group has a different chemical property.

▶ The chemical properties of the amino acids are important because the chemical interactions between amino acids cause a protein fold into a specific three dimensional shape. The protein's shape helps it carry out its specialized role.

The general structure of an amino acid

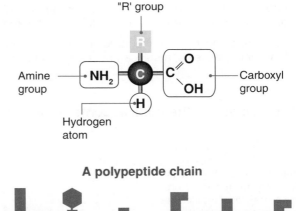

The order of amino acids in a protein is directed by the order of nucleotides in DNA (and therefore mRNA).

A polypeptide chain

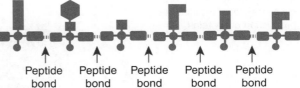

Peptide bond — Peptide bond — Peptide bond — Peptide bond — Peptide bond

The shape of a protein determines its role

▶ The sequence of amino acids determines how a protein will fold up (i.e. the shape it will form).

▶ The shape of a protein determines its role. Proteins generally fall into two groups, globular and fibrous (right).

▶ The shape of a protein is so important to its function, that if the structure of a protein is destroyed (**denatured**) it can no longer carry out its function.

Insulin: a globular protein

Collagen: a fibrous protein

Globular proteins

▶ Globular proteins are round and water soluble. Their functions include:

• Catalytic (e.g. enzymes)

• Regulation (e.g. hormones)

• Transport (e.g. hemoglobin)

• Protective (e.g. antibodies)

Fibrous proteins

▶ Fibrous proteins are long and strong. Their functions include:

• Support and structure (e.g. connective tissue)

• Contractile (e.g. myosin, actin)

1. (a) Name the four components of an amino acid: _____

   _____

   (b) What makes each type of amino acid unique? _____

2. Why are proteins important in organisms? _____

   _____

3. (a) Why is the shape of a protein important? _____

   _____

   (b) What happens to a protein if it loses its shape? _____

   _____

 © 2016 **BIOZONE** International
ISBN: 978-1-927309-46-9
Photocopying Prohibited

CCC   WEB

SF   51   **KNOW**

52 Proteins Have Many Roles in Cells

Key Idea: DNA determines the structure of proteins. Proteins carry out the essential functions of life and have structural, catalytic, and regulatory roles.

▶ In eukaryotic cells, most of a cell's genetic information (DNA) is found in the **nucleus**, a large membrane-bound organelle. The nucleus directs all cellular activities by controlling the synthesis of proteins.

▶ DNA provides instructions that code for the formation of proteins. Proteins carry out most of a cell's work. A cell produces many different types proteins, each carries out a specific task in the cell.

▶ Proteins are involved in the structure, function, and regulation of the body's cells, tissues, and organs. Without functioning proteins, a cell can not carry out its specialized role and the organism may die.

The nucleus is the control center of a cell

The DNA within the nucleus provides instructions to a cell on how to carry out its functions to sustain essential life processes. This includes the production of proteins.

Different sections of DNA, called genes, code for specific proteins. A cell can control the type of protein it produces by only transcribing (rewriting) specific genes as their proteins are required.

The nuclear envelope is formed by a double-layered membrane. It keeps the DNA within the nucleus.

In eukaryotes, production of the protein is completed outside of the nucleus. Synthesis continues on ribosomes, which may be free in the cytoplasm or associated with the rough endoplasmic reticulum (rER).

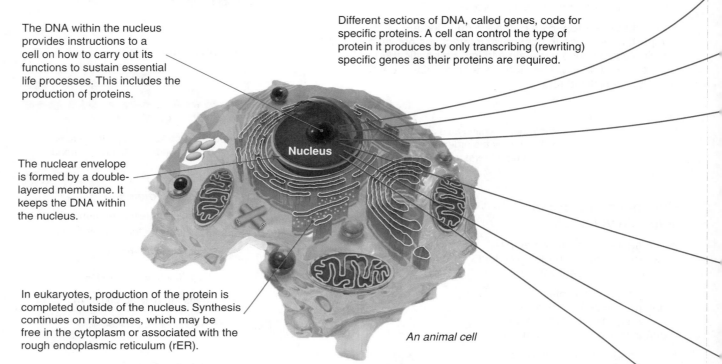

An animal cell

1. Draw a line to match the protein function with its description. Provide examples of each:

Function	Description	Example(s)
Internal defense	Some proteins can function as enzymes, thereby controlling metabolism.	_____
Contractile	Proteins can function as chemical messenger molecules.	_____
Catalytic	Proteins can make up structural components of tissues and organs.	_____
Regulation	Some proteins can act as carrier molecules, to transport molecules from one place to another.	_____
Structural	Some proteins form antibodies that combat disease causing organisms.	_____
Transport	Some proteins form contractile elements in cells and bring about movement.	_____

© 2016 **BIOZONE** International
ISBN: 978-1-927309-46-9
Photocopying Prohibited

Internal defense

Antibodies (also called immunoglobulins) are "Y" shaped proteins that protect the body by identifying and killing disease-causing organisms such as bacteria and viruses.

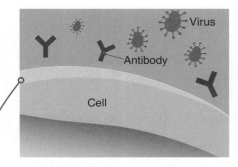

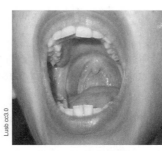

Lusb cc3.0

EXAMPLE: IgA is an antibody found in the gut and airways. It destroys disease-causing organisms growing in these areas and stops them causing an infection.

Movement

Contractile proteins are involved in movement of muscles and form the cytoskeleton of cells.

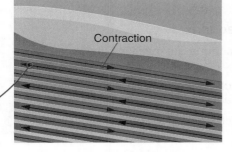

EXAMPLE: Actin and myosin are two proteins involved in the contraction of skeletal muscles.

Catalytic

Thousands of different chemical reactions take place in an organism. Each chemical reaction is catalyzed by enzymes. The ending "ase" identifies a molecule as an enzyme.

Enzyme catalyzes breakdown of substrate

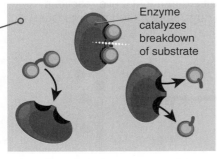

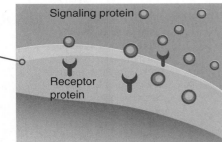

EXAMPLE: Digestion requires enzymes to break down food into smaller components. Amylase will break down the starch in this apple into maltose and glucose.

Regulation

Regulatory proteins such as hormones act as signal molecules to control the timing and occurrence of biological processes and coordinate responses in cells, tissues, and organs.

Signaling protein

Receptor protein

EXAMPLE: The hormone insulin is released after eating in response to high blood glucose and stimulates glucose uptake by cells. When blood glucose falls, the hormone glucagon stimulates processes that release glucose into the blood.

Structural

Structural proteins provide physical support or protection. They are strong, fibrous (thread like) and stringy.

Collagen fiber

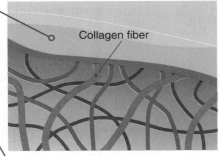

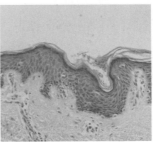

EXAMPLE: Collagen (left) is found in skin and connective tissues, including bones, tendons, and ligaments. Elastin is a highly elastic protein. Elastin helps skin to return to its original position when it is pinched.

Transport

Proteins can carry substances around the body or across membranes. For example, hemoglobin (Hb) transports oxygen (left) and proteins in cell membranes help molecules move into and out of cells.

O_2 released

O_2

Red blood cell

Hb binds O_2

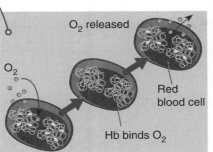

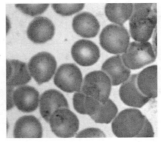

EXAMPLE: Hemoglobin is a four unit protein found in red blood cells (left). It binds oxygen and carries it through the blood, delivering it to cells.

53 Reactions in Cells

Key Idea: Anabolic reactions build complex molecules and structures from simpler ones. Catabolic reactions break down larger molecules into smaller molecules.

▶ Metabolism refers to all of the chemical reactions carried out within a living organism to maintain life. All metabolic reactions are controlled by enzymes.

▶ There are two categories of metabolic reactions: anabolic and catabolic.

Anabolic reactions

▶ **Anabolic reactions** are reactions that result in the production (synthesis) of a more complex molecule from smaller components or smaller molecules. During anabolic reactions, simple molecules are joined to form a larger, more complex molecule.

▶ Anabolic reactions need a net input of energy to proceed. They are called endergonic reactions.

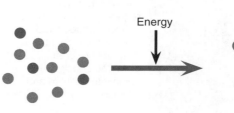

Many small molecules   *One large molecule*

Catabolic reactions

▶ **Catabolic reactions** are reactions that break down large molecules into smaller components.

▶ Catabolic reactions involve a net release of energy. They are called exergonic reactions.

▶ Catabolic reactions are the opposite of anabolic reactions.

▶ The energy released from catabolic reactions can be used to drive other metabolic processes.

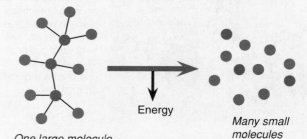

One large molecule   *Many small molecules*

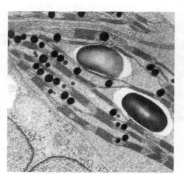

Plants carry out photosynthesis in organelles called chloroplasts (left). Photosynthesis is an anabolic process because it converts carbon dioxide and water into glucose. Energy from the sun is required to drive photosynthesis.

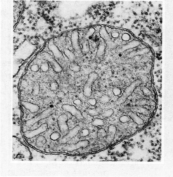

Cellular respiration is an example of a catabolic reaction. Glucose is broken down in a series of reactions to release carbon dioxide, water, and ATP (energy). The energy is used to fuel other activities in the cell. Some stages of cellular respiration take place in the mitochondria (left).

1. What is an anabolic reaction? _____

_____

2. (a) What is a catabolic reaction? _____

_____

(b) Why are catabolic reactions considered to be the opposite to anabolic reactions? _____

_____

_____

3. Identify the following reactions as either catabolic or anabolic:

(a) Protein synthesis: _____   (c) Digestion: _____

(b) ATP conversion to ADP: _____   (d) DNA synthesis: _____

© 2016 **BIOZONE** International
ISBN: 978-1-927309-46-9
Photocopying Prohibited

KNOW

54 Enzymes Catalyze Reactions in Cells

Key Idea: Enzymes are biological catalysts. They speed up biological reactions. They are not consumed in the reaction but released to work again.

What are enzymes?

▶ **Enzymes** are proteins. They control all the metabolic reactions that take place in a cell.

▶ Enzymes are called **biological catalysts** because they speed up biochemical reactions.

▶ During the reaction the enzyme itself remains unchanged, and is not used up during the reaction.

▶ Each enzyme controls a very specific metabolic reaction or series of related metabolic reactions.

▶ Enzymes may break down a single substrate molecule (catabolism), or join two or more substrate molecules together (anabolism).

▶ Extremes of temperature or pH can alter the enzyme's active site where catalysis occurs. This can lead to loss of function and is called **denaturation**.

The **active site** is the region of an enzyme where substrate molecules bind and undergo a chemical reaction.

The chemical that an enzyme acts on is the **substrate**. An enzyme acts on a specific substrate or group of related substrates.

How enzymes work

▶ An early model to explain enzyme activity described the enzyme and its substrate as a **lock and key**, where the substrate fitted neatly into the active site of the enzyme. Evidence showed this model to be flawed and it has since been modified to recognize the flexible nature of enzymes (the induced fit model).

▶ The **induced fit model** for enzyme action is shown below. The shape of the enzyme changes when the substrate fits into the active site. The substrate becomes bound to the enzyme by weak chemical bonds. This weakens bonds within the substrate itself, allowing the reaction to proceed more readily.

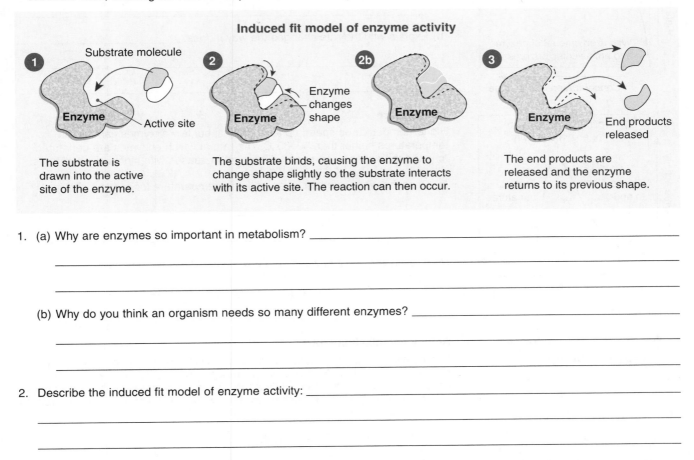

Induced fit model of enzyme activity

1 Substrate molecule / Enzyme / Active site
The substrate is drawn into the active site of the enzyme.

2 Enzyme / Enzyme changes shape
2b Enzyme
The substrate binds, causing the enzyme to change shape slightly so the substrate interacts with its active site. The reaction can then occur.

3 Enzyme / End products released
The end products are released and the enzyme returns to its previous shape.

1. (a) Why are enzymes so important in metabolism? _____

_____

_____

(b) Why do you think an organism needs so many different enzymes? _____

_____

_____

2. Describe the induced fit model of enzyme activity: _____

_____

_____

_____

 © 2016 **BIOZONE** International
ISBN: 978-1-927309-46-9
Photocopying Prohibited

CCC   WEB

SF 54   **KNOW**

55 Enzymes Have Optimal Conditions to Work

Key Idea: Enzymes have a narrow range of conditions within which they operate most efficiently. Outside this range, activity decreases and the enzyme may lose its structure.

▶ Enzymes usually have an set of conditions (e.g. pH and temperature) where their activity is greatest. This is called their **optimum**. At low temperatures, the activity of most enzymes is very slow, or does not proceed at all. Enzyme activity increases with increasing temperature, but falls off after the optimum temperature is exceeded and the enzyme is denatured. Extremes in pH can also cause denaturation.

▶ Within their normal operating conditions, enzyme reaction rates are influenced by enzyme and substrate concentration in a predictable way (below). In the graphs below, the rate of reaction or degree of enzyme activity is plotted against each of four factors that affect enzyme function.

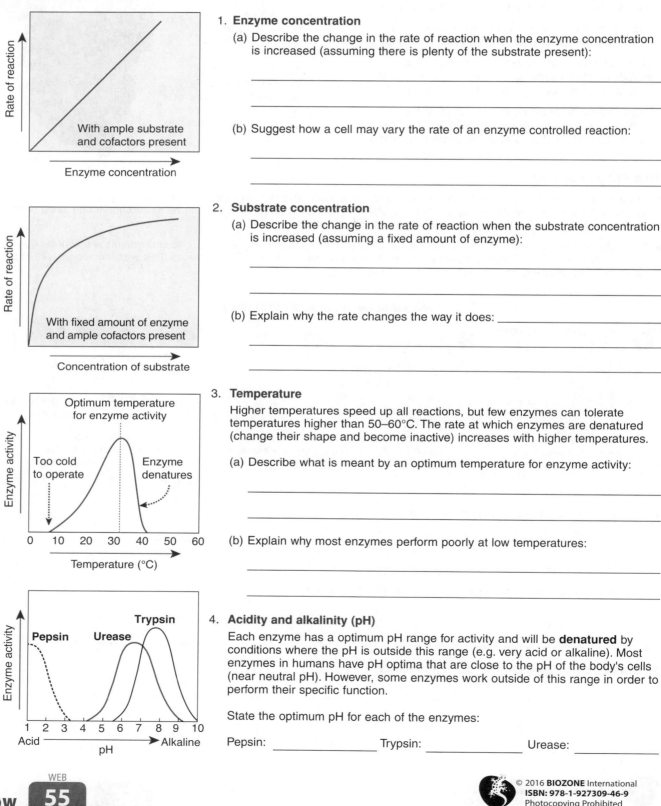

1. **Enzyme concentration**

   (a) Describe the change in the rate of reaction when the enzyme concentration is increased (assuming there is plenty of the substrate present):

   _____

   _____

   (b) Suggest how a cell may vary the rate of an enzyme controlled reaction:

   _____

   _____

2. **Substrate concentration**

   (a) Describe the change in the rate of reaction when the substrate concentration is increased (assuming a fixed amount of enzyme):

   _____

   _____

   (b) Explain why the rate changes the way it does: _____

   _____

   _____

3. **Temperature**

   Higher temperatures speed up all reactions, but few enzymes can tolerate temperatures higher than 50–60°C. The rate at which enzymes are denatured (change their shape and become inactive) increases with higher temperatures.

   (a) Describe what is meant by an optimum temperature for enzyme activity:

   _____

   _____

   (b) Explain why most enzymes perform poorly at low temperatures:

   _____

   _____

4. **Acidity and alkalinity (pH)**

   Each enzyme has a optimum pH range for activity and will be **denatured** by conditions where the pH is outside this range (e.g. very acid or alkaline). Most enzymes in humans have pH optima that are close to the pH of the body's cells (near neutral pH). However, some enzymes work outside of this range in order to perform their specific function.

   State the optimum pH for each of the enzymes:

   Pepsin: _____ Trypsin: _____ Urease: _____

© 2016 **BIOZONE** International
ISBN: 978-1-927309-46-9
Photocopying Prohibited

56 Investigating Catalase Activity

Key Idea: Catalase activity can be measured in germinating seeds. Activity changes with amount of enzyme present, which varies with stage of germination.

▶ Enzyme activity can be measured easily in simple experiments. This activity describes an experiment in which germinating seeds of different ages were tested for their level of catalase activity using hydrogen peroxide solution as the substrate and a simple apparatus to measure oxygen production (see background).

The aim and hypothesis

To investigate the effect of germination age on the level of catalase activity in mung beans. The students hypothesized that if metabolic activity increased with germination age, catalase activity would also increase with germination age.

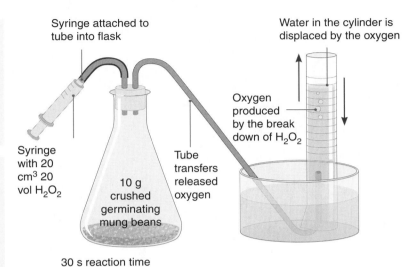

Syringe attached to tube into flask

Syringe with 20 cm³ 20 vol H_2O_2

10 g crushed germinating mung beans

Tube transfers released oxygen

Water in the cylinder is displaced by the oxygen

Oxygen produced by the break down of H_2O_2

30 s reaction time

Background

Germinating seeds are metabolically very active. The metabolism produces reactive oxygen species, including hydrogen peroxide (H_2O_2). H_2O_2 helps germination by breaking dormancy, but it is also toxic.

To counter the toxic effects of H_2O_2 and prevent cellular damage, germinating seeds produce catalase, an enzyme that breaks down H_2O_2 to water and oxygen.

A class was divided into six groups with each group testing the seedlings of each age. Each group's set of results (for 0.5, 2, 4, 6, and 10 days) therefore represents one trial.

The apparatus and method

Ten grams of germinating mung bean seeds (0.5, 2, 4, 6, or 10 days old) were crushed and placed in a conical flask. There were six trials at each of the five seedling ages. With each trial, 20 cm³ of 20 vol H_2O_2 was added to the flask at time 0 and the reaction was run for 30 seconds.

The oxygen released was collected via a tube into an inverted measuring cylinder. The volume of oxygen produced is measured by the amount of water displaced from the cylinder. The results are presented in the table below:

NEED HELP? See Activities 6, 23, 24

Stage of germination (days)	Volume of oxygen collected after 30 s (cm³)						Mean	Standard deviation	Mean rate (cm³ s⁻¹ g⁻¹)
Group (trial) #	1	2	3	4	5	6			
0.5	9.5	10	10.7	9.5	10.2	10.5			
2	36.2	30	31.5	37.5	34	40			
4	59	66	69	60.5	66.5	72			
6	39	31.5	32.5	41	40.3	36			
10	20	18.6	24.3	23.2	23.5	25.5			

1. Write the equation for the catalase reaction with hydrogen peroxide: _____

2. Complete the table above to summarize the data from the six trials:

   (a) Calculate the mean volume of oxygen for each stage of germination and enter the values in the table.

   (b) Calculate the standard deviation for each mean and enter the values in the table (you may use a spreadsheet).

   (c) Calculate the mean rate of oxygen production in cm³ per second per gram. For the purposes of this exercise, assume that the weight of germinating seed in every case was 10.0 g.

© 2016 **BIOZONE** International
ISBN: 978-1-927309-46-9
Photocopying Prohibited

PRACTICES   PRACTICES   PRACTICES   WEB

56   **KNOW**

3. (a) What sort of graph would you use to plot the results of this experiment? _____

    (b) Explain your choice: _____

    _____

    (c) Use the tabulated data to plot the results on the grid provided below. Include the standard deviation as error bars above and below each mean.

4. (a) Describe the trend in the data: _____

    _____

    _____

    (b) Explain the relationship between stage of germination and catalase activity shown in the data: _____

    _____

    _____

    (c) Do the results support the students' hypothesis? _____

    _____

5. Describe any potential sources of errors in the apparatus or the procedure: _____

    _____

    _____

    _____

6. Describe two things that might affect the validity of findings in this experimental design: _____

    _____

    _____

© 2016 **BIOZONE** International
ISBN: 978-1-927309-46-9
Photocopying Prohibited

57 Organ Systems Work Together

Key Idea: The different organ systems of an organism have specific roles and interact to bring about the efficient functioning of the body.

▶ An **organ system** is a group of organs that work together to perform a certain group of tasks.

▶ Although each system has a specific job (e.g. digestion, reproduction, internal transport, or gas exchange) the organ systems must interact to maintain the functioning of the organism.

▶ There are 11 organ (body) systems in humans. Plants have fewer organ systems.

▶ The example on this page shows how the muscular and skeletal organ systems in humans work together to achieve movement in the arm.

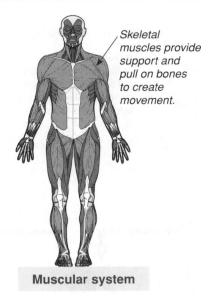

Skeletal muscles provide support and pull on bones to create movement.

Muscular system

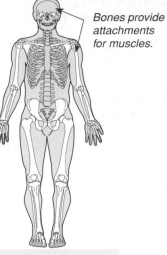

Bones provide attachments for muscles.

Skeletal system

Bones act with muscles to form levers that enable movement

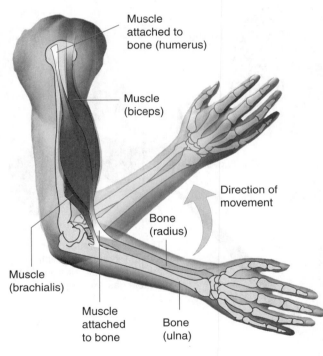

Muscle attached to bone (humerus)

Muscle (biceps)

Muscle (brachialis)

Muscle attached to bone

Bone (radius)

Bone (ulna)

Direction of movement

While you are sitting at your desk, lay your forearm on the desk, palm facing upwards. Now put your other hand on your upper arm and raise your resting arm towards you, bending at the elbow. You would have felt the muscles in your upper arm move. This movement required two organ systems, the skeletal and muscular systems, to work together.

What you felt was the muscle contracting (shortening), causing the bone attached to it to move at its joint, creating movement. A pair of muscles, each with opposing actions, can work in opposition to create movement of a body part.

1. What is an organ system? _____

_____

2. (a) What is the role of the skeletal system in movement? _____

_____

(b) What is the role of the muscular system in movement? _____

_____

_____

PRACTICES   CCC

 KNOW

58 Circulation and Gas Exchange Interactions

Key Idea: The circulatory and respiratory systems interact to provide the body's tissues with oxygen and remove carbon dioxide.

Circulatory system

Function

Delivers oxygen (O_2) and nutrients to all cells and tissues. Removes carbon dioxide (CO_2) and other waste products of metabolism. CO_2 is transported to the lungs.

Components

▶ Heart

▶ Blood vessels:
- Arteries
- Veins
- Capillaries

▶ Blood

Interaction between systems

In vertebrates, the respiratory system and cardiovascular system interact to supply oxygen and remove carbon dioxide from the body.

Respiratory system

Function

Provides surface for gas exchange. Moves fresh air into and stale air out of the body.

Components

▶ Airways:
- Pharynx
- Larynx
- Trachea

▶ Lungs:
- Bronchi
- Bronchioles
- Alveoli

▶ Diaphragm

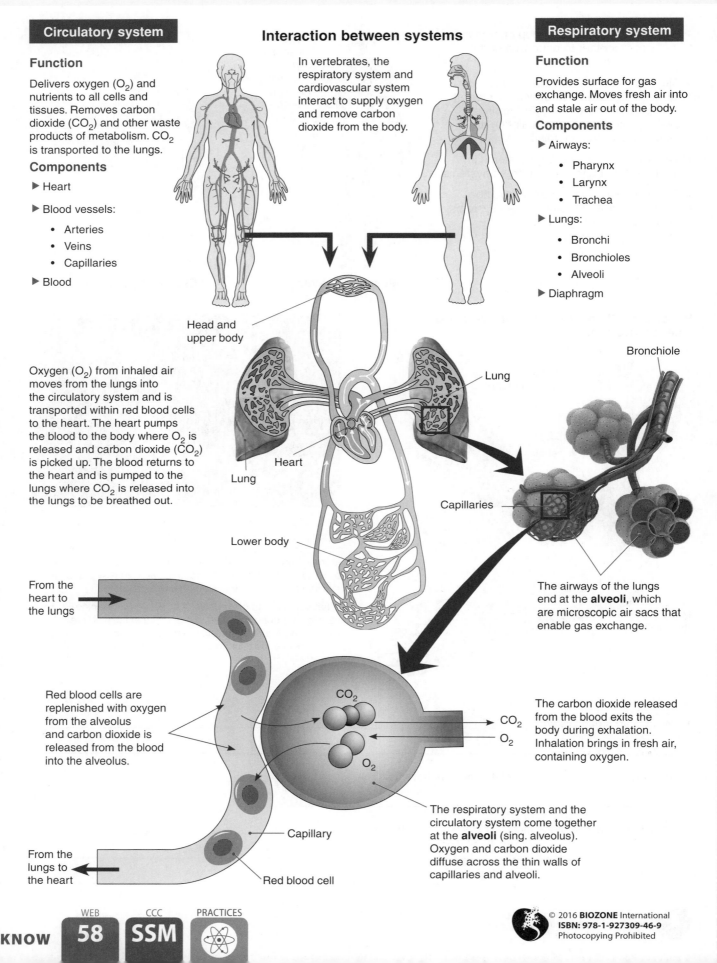

Head and upper body

Oxygen (O_2) from inhaled air moves from the lungs into the circulatory system and is transported within red blood cells to the heart. The heart pumps the blood to the body where O_2 is released and carbon dioxide (CO_2) is picked up. The blood returns to the heart and is pumped to the lungs where CO_2 is released into the lungs to be breathed out.

Lung

Heart

Lung

Lower body

Bronchiole

Lung

Capillaries

The airways of the lungs end at the **alveoli**, which are microscopic air sacs that enable gas exchange.

From the heart to the lungs

Red blood cells are replenished with oxygen from the alveolus and carbon dioxide is released from the blood into the alveolus.

CO_2

CO_2

O_2

O_2

The carbon dioxide released from the blood exits the body during exhalation. Inhalation brings in fresh air, containing oxygen.

Capillary

From the lungs to the heart

Red blood cell

The respiratory system and the circulatory system come together at the **alveoli** (sing. alveolus). Oxygen and carbon dioxide diffuse across the thin walls of capillaries and alveoli.

WEB   CCC   PRACTICES

KNOW 58 SSM

Responses to exercise

▶ During exercise, your body needs more oxygen to meet the extra demands placed on the muscles, heart, and lungs. At the same time, more carbon dioxide must be expelled. To meet these increased demands, blood flow must increase. This is achieved by increasing the rate of heart beat. As the heart beats faster, blood is circulated around the body more quickly, and exchanges between the blood and tissues increase.

▶ The arteries and veins must be able to resist the extra pressure of higher blood flow and must expand (dilate) to accommodate the higher blood volume. If they didn't, they could rupture (break). During exercise, the muscular, cardiovascular, and nervous systems interact to maintain the body's systems in spite of increased demands (right).

Muscular system
Increased activity increases demand for oxygen and nutrients.

Nervous system

Heart
Heart beats faster and rate of blood flow increases

Blood vessels
Arteries dilate (widen) to accommodate increased blood flow

Delivery of blood to capillaries of working muscle increases

A thick layer of elastic tissue and smooth muscle. When the smooth muscle relaxes, the artery expands to allow more blood to flow.

Capillaries dilate during exercise to increase the rate of exchanges of gases, nutrients, and wastes between the blood and the tissues.

Elastic outer layer prevents the artery over-expanding.

Endothelium is in contact with the blood

Muscular activity helps return blood to the heart

Valves stop back-flow of blood

Artery
The strong stretchy structure of arteries enables them to respond to increases in blood flow and pressure as more blood is pumped from the heart.

Vein
Veins return blood to the heart. They are less muscular than arteries, but valves and the activity of skeletal muscles, especially during exercise, help venous return.

1. In your own words, describe how the circulatory system and respiratory system work together to provide the body with oxygen and remove carbon dioxide:

_____

_____

_____

_____

_____

_____

2. (a) What happens to blood flow during exercise? _____

   (b) How do body systems interact to accommodate the extra blood flow needed when a person exercises?

   _____

   _____

   _____

   _____

© 2016 **BIOZONE** International
ISBN: 978-1-927309-46-9
Photocopying Prohibited

59 Circulation and Digestive Interactions

Key Idea: The circulatory and digestive systems interact to provide the body's tissues with nutrients.

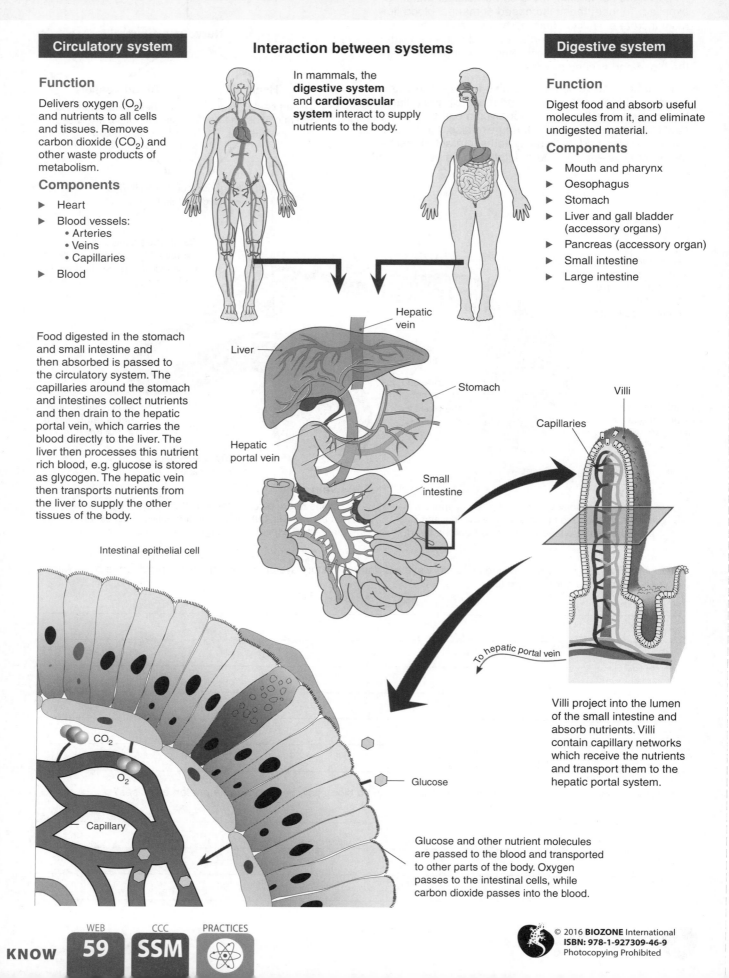

Circulatory system

Function

Delivers oxygen (O_2) and nutrients to all cells and tissues. Removes carbon dioxide (CO_2) and other waste products of metabolism.

Components

▶ Heart
▶ Blood vessels:
  • Arteries
  • Veins
  • Capillaries
▶ Blood

Food digested in the stomach and small intestine and then absorbed is passed to the circulatory system. The capillaries around the stomach and intestines collect nutrients and then drain to the hepatic portal vein, which carries the blood directly to the liver. The liver then processes this nutrient rich blood, e.g. glucose is stored as glycogen. The hepatic vein then transports nutrients from the liver to supply the other tissues of the body.

Interaction between systems

In mammals, the **digestive system** and **cardiovascular system** interact to supply nutrients to the body.

Hepatic vein

Liver

Stomach

Hepatic portal vein

Small intestine

Digestive system

Function

Digest food and absorb useful molecules from it, and eliminate undigested material.

Components

▶ Mouth and pharynx
▶ Oesophagus
▶ Stomach
▶ Liver and gall bladder (accessory organs)
▶ Pancreas (accessory organ)
▶ Small intestine
▶ Large intestine

Villi

Capillaries

To hepatic portal vein

Villi project into the lumen of the small intestine and absorb nutrients. Villi contain capillary networks which receive the nutrients and transport them to the hepatic portal system.

Intestinal epithelial cell

CO_2

O_2

Capillary

Glucose

Glucose and other nutrient molecules are passed to the blood and transported to other parts of the body. Oxygen passes to the intestinal cells, while carbon dioxide passes into the blood.

© 2016 **BIOZONE** International
ISBN: 978-1-927309-46-9
Photocopying Prohibited

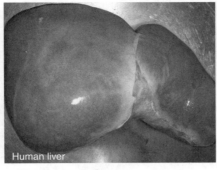

Human liver

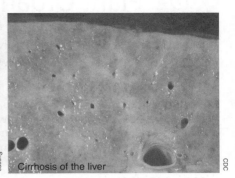

Cirrhosis of the liver

Blood flow to the digestive tract increases steadily after a meal and remains elevated for about 2.5 hours, reaching a maximum after about 30 minutes. During exercise, blood flow in the digestive tract is reduced as it is redirected to the muscles.

Nutrients, e.g. minerals, sugars, and amino acids, are transported in the blood plasma to the liver. The liver receives nutrient-rich deoxygenated blood from the digestive system via the hepatic portal vein and oxygen rich blood from the hepatic artery.

Scarring of the liver tissue, or cirrhosis, can result in portal hypertension (high blood pressure). The scarred tissue obstructs blood flow in the liver. This causes pressure to build up in upstream blood vessels, resulting in swelling and possible hemorrhage.

1. How are nutrients transported in the blood? _____

_____

2. Explain how a liver cirrhosis affects the circulatory system: _____

_____

_____

_____

_____

3. (a) At which two points in the body do the digestive and circulatory systems directly interact? _____

_____

   (b) Explain what is happening at these points: _____

_____

_____

_____

4. (a) What happens to blood flow to the digestive tract after a meal? _____

_____

   (b) Explain why it is often recommended that a person should exercise within 2.5 hours of eating, or eat within half an hour after exercising to gain a most benefit from the exercise (in terms of muscle development):

_____

_____

_____

_____

5. In your own words, describe how the circulatory and digestive systems work together to provide the body with nutrients:

_____

_____

_____

_____

_____

 © 2016 BIOZONE International
ISBN: 978-1-927309-46-9
Photocopying Prohibited

60 Plant Organ Systems

Key Idea: The plant body is divided into the shoot system (stems, leaves, and other above-ground parts) and the below-ground root system.

Plants have fewer organ systems than animals because they are simpler and have lower energy demands. The two primary plant organ systems are the shoot system and the root system.

Shoot system

The above-ground parts of the plant: including organs such as leaves, buds, stems, and the flowers and fruit (or cones) if present. All parts of the shoot system produce hormones.

The shoot and root systems of plants are connected by transport tissues (xylem and phloem) that are continuous throughout the plant.

Leaves

► Manufacture food via photosynthesis.

► Exchange gases with the environment.

► Store food and water.

Stems

► Transport water and nutrients between roots and leaves.

► Support and hold up the leaves, flowers and fruit.

► Produce new tissue for photosynthesis and support.

► Store food and water.

Structures for sexual reproduction

► Reproductive structures are concerned with passing on genes to the next generation.

► Flowers or cones are the reproductive structures of seed plants (angiosperms and gymnosperms).

► Fruits provide flowering plants with a way to disperse the seeds.

Root system

The below-ground parts of the plant, including the roots and root hairs.

Roots

► Anchor the plant in the soil

► Absorb and transport minerals and water

► Store food

► Produce hormones

► Produce new tissue for anchorage and absorption.

1. Describe how each of the following systems provides for the essential functions of life for the plant:

   (a) Root system: _____

   _____

   (b) Shoot system: _____

   _____

2. In the following list of plant functions, circle in blue the functions that are shared by the root and shoot system, circle in red those unique to the shoot system, and circle in black those unique to the root system:

   Photosynthesis, transport, absorption, anchorage, storage, sexual reproduction, hormone production, growth

CCC   PRACTICES

KNOW SSM

61 Interacting Systems in Plants

Key Idea: The shoot and root systems of plants interact to balance water uptake and loss, so that the plant can maintain the essential functions of life.

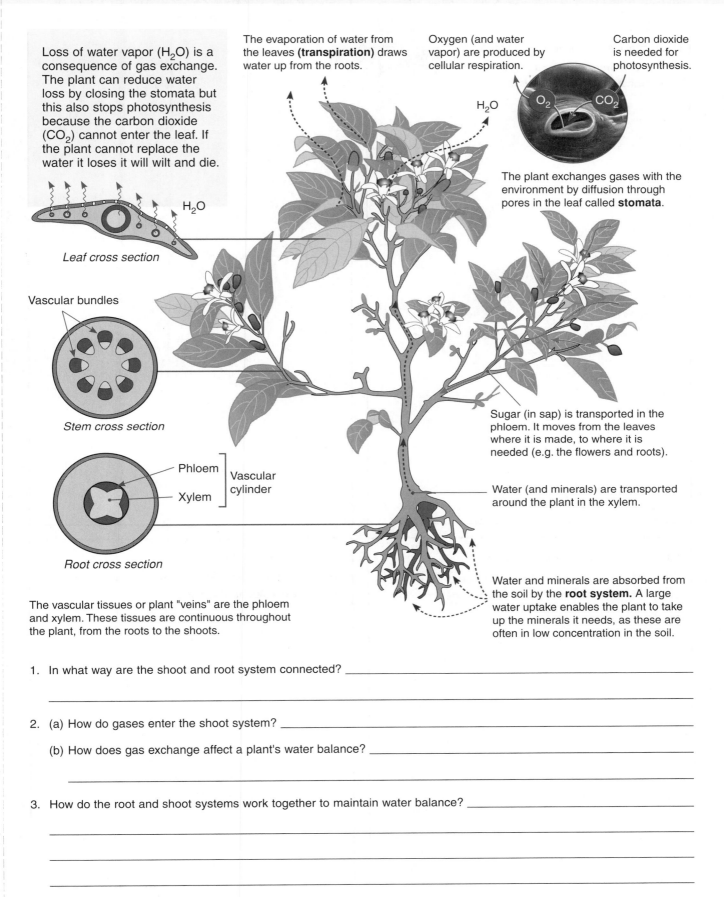

Loss of water vapor (H_2O) is a consequence of gas exchange. The plant can reduce water loss by closing the stomata but this also stops photosynthesis because the carbon dioxide (CO_2) cannot enter the leaf. If the plant cannot replace the water it loses it will wilt and die.

H_2O

Leaf cross section

The evaporation of water from the leaves **(transpiration)** draws water up from the roots.

H_2O

Oxygen (and water vapor) are produced by cellular respiration.

O_2 CO_2

Carbon dioxide is needed for photosynthesis.

The plant exchanges gases with the environment by diffusion through pores in the leaf called **stomata**.

Vascular bundles

Stem cross section

Phloem
Xylem
} Vascular cylinder

Root cross section

The vascular tissues or plant "veins" are the phloem and xylem. These tissues are continuous throughout the plant, from the roots to the shoots.

Sugar (in sap) is transported in the phloem. It moves from the leaves where it is made, to where it is needed (e.g. the flowers and roots).

Water (and minerals) are transported around the plant in the xylem.

Water and minerals are absorbed from the soil by the **root system.** A large water uptake enables the plant to take up the minerals it needs, as these are often in low concentration in the soil.

1. In what way are the shoot and root system connected? _____

_____

2. (a) How do gases enter the shoot system? _____

   (b) How does gas exchange affect a plant's water balance? _____

   _____

3. How do the root and shoot systems work together to maintain water balance? _____

_____

_____

_____

PRACTICES   CCC   WEB

SSM   61   **KNOW**

62 Chapter Review

Summarize what you know about this topic so far under the headings provided. You can draw diagrams or mind maps, or write notes to organize your thoughts. Use the checklist in the introduction and the hints to help you:

Cells provide essential life functions

HINT: Cell organelles and their relationship to cell processes.

DNA codes for proteins

HINT: Describe the structure and role of DNA and RNA. Include information about the structure of proteins and its relationship to the genetic code.

Interactions between organ systems

HINT: What is the role of specialized cells in the body? How do organ systems work together?

Proteins carry out the work of cells

HINT: Globular and fibrous proteins and their different roles in the body.

© 2016 BIOZONE International
ISBN: 978-1-927309-46-9
Photocopying Prohibited

REVISE

63 KEY TERMS AND IDEAS: Did You Get It?

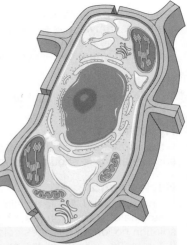

1. (a) Is the cell on the right an animal or plant cell?

   _____

   (b) List the features of the cell that support your answer:

   _____

   _____

   _____

   _____

2. Organelles are the cell's "organs": They carry out the cell's work. Draw lines to match the organelle on the left with its description in the right hand column.

Cell wall	A bilayered structure of phospholipids surrounding a cell. It controls the movement of substances into and out of the cell.
Chloroplast	A structure present in plant cells but not animal cells. It is found outside the plasma membrane and gives rigidity to the cell.
Nucleus	Membrane-bound area within a eukaryotic cell where most of a cell's DNA is found.
Mitochondrion	These structures are involved in making proteins in a cell.
Plasma membrane	An organelle found in plants which contains chlorophyll and is the site of photosynthesis.
Ribosome	This organelle is involved in generating the cell's usable energy molecule, ATP.

3. (a) What type of metabolic reaction is taking place in the diagram right?

   _____

   (b) What is occurring during this reaction? _____

   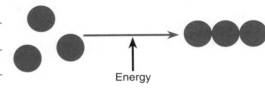

   _____

   _____

   (c) Give an example of this type of metabolic reaction: _____

4. A grasshopper has the following percentages of nucleotides in its DNA: A = 29.3, G = 20.5, C = 20.7, T = 29.3, %GC = 41.2, %AT = 58.6. For a rat, the percentages are A = 28.6, G = 21.4, C = 20.5, T = 28.4, %GC = 42.9, %AT = 57.0. How do these numbers demonstrate Chargaff's rules?

   _____

   _____

5. Fill in the missing words in the paragraph below. Use the word list below to help you.
   Guanine, nucleotides, protein, shape, double-helix, base, DNA, phosphate, denatured, cytosine, base pairing

   All living cells contain genetic material called _____, which stores and transmits the information an organism needs

   to develop, function, and reproduce. DNA is very large. It is made up of building blocks called _____ joined

   together. A nucleotide has three parts, a _____ , a sugar, and a _____ group. There are four

   different types of nucleotides in DNA, adenine,_____ , _____ , and thymine. They pair together

   in a very specific way called the _____ _____ rule. This pairing contributes to DNA's characteristic

   _____ shape. Segments of DNA, called genes, code for a specific _____. Proteins are very

   important because they control every aspect of an organism's structure and function. The _____ of a protein

   determines its functional role. If a protein loses its shape, becomes _____ and can no longer perform its role.

TEST

64 Summative Assessment

Amylase is a digestive enzyme that hydrolyzes (breaks down) starch into the sugars maltose and glucose. In mammals, amylase is secreted by the salivary glands in the mouth into the saliva and by the pancreas into the small intestine. The mouth has a near neutral pH (around 7.0) and a temperature about 37°C. Like all enzymes, amylase works best under certain conditions. In the experiment below, students investigated how pH affected amylase activity.

Aim
To determine the optimum pH for salivary amylase.

Hypothesis
If the normal pH for saliva is 6.5-7.5, then the optimum pH for salivary amylase should be approximately pH 7.

Background
Iodine solution (I_2/KI) is a yellow/orange color, but in the presence of starch, it turns a blue/black color. When the iodine solution no longer changes color after the sample is added (i.e. remains yellow), all the starch has been hydrolyzed.

Method
The experiment was performed at room temperature. A single drop of 0.1 M iodine solution was placed into the wells of spotting plates. 2 cm^3 of 1% amylase solution and 1 cm^3 of a buffered solution, pH 4, were added to the test tube. The solutions were mixed and 2 cm^3 of a 1% starch solution was added. A timer was immediately started. After 10 seconds a plastic pipette was used to remove a small amount of solution. A single drop was added to the first well of the **spotting plate** (right) and the remaining solution inside the pipette returned to the test tube. This action was repeated at 10 second intervals, adding a drop of the reaction solution into a new well until the iodine solution no longer changed color (remained yellow/orange). The experiment was repeated using buffer solutions of pH 5, 6, 7, and 8.

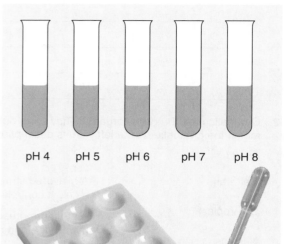

pH 4   pH 5   pH 6   pH 7   pH 8

Spotting plate: each well contains a single drop of 0.1 M iodine solution (iodine dissolved in a solution of potassium iodide). Multiple spotting plates were set up to accommodate the number of tests required.

Results
The table below shows how many drops it took until there was no color change (the iodine solution remained yellow).

pH	Number of drops until no color change occurred	Number of seconds until no color change occurred	Reaction rate (s^{-1})
4	19	190	
5	12	120	
6	10	100	
7	6	60	
8	29	290	

1. Complete the table (above) by calculating the reaction rate for each pH (1 ÷ seconds):

2. (a) Graph the reaction rate vs pH on the grid.

   (b) Identify the pH where amylase activity was the highest: _____

   (c) Is this what you had expected? Explain: _____

   _____

© 2016 **BIOZONE** International
ISBN: 978-1-927309-46-9
Photocopying Prohibited

3. The students repeated the experiment at pH 1. Each sample turned blue/black when added to the iodine even after five minutes of sampling. Explain what has happened here:

_____

_____

_____

In a second experiment, students wanted to determine the **temperature optimum** for salivary amylase.

4. (a) Predict the optimum temperature for salivary amylase: _____

   (b) Explain your reasoning: _____

   _____

   _____

5. (a) What pH should the experiment be carried out at? _____

   (b) Why did you choose this pH? _____

6. In the space below, design an experiment to determine the optimum temperature for salivary amylase: _____

_____

_____

_____

_____

_____

_____

_____

_____

_____

_____

_____

_____

_____

_____

_____

7. A group of students tested amylase activity at 60°C. Samples were taken at 1 minute intervals and tested for the presence of starch using iodine solution. Their results, all blue/black, are shown on the right below.

   (a) Has the starch been broken down? _____

   (b) Explain your answer in (a): _____

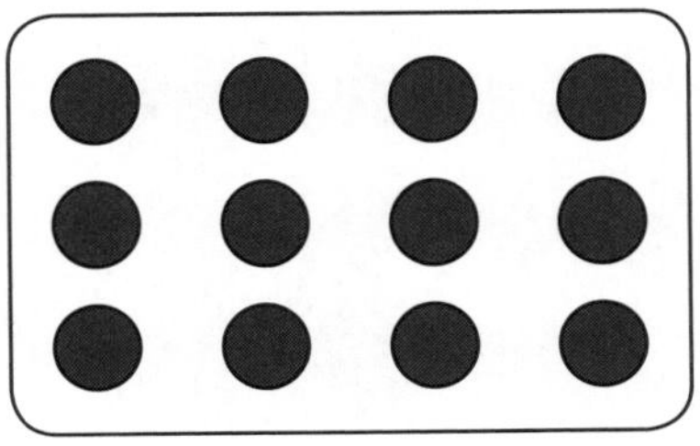

   _____

   _____

   _____

8. Use your knowledge of enzymes to explain what would happen to enzyme activity if the experiment was carried out below 10°C:

_____

_____

© 2016 **BIOZONE** International
ISBN: 978-1-927309-46-9
Photocopying Prohibited

Feedback Mechanisms

Key terms

diabetes mellitus

feedback
mechanisms

homeostasis

negative feedback

positive feedback

thermoregulation

transpiration

Disciplinary core ideas

Show understanding of these core ideas

Activity number

Feedback mechanisms maintain the internal environment within certain limits

☐ 1 Homeostatic mechanisms help the body maintain a constant internal environment, even when external conditions are changing. — 65

☐ 2 The body's systems must maintain homeostasis so essential life process can be carried out to maintain life. — 66

☐ 3 During exercise, the circulatory and respiratory systems are mainly responsible for maintaining homeostasis. — 75 76 77

Feedback mechanisms can be positive or negative

☐ 4 Positive feedback mechanisms amplify a response, usually to achieve a specific outcome. Labor is an example of a physiological process involving positive feedback in which the outcome is the delivery of the baby. — 68

☐ 5 Negative feedback mechanisms have a stabilizing effect, are self correcting, and encourage a return to the steady state. Negative feedback regulates most natural systems, e.g. in physiology and ecology. — 67

☐ 6 Thermoregulation (regulation of body temperature) is controlled by negative feedback. Failure to maintain a constant body temperature can be fatal. — 69 70 71 72

☐ 7 Blood sugar levels are tightly regulated by negative feedback mechanisms. Diabetes mellitus is a life-threatening disease which can occur when normal regulatory controls no longer work. — 73 74

☐ 8 Plants lose water by transpiration. They must take up enough water to balance the water loss so they can continue to carry out essential life processes. — 78 79

Crosscutting concepts

Understand how these fundamental concepts link different topics

Activity number

☐ 1 **SC** ▸ Negative feedback can stabilize a system, whereas positive feedback can have a destabilizing effect. — 65-68 70 71 73-79

Science and engineering practices

Demonstrate competence in these science and engineering practices

Activity number

☐ 1 Use an evidence-based model to show how positive feedback operates. — 68

☐ 2 Use a model based on evidence to show how negative feedback mechanisms maintain homeostasis. — 67 70 71 73

☐ 3 Carry out an investigation to show how the body maintains homeostasis, for example, during exercise. — 76

☐ 4 Carry out an investigation and analyze data to show how plants maintain water balance in changing environmental conditions. — 79

☐ 5 Use mathematics and computational tools to support explanations of how organisms maintain homeostasis. — 77

65 Homeostasis

Key Idea: Homeostasis is the ability to maintain a constant internal environment despite changes in the external environment.

What is homeostasis?

Homeostasis literally means "constant state". Organisms maintain homeostasis, i.e. a relatively constant internal environment, even when the external environmental is changing. This takes energy.

For example, when you exercise (right), your body must keep your body temperature constant at about 37.0 °C despite the increased heat generated by activity. Similarly, you must regulate blood sugar levels and blood pH, water and electrolyte balance, and blood pressure. Your body's organ systems carry out these tasks.

To maintain homeostasis, the body must detect changes in the environment (through receptors), process this sensory information, and respond to it appropriately. The response provides new feedback to the receptor. These three components are illustrated below.

How homeostasis is maintained

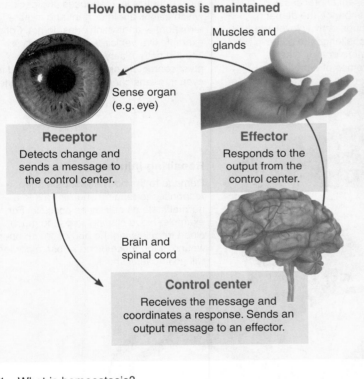

Muscles and glands

Sense organ (e.g. eye)

Receptor
Detects change and sends a message to the control center.

Effector
Responds to the output from the control center.

Brain and spinal cord

Control center
Receives the message and coordinates a response. Sends an output message to an effector.

The analogy of a thermostat on a heater is a good way to understand how homeostasis is maintained. A heater has sensors (a receptor) to monitor room temperature. It also has a control center to receive and process the data from the sensors. Depending on the data it receives, the control center activates the effector (heating unit), switching it on or off. When the room is too cold, the heater switches on. When it is too hot, the heater switches off. This maintains a constant temperature.

1. What is homeostasis? _____

_____

_____

2. What is the role of the following components in maintaining homeostasis:

   (a) Receptor: _____

   _____

   (b) Control center: _____

   _____

   (c) Effector: _____

   _____

 © 2016 **BIOZONE** International
ISBN: 978-1-927309-46-9
Photocopying Prohibited

66 Keeping in Balance

Key Idea: Essential life processes, such as growth, require the body's systems to be kept in balance. Many organ systems work together to maintain homeostasis.

Why is homeostasis important?

▶ An organism must constantly regulate its internal environment in order to carry out essential life processes, such as growing and responding to the environment. Changes outside of normal levels for too long can stop the body systems working properly, and can result in illness or death.

▶ Homeostasis relies on monitoring all the information received from the internal and external environment and coordinating appropriate responses. This often involves many different organ systems working together.

▶ Most of the time an organism's body systems are responding to changes at the subconscious level, but sometimes homeostasis is achieved by changing a behavior (e.g. finding shade if the temperature is too high).

Some examples of how the body keeps in balance

Regulating respiratory gases

Cellular respiration, which produces usable energy as ATP, takes place in all cells. It requires a constant supply of oxygen and produces carbon dioxide as a waste product. The changing demands of the body for supply of oxygen and removal of CO_2 (as when exercising) are met by the gas exchange and circulatory systems of the body.

Maintaining nutrient supply

Food and drink provide the energy (fuel) and nutrients the body needs to carry out the essential processes of life. Factors that change the demand for nutrients include activity level and environmental challenges (e.g. the body requires more energy when active or in extremes of temperature).

Coordinating responses

The body is constantly bombarded by stimuli from the environment. The brain must prioritize its responses and decide which stimuli are important and require a response, and which ones do not. For example, if a vehicle suddenly swerves into the path of a cyclist in a race, he must coordinate his responses to avoid it, even it means changing his chosen route.

Repairing injuries

Damage to the body's tissues triggers responses to repair it and return it to a normal state as quickly as possible. For example, blood clotting stops too much blood from leaving the body from an open wound. If too much blood is lost, a person will go into fatal shock.

Maintaining fluid and electrolyte balance

Fluid and salts are taken in with food and drink but imbalances quickly lead to cellular disruption and death if not treated. Water balance is maintained by ensuring that the amount of fluid consumed and generated by metabolism equals the amount of water lost in urine and feces, sweat, and breathing. Maintaining the body's levels of fluid and salts is the job of the kidneys.

Protecting the body against disease

Like all organisms, we are under constant attack from disease-causing organisms (pathogens) which can cause damage to the body's systems. The body's skin and immune system produce chemicals and cells that act to prevent the entry of pathogens and limit the damage they cause if they do enter. The cardiovascular system circulates these components through the body.

WEB   CCC

KNOW   **66**   **SC**

The body's organ systems work together to maintain homeostasis

Organ systems work together to maintain the environment necessary for the functioning of the body's cells. A constant internal environment allows an organism to be somewhat independent of its external environment, so that it can move about even as its environment changes. The simplified example below illustrates how organ systems interact to exchange material with each other to maintain a constant internal environment.

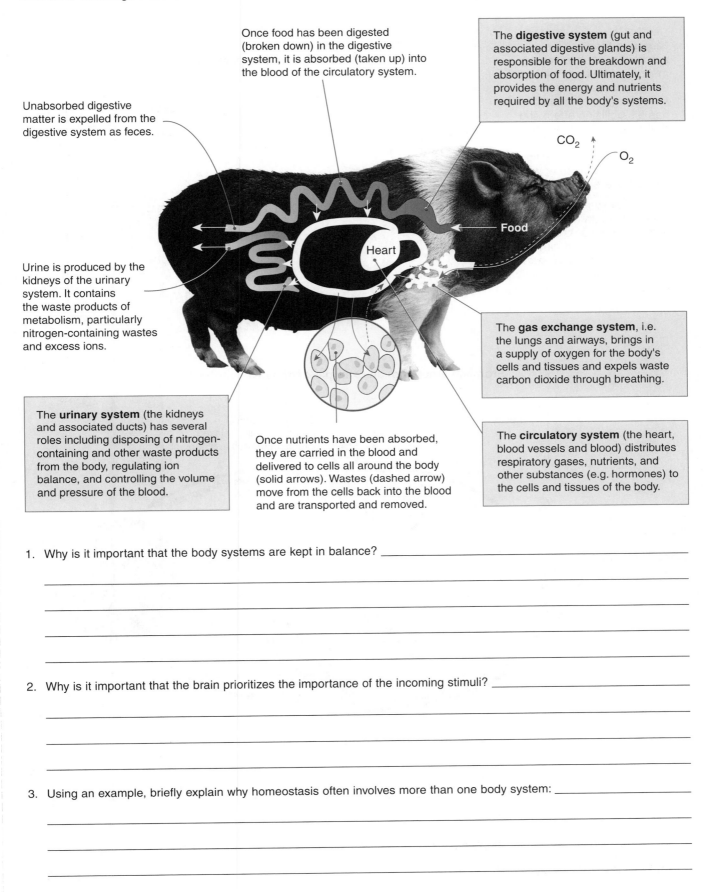

Once food has been digested (broken down) in the digestive system, it is absorbed (taken up) into the blood of the circulatory system.

The **digestive system** (gut and associated digestive glands) is responsible for the breakdown and absorption of food. Ultimately, it provides the energy and nutrients required by all the body's systems.

Unabsorbed digestive matter is expelled from the digestive system as feces.

CO_2

O_2

Food

Heart

Urine is produced by the kidneys of the urinary system. It contains the waste products of metabolism, particularly nitrogen-containing wastes and excess ions.

The **gas exchange system**, i.e. the lungs and airways, brings in a supply of oxygen for the body's cells and tissues and expels waste carbon dioxide through breathing.

The **urinary system** (the kidneys and associated ducts) has several roles including disposing of nitrogen-containing and other waste products from the body, regulating ion balance, and controlling the volume and pressure of the blood.

Once nutrients have been absorbed, they are carried in the blood and delivered to cells all around the body (solid arrows). Wastes (dashed arrow) move from the cells back into the blood and are transported and removed.

The **circulatory system** (the heart, blood vessels and blood) distributes respiratory gases, nutrients, and other substances (e.g. hormones) to the cells and tissues of the body.

1. Why is it important that the body systems are kept in balance? _____

_____

_____

_____

_____

2. Why is it important that the brain prioritizes the importance of the incoming stimuli? _____

_____

_____

_____

3. Using an example, briefly explain why homeostasis often involves more than one body system: _____

_____

_____

_____

67 Negative Feedback Mechanisms

Key Idea: Negative feedback mechanisms detect changes in the internal environment away from the normal and act to return the internal environment back to a steady state.

Negative feedback is a control system which maintains the body's internal environment at a steady state. Negative feedback has a stabilizing effect and discourages variations from a set point. It works by returning internal conditions back to a steady state when variations are detected (right).

Most body systems achieve homeostasis through negative feedback. Body temperature, blood glucose levels, and blood pressure are all controlled by negative feedback mechanisms.

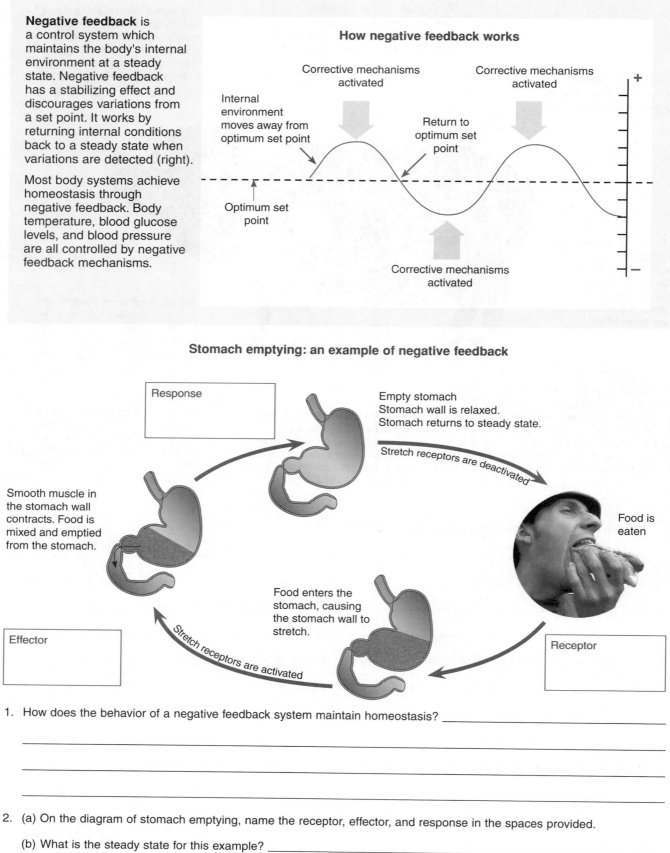

How negative feedback works

Stomach emptying: an example of negative feedback

1. How does the behavior of a negative feedback system maintain homeostasis? _____

_____

_____

_____

2. (a) On the diagram of stomach emptying, name the receptor, effector, and response in the spaces provided.

(b) What is the steady state for this example? _____

© 2016 **BIOZONE** International
ISBN: 978-1-927309-46-9
Photocopying Prohibited

68 Positive Feedback Mechanisms

Key Idea: Positive feedback mechanisms amplify a physiological response in order to achieve a particular outcome.

Positive feedback mechanisms amplify (increase) or speed up a physiological response, usually to achieve a particular outcome. Examples of positive feedback include fruit ripening, fever, blood clotting, childbirth (labor) and lactation (production of milk). A positive feedback mechanism stops when the end result is achieved (e.g. the baby is born, a pathogen is destroyed by a fever, or ripe fruit falls off a tree).

Positive feedback is less common than negative feedback because it creates an escalation in response, which is unstable. This response can be dangerous (or even cause death) if it is prolonged.

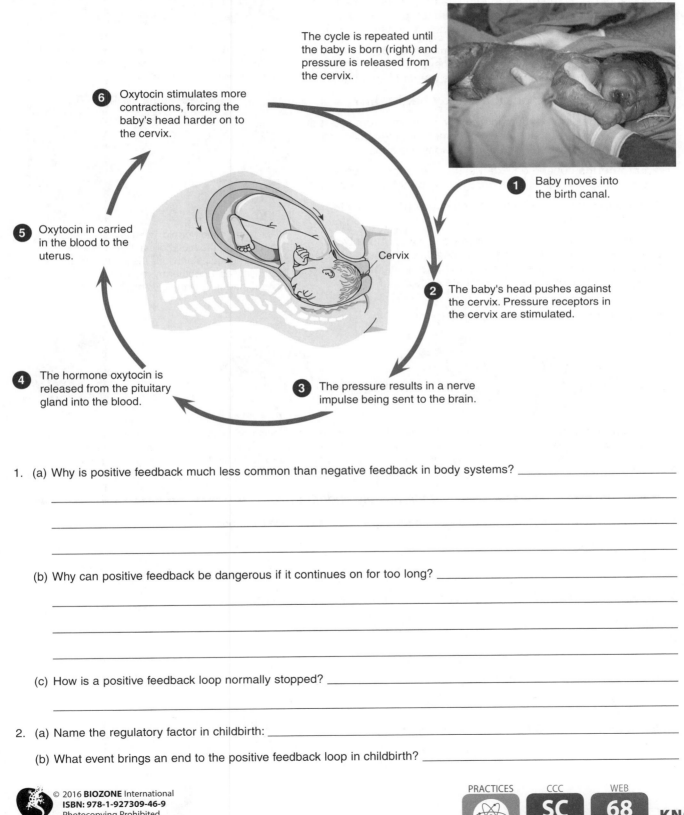

The cycle is repeated until the baby is born (right) and pressure is released from the cervix.

6 Oxytocin stimulates more contractions, forcing the baby's head harder on to the cervix.

5 Oxytocin in carried in the blood to the uterus.

4 The hormone oxytocin is released from the pituitary gland into the blood.

Cervix

1 Baby moves into the birth canal.

2 The baby's head pushes against the cervix. Pressure receptors in the cervix are stimulated.

3 The pressure results in a nerve impulse being sent to the brain.

1. (a) Why is positive feedback much less common than negative feedback in body systems? _____

    _____

    _____

    _____

   (b) Why can positive feedback be dangerous if it continues on for too long? _____

    _____

    _____

    _____

   (c) How is a positive feedback loop normally stopped? _____

    _____

2. (a) Name the regulatory factor in childbirth: _____

   (b) What event brings an end to the positive feedback loop in childbirth? _____

PRACTICES   CCC   WEB

SC   68   **KNOW**

69 Sources of Body Heat

Key Idea: Animals have an optimal body temperature for their essential life processes. Ectotherms obtain heat from the environment. Endotherms generate heat from metabolism.

Why is body heat important?

The essential processes of life are regulated by enzymes and so require a certain optimum temperature in order to operate most efficiently. This optimal temperature varies depending on the organism. A bacterium living in a geothermal hot pool will have a very different optimal temperature to a soil bacterium in a woodland! Below the optimum temperature, metabolic reactions proceed very slowly. Above the optimum temperature, the enzymes may be damaged and the reaction doesn't proceed.

The average body temperature of mammals is ~38°C. For birds, it is ~40°C. Most snakes and lizards operate best in the range 24-35°C, although some operate in the mammalian range.

Where do animals get their body heat from?

So where does the heat come from? It can come from external sources, such as the sun, or it can come from metabolic activity. These two strategies divide animals into two groups.

▶ **Ectotherms** depend on external sources of heat (from the environment) for their heat energy (e.g. heat from the sun).

▶ **Endotherms** generate most of their body heat from internal metabolic processes.

In reality, animals often fall somewhere between these two extremes. For example, some large insects use muscular activity to raise the temperature of their thorax prior to flight and some snakes use metabolic heat to incubate eggs.

Reptiles are ectotherms. These captive turtles are basking under a lamp to raise their body temperature for activity. In the wild, a low body temperature reduces their ability to forage and escape predators.

All birds are endothermic. Even in cold Antarctic temperatures, the metabolic activity of these Emperor penguins provides the warmth to sustain life processes. Endothermy requires a lot of energy, so endotherms cannot go without food for long.

The ectotherm-endotherm continuum

Most (but not all) fish are fully ectothermic and rely solely on the environment for their body heat.

Snakes use heat energy from the environment to increase their body temperature for activity.

Some large insects, such as bumblebees, may raise their temperature for short periods through muscular activity.

Mammals and birds achieve high body temperatures through metabolic activity and reduction of heat losses.

→ *Increasingly endothermic* →

1. Distinguish between **ectotherms** and **endotherms** in terms of their sources of body heat: _____

_____

_____

2. (a) Why are the movements of many ectotherms slow in the early morning? _____

_____

_____

(b) Why could this be a disadvantage?_____

_____

(c) What might be an advantage of ectothermy? _____

_____

 © 2016 **BIOZONE** International
ISBN: 978-1-927309-46-9
Photocopying Prohibited

70 | Thermoregulation

Key Idea: Thermoregulation is the regulation of body temperature independently of changes in the environmental temperature. For homeotherms, this represents a large energetic cost.

Thermoregulation refers to the regulation of body temperature in the face of changes in the temperature of the environment. Animals show two extremes of body temperature tolerance:

▶ **Homeotherms** (all birds and mammals) maintain a constant body temperature independently of environmental variation.

▶ **Poikilotherms** allow their body temperature to vary with the temperature of the environment. Most fish and all amphibians cannot regulate body temperature at all, but most reptiles use behavior both to warm up and to avoid overheating (below).

We have seen that animals are classed as ectotherms or endotherms depending on their sources of heat energy. Most endotherms are also strict homeotherms (always thermoregulate) and most, but not all, ectotherms allow their body temperatures to vary with environmental fluctuations. Thermoregulation relies on physical, physiological, and behavioral mechanisms.

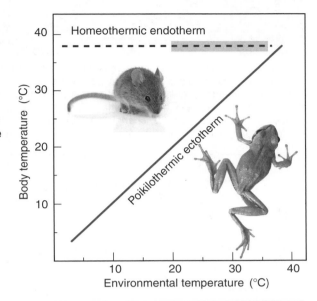

Homeothermic endotherm

Body temperature (°C): 10, 20, 30, 40

Poikilothermic ectotherm

Environmental temperature (°C): 10, 20, 30, 40

Mechanisms of thermoregulation

Homeothermic endotherm (mammal)

Always thermoregulate, largely by physiological mechanisms. This requires a large amount of energy, especially when outside the range of their normal body temperature.

Wool, hair, or fur traps air next to the skin providing an insulating layer to reduce heat loss and slow heat gain.

Panting and sweating cool through evaporation. Mammals usually sweat or pant but not both.

Heat can be generated by shivering.

In cold weather, many mammals cluster together to retain body heat.

Poikilothermic ectotherm (reptile)

Thermoregulate at the extremes of their temperature range. The energetic costs of this are much lower than for homeotherms because the environment provides heat energy for warming.

Increasing blood flow to the surface can help lose heat quickly.

Basking in the sun is common in lizards and snakes. The sun warms the body up and they seek shade to cool down.

Some lizards reduce points of contact with hot ground (e.g. standing on two legs instead of four) reducing heat uptake via conduction.

1. What is **thermoregulation**? _____

_____

2. (a) The graph above shows body temperature variations in a mammal and an amphibian with change in environmental temperature. Compare and explain their different responses:

_____

_____

(b) The blue shaded area on the graph marks the region where the energy cost of thermoregulation is lowest for the mouse. Why is the mouse using less energy to thermoregulate in this temperature range?

_____

_____

© 2016 **BIOZONE** International
ISBN: 978-1-927309-46-9
Photocopying Prohibited

PRACTICES   CCC   WEB

SC   70   **KNOW**

Liolaemus

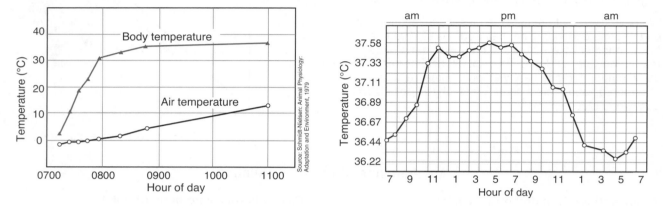

Core body temperature falls during sleep

The Peruvian mountain lizard (*Liolaemus*) emerges in the morning when the air temperature is below freezing. It exposes itself to the sun, rapidly heating up to a body temperature that enables it to be fully active (below). Once warm, it maintains its preferred body temperature of around 35°C by changing posture and orientation to the sun and thereby controlling the amount of heat absorbed.

Although human core body temperature is relatively stable around 36-37°C, it does vary slightly over a 24 hour period (below). These variations occur in response to the body's internal rhythm and not the environment. Assuming a person sleeps at night and is awake during the day, the lowest body temperatures occur in the early morning and the highest in the afternoon.

Source: Schmidt-Nielsen: Animal Physiology: Adaptation and Environment, 1979

3. Most of thermoregulation in homeotherms occurs through physiological processes, e.g. sweating. Describe another physiological mechanism that mammals use to thermoregulate and describe its effect:

_____

_____

4. Thermoregulation can be aided by both physical features (structure) and behavior. Describe an example of each:

   (a) Behavior: _____

   _____

   (b) Physical features: _____

   _____

5. Describe the main difference between the thermoregulatory mechanisms of a mammal and a lizard such as *Liolaemus*:

   _____

   _____

6. Both humans and the Peruvian mountain lizard show a daily variation in core body temperature. Describe two main differences between these daily variations:

   _____

   _____

   _____

   _____

7. What might be the advantage to humans to allow a slight fall in core body temperature when sleeping?

   _____

   _____

© 2016 BIOZONE International
ISBN: 978-1-927309-46-9
Photocopying Prohibited

71 Thermoregulation in Humans

Key Idea: The hypothalamus regulates body temperature in humans. It coordinates nervous and hormonal responses to keep the body temperature within its normal range.

The hypothalamus regulates temperature

▶ In humans, the temperature regulation center is a region of the brain called the hypothalamus. It has thermoreceptors that monitor core body temperature and has a 'set-point' temperature of 36.7°C (98.6°F).

▶ The hypothalamus acts like a thermostat. It registers changes in the core body temperature and also receives information about temperature change from thermoreceptors in the skin. It then coordinates the appropriate nervous and hormonal responses to counteract the changes and restore normal body temperature, as shown in the diagram below.

▶ When normal temperature is restored, the corrective mechanisms are switched off. This is an example of a negative feedback regulation.

▶ Infection can reset the set-point of the hypothalamus to a higher temperature. Homeostatic mechanisms then act to raise the body temperature to the new set point, resulting a fever (right).

Fever is an important defense against infection, but if the body temperature rises much above 42°C, a dangerous positive feedback loop can begin, making the body produce heat faster than it can get rid of it.

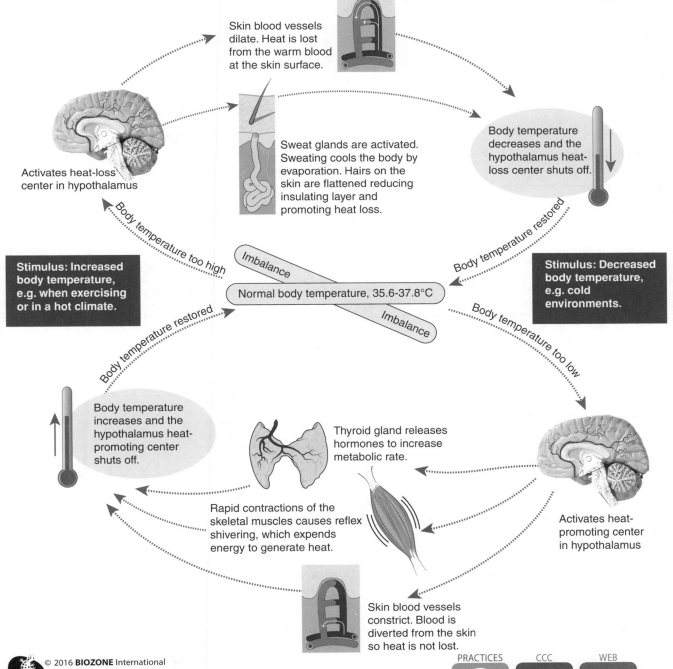

Skin blood vessels dilate. Heat is lost from the warm blood at the skin surface.

Sweat glands are activated. Sweating cools the body by evaporation. Hairs on the skin are flattened reducing insulating layer and promoting heat loss.

Activates heat-loss center in hypothalamus

Body temperature decreases and the hypothalamus heat-loss center shuts off.

Body temperature too high

Body temperature restored

Stimulus: Increased body temperature, e.g. when exercising or in a hot climate.

Imbalance

Normal body temperature, 35.6-37.8°C

Imbalance

Stimulus: Decreased body temperature, e.g. cold environments.

Body temperature restored

Body temperature too low

Body temperature increases and the hypothalamus heat-promoting center shuts off.

Thyroid gland releases hormones to increase metabolic rate.

Rapid contractions of the skeletal muscles causes reflex shivering, which expends energy to generate heat.

Activates heat-promoting center in hypothalamus

Skin blood vessels constrict. Blood is diverted from the skin so heat is not lost.

PRACTICES

CCC

WEB

SC 71 **KNOW**

Thermoregulation in newborns

▶ Newborn babies cannot fully thermoregulate until six months of age. They can become too cold or too hot very quickly.

▶ Newborns minimize heat loss by reducing the blood supply to the periphery (skin, hands, and feet). This helps to maintain the core body temperature. Increased brown fat activity and general metabolic activity generates heat. Newborns are often dressed in a hat to reduce heat loss from the head, and tightly wrapped to trap heat next to their bodies.

▶ Newborns lower their temperature by increasing peripheral blood flow. This allows heat to be lost, cooling the core temperature. Newborns can also reduce their body temperature by sweating, although their sweat glands are not fully functional until four weeks after birth.

Newborns cannot shiver to produce heat.

Heat losses from the head are high because the head is very large compared to the rest of the body.

A baby's body surface is three times greater than an adult's. There is greater surface area for heat to be lost from.

Newborns have thin skin, and blood vessels that run close to the skin, these features allow heat to be lost easily.

Newborns have very little white fat beneath their skin to insulate them against heat loss.

1. (a) Where is the temperature regulation center in humans located? _____

   (b) How does it carry out this role? _____
   _____

2. Describe the role of the following in maintaining a constant body temperature in humans:

   (a) The skin: _____
   _____

   (b) The muscles: _____
   _____

   (c) The thyroid gland: _____
   _____

3. How is negative feedback involved in keeping body temperature within narrow limits? _____
   _____
   _____
   _____

4. (a) Why does infection result in an elevated core body temperature? _____
   _____

   (b) What is the purpose of this? _____
   _____

   (c) Explain why a prolonged fever can be fatal: _____
   _____

5. (a) What features of a newborn cause it to lose heat quickly? _____
   _____

   (b) What mechanisms do newborns have to control body temperature? _____
   _____
   _____

© 2016 BIOZONE International
ISBN: 978-1-927309-46-9
Photocopying Prohibited

72 Body Shape and Heat Loss

Key Idea: Body shape influences how quickly heat is lost from the body's surface. Death can occur if body temperature deviates too far either side of the set point.

Body shape influences heat loss

▶ Body shape influences how heat is retained or lost. Animals with a lower surface area to volume ratio will lose less body heat per unit of mass than animals with a high surface area to volume ratio.

▶ In humans there is a negative relationship between surface area and latitude. Indigenous people living near the equator tend to have a higher surface area to volume ratio so they can lose heat quickly. In contrast, people living in higher latitudes near the poles have a lower surface area to volume ratio so that they can conserve heat.

▶ People from low latitudes (equatorial regions) have a taller, more slender body and proportionately longer limbs. Those at high latitudes are stockier, with shorter limbs.

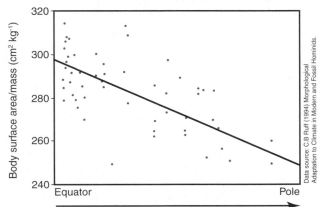
Relationship between ratio of surface area to body mass and latitude

Body surface area/mass (cm² kg⁻¹)

Increasing latitude

Data source: C.B Ruff (1994) Morphological Adaptation to Climate in Modern and Fossil Hominids.

The Inuit people of Arctic regions are stocky, with relatively short extremities. This body shape is well suited to reducing the surface area over which heat can be lost from the body.

The indigenous peoples of equatorial Africa, such as these young Kenyan men, and tall and slender, with long limbs. This body shape increases the surface area over which heat can be lost.

The relationship holds true for fossil hominids. Neanderthals, which inhabited Eurasia during the last glacial, had robust, stocky bodies relative to modern humans as this reconstruction shows.

1. Why would having a reduced surface area to volume ratio be an advantage in a cold climate? _____

_____

_____

2. Hypothermia is a condition that occurs when the body cannot generate enough heat and the core body temperature drops below 35°C. Prolonged hypothermia is fatal. Exposure to cold water results in hypothermia more quickly than exposure to the same temperature of air because water is much more effective than air at conducting heat away from the body.
In the graph (right), hypothermia resulting in death is highly likely in region 1 and highly unlikely in region 2.

(a) Which body shape has best survival at 15°C?

_____

(b) Explain your choice: _____

_____

_____

_____

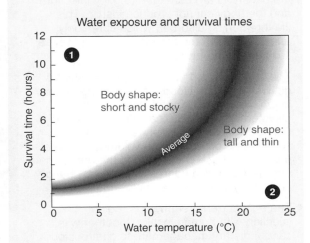
Water exposure and survival times

Survival time (hours)

Body shape: short and stocky

Average

Body shape: tall and thin

Water temperature (°C)

KNOW

73 Controlling Blood Glucose

Key Idea: Blood glucose levels are regulated by negative feedback involving two hormones, insulin and glucagon. Lack of insulin results in a disorder called type 1 diabetes.

The importance of blood glucose

▶ **Glucose** is the body's main energy source. It is chemically broken down during cellular respiration to generate ATP, which is used to power metabolism. Glucose is the main sugar circulating in blood, so it is often called blood sugar.

▶ Blood glucose levels are tightly controlled because cells must receive an adequate and regular supply of fuel. Prolonged high or low blood glucose causes serious physiological problems and even death. Normal activities, such as eating and exercise, alter blood glucose levels, but the body's control mechanisms regulate levels so that fluctuations are minimized and generally occur within a physiologically acceptable range. For humans this is 60-110 mg dL^{-1}, indicated by the shaded area in the graph below.

Insulin enables cells to take up glucose

After a meal is eaten, food is broken down by the digestive system, and the components of the food, including glucose, are absorbed into the bloodstream and transported around the body.

The rise in blood glucose after a meal stimulates the release of the hormone **insulin** from the pancreas. Insulin stimulates cells to take up glucose, the fuel they need to carry out their functions.

When the cells take up glucose, the amount in the blood is reduced and blood glucose level returns to normal. Between meals, the liver can release glucose from stored glycogen to keep blood glucose stable.

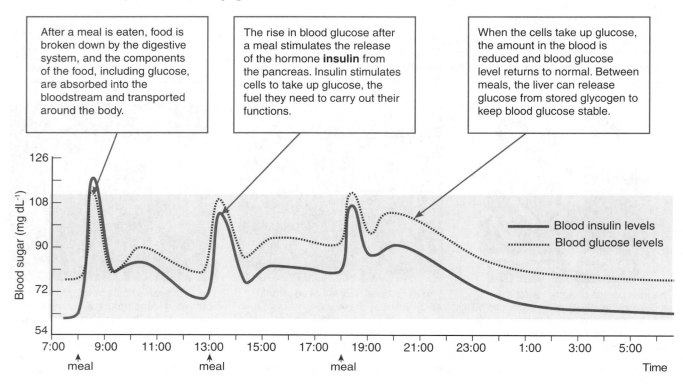

What happens if your body does not produce insulin?

▶ In some people, the insulin-producing cells of the pancreas are damaged (e.g. by infection) and the body cannot produce insulin. This life threatening disorder, which commonly affects children and teenagers, is called **type 1 diabetes mellitus**.

▶ The cells of type 1 diabetics cannot take up glucose from the blood, so glucose remains in the blood and blood glucose levels are elevated (inset right). The kidneys try to rid the body of the apparently 'excess' glucose so sufferers produce large volumes of 'sweet' urine as glucose is excreted in the urine. They feel tired and weak and are constantly hungry and thirsty. Fats are metabolized for fuel.

▶ The only current treatment is regular injection with human insulin, together with careful dietary management to control blood glucose levels.

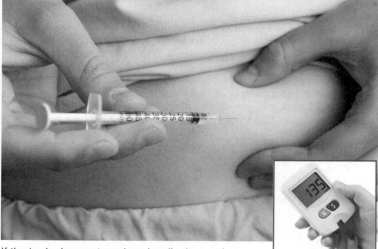

If the body does not produce insulin, it must be provided by injection. A simple blood test (right) is used to monitor blood glucose.

© 2016 **BIOZONE** International
ISBN: 978-1-927309-46-9
Photocopying Prohibited

Controlling blood glucose levels

Blood glucose (BG) is controlled by two hormones produced by special endocrine cells in the pancreas. The hormones work antagonistically (oppose each other) and levels are tightly controlled by **negative feedback**.

▶ **Insulin** lowers blood glucose by promoting glucose uptake by cells and glycogen storage in the liver.

▶ **Glucagon** increases blood glucose by promoting release of glucose from the breakdown of glycogen in the liver.

▶ When normal blood glucose levels are restored, negative feedback stops hormone secretion.

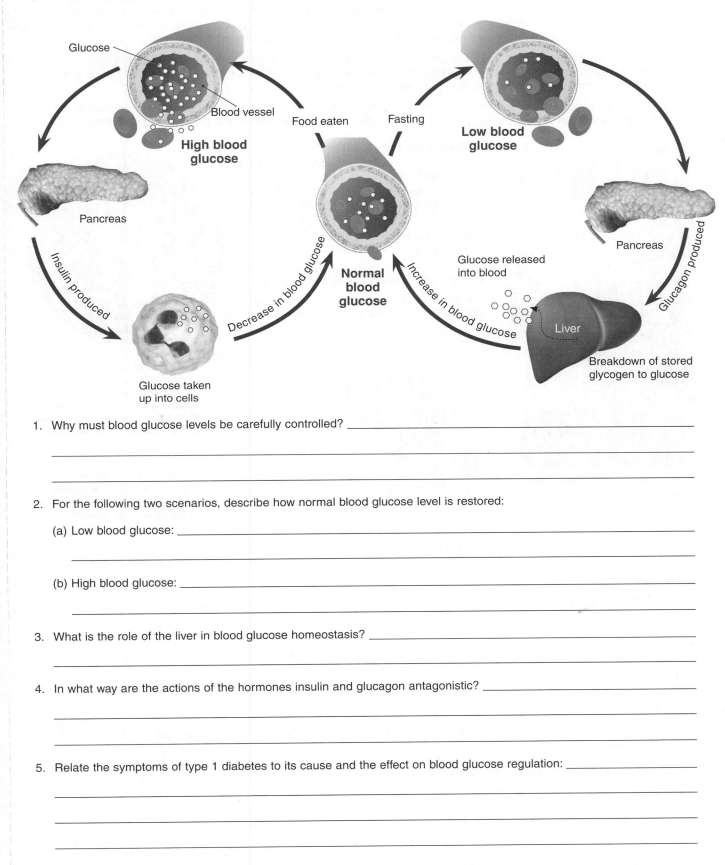

1. Why must blood glucose levels be carefully controlled? _____

_____

_____

2. For the following two scenarios, describe how normal blood glucose level is restored:

   (a) Low blood glucose: _____

   _____

   (b) High blood glucose: _____

   _____

3. What is the role of the liver in blood glucose homeostasis? _____

_____

4. In what way are the actions of the hormones insulin and glucagon antagonistic? _____

_____

_____

5. Relate the symptoms of type 1 diabetes to its cause and the effect on blood glucose regulation: _____

_____

_____

_____

74 Type 2 Diabetes

Key Idea: Diabetes mellitus is a condition where blood glucose levels are too high. In type 2 diabetes, insulin is present but the body is resistant to its effects.

▶ As we have seen, high blood glucose levels are the result of a condition called diabetes mellitus. In type 1 diabetes, the body cannot produce insulin and so the body's cells cannot take up glucose at all. However, a second form of diabetes occurs when the body is resistant to insulin's effects. The pancreas produces insulin, but the body's cells stop responding to normal regulatory insulin levels and glucose levels in the blood remain high.

▶ Symptoms are similar to type 1 diabetes and are mild at first. The body's cells do not respond appropriately to the insulin present and blood glucose levels become elevated. Type 2 diabetes typically affects older people but is increasingly common in young adults and overweight children.

▶ Type 2 diabetes is a disease that becomes progressively worse over time, eventually damaging blood vessels and leading to heart disease and stroke. Treatment is through dietary management and prescribed anti-diabetic drugs.

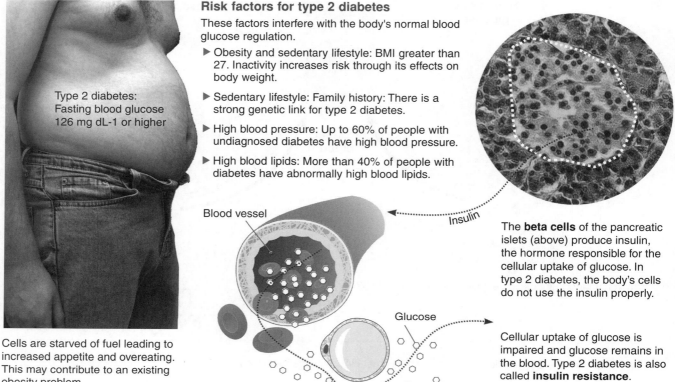

Type 2 diabetes:
Fasting blood glucose
126 mg dL-1 or higher

Risk factors for type 2 diabetes
These factors interfere with the body's normal blood glucose regulation.

▶ Obesity and sedentary lifestyle: BMI greater than 27. Inactivity increases risk through its effects on body weight.

▶ Sedentary lifestyle: Family history: There is a strong genetic link for type 2 diabetes.

▶ High blood pressure: Up to 60% of people with undiagnosed diabetes have high blood pressure.

▶ High blood lipids: More than 40% of people with diabetes have abnormally high blood lipids.

Blood vessel

Insulin

Glucose

The **beta cells** of the pancreatic islets (above) produce insulin, the hormone responsible for the cellular uptake of glucose. In type 2 diabetes, the body's cells do not use the insulin properly.

Cells are starved of fuel leading to increased appetite and overeating. This may contribute to an existing obesity problem.

Cellular uptake of glucose is impaired and glucose remains in the blood. Type 2 diabetes is also called **insulin resistance**.

1. How does type 2 diabetes differ from type 1 diabetes with respect to:

    (a) Cause: _____

    _____

    (b) Symptoms: _____

    _____

    (c) Treatment: _____

    _____

2. Antidiabetic drugs can work in several ways. For each of the statements below, state how the drug would help restore blood glucose homeostasis:

    (a) Drug increases sensitivity of cells to insulin: _____

    _____

    (b) Drug increases pancreatic secretion of insulin: _____

    _____

WEB
74

CCC
SC

 © 2016 **BIOZONE** International
ISBN: 978-1-927309-46-9
Photocopying Prohibited

75 Homeostasis During Exercise

Key Idea: The circulatory and respiratory systems are primarily responsible for maintaining homeostasis during exercise.

▶ During exercise, greater metabolic demands are placed on the body, and it must work harder to maintain homeostasis.

▶ Maintaining homeostasis during exercise is principally the job of the circulatory and respiratory systems, although the skin, kidneys, and liver are also important.

Working muscles need more ATP than muscles at rest.

Increased body temperature

During exercise, the extra heat produced by muscle contraction must be dispersed to prevent overheating. Thermoregulatory mechanisms, such as sweating and increased blood flow to the skin, release excess heat into the surrounding environment and help cool the body.

Increased heart rate

An increased heart rate circulates blood around the body more quickly. This increases the rate at which exchanges can be made between the blood and the working tissues. Oxygen and glucose are delivered and metabolic wastes (e.g. carbon dioxide) are removed.

Increased breathing rate

Exercise increases the body's demand for energy (ATP). Oxygen is required for cellular respiration and ATP production. Increasing the rate of breathing delivers more oxygen to working tissues and enables them to make the ATP they need to keep working. An increased breathing rate also increases the rate at which carbon dioxide is expelled from the body.

Increased glucose production

During exercise, working muscles quickly take up and use the freely available blood glucose. Glucose is mobilized from glycogen stores in the liver and supplies the body with fuel to maintain ATP production.

1. The graph (right) compares the change in cardiac output (a measure of total blood flow in L) during rest and during exercise. The color of the bars indicates the proportion of blood flow in skeletal muscle relative to other body parts.

   ■ Blood flow to muscle
   ☐ Blood flow to other body parts

   Cardiac output (L)

   Total 5.5 L: muscle 0.9 L

   Resting

   Total 22.5 L: muscle 17 L

   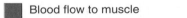

   Heavy exercise

   (a) What percentage of the blood goes to the muscles at rest?

   _____

   _____

   (b) What percentage of the blood goes to the muscles during exercise?

   _____

   _____

2. (a) What happens to the total blood flow during heavy exercise compared to at rest? _____

   _____

   _____

   (b) Why does this occur? _____

   _____

   (c) What would be happening to breathing rate during this time? _____

© 2016 **BIOZONE** International
ISBN: 978-1-927309-46-9
Photocopying Prohibited

CCC

SC   **KNOW**

76 Effect of Exercise on Breathing and Heart Rate

Key Idea: The body's response to exercise can be measured by monitoring changes in heart rate and breathing rate before and after a controlled physical effort.

In this practical, you will work in groups of three to see how exercise affects breathing and heart rates. Choose one person to carry out the exercise and one person each to record heart rate and breathing rate.

Heart rate (beats per minute) is obtained by measuring the pulse (right) for 15 seconds and multiplying by four.

Breathing rate (breaths per minute) is measured by counting the number of breaths taken in 15 seconds and multiplying it by four.

CAUTION: The person exercising should have no known pre-existing heart or respiratory conditions.

Gently press your index and middle fingers, not your thumb, against the carotid artery in the neck (just under the jaw) or the radial artery (on the wrist just under the thumb) until you feel a pulse.

Measuring the carotid pulse

Measuring the radial pulse

Procedure

Resting measurements

Have the person carrying out the exercise sit down on a chair for 5 minutes. They should try not to move. After 5 minutes of sitting, measure their heart and breathing rates. Record the resting data on the table (right).

Exercising measurements

Choose an exercise to perform. Some examples include step ups onto a chair, skipping rope, jumping jacks, and running in place.

Begin the exercise, and take measurements after 1, 2, 3, and 4 minutes of exercise. The person exercising should stop just long enough for the measurements to be taken. Record the results in the table.

Post exercise measurements

After the exercise period has finished, have the exerciser sit down in a chair. Take their measurements 1 and 5 minutes after finishing the exercise. Record the results on the table, right.

	Heart rate (beats minute^{-1})	Breathing rate (breaths minute^{-1})
Resting		
1 minute		
2 minutes		
3 minutes		
4 minutes		
1 minute after		
5 minutes after		

1. (a) Graph your results on separate piece of paper. You will need to use one vertical axis for heart rate and another for breathing rate. When you have finished answering the questions below, attach it to this page.

   (b) Analyze your graph and describe what happened to heart rate and breathing rate during exercise:

   _____

   _____

2. (a) Describe what happened to heart rate and breathing rate after exercise: _____

   _____

   _____

   (b) Why did this change occur? _____

   _____

   _____

3. Design an experiment to compare heart and breathing rates pre- and post exercise in students that do and do not regularly participate in a sport. Attach your design to this page. What do you predict will be the result and why?

PRAC

© 2016 **BIOZONE** International
ISBN: 978-1-927309-46-9
Photocopying Prohibited

77 Is the Effect of Exercise on Heart Rate Significant?

Key Idea: A simple statistic called the 95% confidence interval can be used to evaluate whether the effect of exercise on heart rate is significant.

We have seen from the previous activity that exercise affects heart rate. How do we know that the response we see is not just representative of the normal variation in a person's heart rate over time? We can evaluate the effect objectively by examining a larger group of people (a larger sample of the population) and comparing the pre- and post exercise heart rates objectively using a statistical measure called the 95% confidence interval of the mean (95% CI). The 95% CI is usually plotted either side of the mean (mean ± 95%CI). This tells you that, on average, 95 times out of 100, the true population mean will lie between these values.

x heart rate (beats per minute)	
Pre-exercise (A)	**Post exercise (B)**
72	102
116	175
79	96
97	100
90	132
67	158
115	152
82	141
95	113
82	136
77	130
$\bar{x}_A =$	$\bar{x}_B =$
$s_A =$	$s_B =$
$SE_A =$	$SE_B =$
$95\%CI_A =$	$95\%CI_B =$
$n_A = 11$	$n_B = 11$
$df = 10$	$df = 10$
$t_{(P=0.05)} = 2.228$	$t_{(P=0.05)} = 2.228$

▶ Students investigated how exercise affects heart rate by comparing their resting heart rate before exercise to their post exercise heart rate immediately after exercising.

▶ The students sat quietly for five minutes, then measured and recorded their heart rate to obtain their pre-exercise heart rate. They then performed star jumps as fast as possible for one minute. Immediately at the end of this, they measured their heart rate again. This was their post exercise heart rate. The class results are presented left:

Calculating 95% CI for pre- and post exercise data

Step 1: Calculate the sample mean
Calculate the sample mean for pre- and post exercise (A and B) and enter the values in the table left. If working longhand, attach your working to this page.

Step 2: Calculate standard deviation (*s*)
Calculate the sample standard deviation for pre- and post exercise and enter the values in the table left. If working longhand, attach your working to this page.

Step 3: Calculate standard error (SE)
Standard error is simply the standard deviation, divided by the square root of the sample size (n). Calculate the standard error of the mean for pre- and post exercise and enter the values in the table left. If working longhand, show your working below:

$$SE = \frac{s}{\sqrt{n}}$$

NEED HELP?
See Activity 24

Step 4: Calculate 95% confidence interval and plot your data
The 95% CI of the mean is given by the standard error multiplied by the value of *t* at P = 0.05 (from a *t* table) for the appropriate degrees of freedom (df) for your sample (this is given by *n* – 1). Calculate the 95% CI of the mean for pre- and post exercise and enter the values in the table. The *t* value at P = 0.05 has been provided for you.

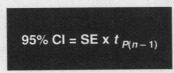
$$95\% \text{ CI} = SE \times t_{P(n-1)}$$

Plot the **mean values** for pre- and post exercise data as a column graph on the grid below. Add the 95% CI to each of the two columns as error bars, above and below the mean. Allow enough vertical scale to accommodate your error bars.

1. Complete steps 1-4 above.

2. (a) Do the error bars indicating the 95% confidence intervals of the two means overlap?

_____

(b) What does this tell you about the difference between the two data sets?

_____

_____

(c) What can you conclude about the experiment from this? _____

_____

_____

_____

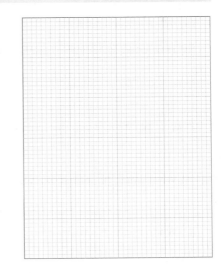

© 2016 **BIOZONE** International
ISBN: 978-1-927309-46-9
Photocopying Prohibited

PRACTICES   PRACTICES   CCC   WEB

SC   77   **DATA**

78 Homeostasis in Plants

Key Idea: In plants, evaporative water loss from stomata drives a transpiration stream that ensures plants have a constant supply of water to support essential life processes.

Maintaining water balance

Like animals, plants need water for life processes. Water gives cells turgor, transports dissolved substances, and is a medium in which metabolic reactions can take place. Maintaining water balance is an important homeostatic function in plants.

Vascular plants obtain water from the soil. Water enters the plant via the roots, and is transported throughout the plant by a specialized tissue called xylem. Water is lost from the plant by evaporation. This evaporative water loss is called **transpiration**.

Transpiration has several important functions:

▶ Provides a constant supply of water needed for essential life processes (such as photosynthesis).

▶ Cools the plant by evaporative water loss.

▶ Helps the plant take up minerals from the soil.

However, if too much water is lost by transpiration, a plant will become dehydrated and may die.

The role of stomata

Water loss occurs mainly through **stomata** (pores in the leaf). The rate of water loss can be regulated by specialized guard cells either side of the stoma, which open or close the pore.

▶ Stomata open: transpiration rate increases.

▶ Stomata closed: transpiration rates decrease.

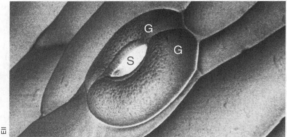

Guard cells (G) control the size of the stoma (S).

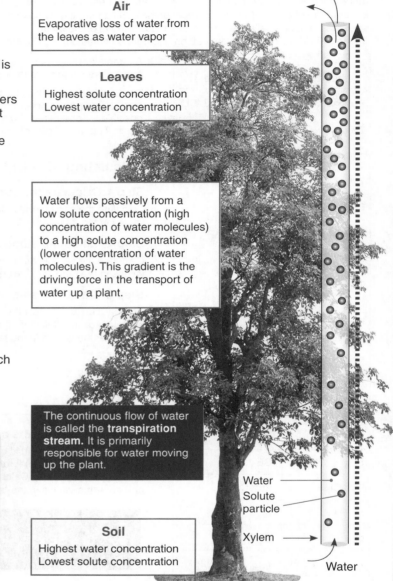

Air
Evaporative loss of water from the leaves as water vapor

Leaves
Highest solute concentration
Lowest water concentration

Water flows passively from a low solute concentration (high concentration of water molecules) to a high solute concentration (lower concentration of water molecules). This gradient is the driving force in the transport of water up a plant.

The continuous flow of water is called the **transpiration stream.** It is primarily responsible for water moving up the plant.

Soil
Highest water concentration
Lowest solute concentration

Water
Solute particle
Xylem
Water

1. (a) What is transpiration? _____

   (b) How does transpiration provide water for essential life processes in plants? _____

   _____

   _____

2. How do plants regulate the amount of water lost from the leaves? _____

   _____

3. (a) What would happen if too much water was lost by transpiration? _____

   _____

   (b) When might this happen? _____

   _____

 © 2016 **BIOZONE** International
ISBN: 978-1-927309-46-9
Photocopying Prohibited

79 Measuring Transpiration in Plants

Key Idea: Physical factors in the environment such as humidity, temperature, light level, and air movement affect transpiration. Transpiration can be measured with a potometer.

Measuring transpiration

▶ Transpiration rate (water loss per unit of time) can be measured using a simple instrument called a potometer. A basic potometer can easily be moved around so that transpiration rate can be measured under different environmental conditions.

▶ Potometers are commonly used to investigate the effect of the following environmental conditions on transpiration rate: Humidity

- Water supply
- Temperature
- Light level
- Air movement

▶ Many plants have adaptations to minimize water loss. Potometers can be used to compare transpiration rates in plants with different adaptations. For example comparing transpiration rates in plants with narrow leaves compared to rates in plants with broad leaves.

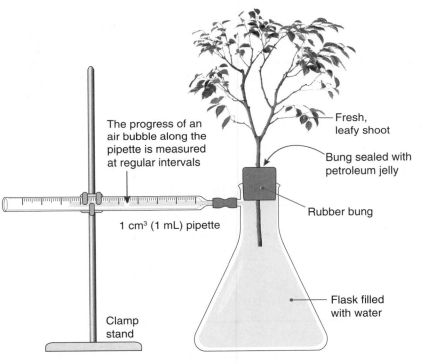

The progress of an air bubble along the pipette is measured at regular intervals

1 cm³ (1 mL) pipette

Clamp stand

Fresh, leafy shoot

Bung sealed with petroleum jelly

Rubber bung

Flask filled with water

How a potometer works

A bubble potometer, like the one shown on the left, measures the rate of water loss indirectly.

Transpiration rate is measured by measuring how much the bubble moves over a period of time. The movement of the bubble is assumed to be caused by the plant taking up water to replace the water lost by transpiration.

In transpiration experiments, it is important that the system is sealed (watertight). You can then be confident that the water losses you are recording are the result of transpiration, not leakage from the system.

Transpiration experiment

▶ This activity describes the results of a plant transpiration experiment investigating the effect of specific environmental conditions on transpiration rate.

▶ The environmental conditions investigated were ambient (standard room conditions), wind (fan), humidity (mist), and bright light (lamp).

▶ A potometer was used to measure transpiration.

▶ Once set up, the apparatus was equilibrated for 10 minutes, and the position of the air bubble in the pipette was recorded. This is the time 0 reading.

▶ The plant was then exposed to one of the environmental conditions described above. Students recorded the bubble position every two minutes over a 20 minute period. Results are given in Table 1.

PRACTICES   PRACTICES   CCC   WEB

SC   79   **PRAC**

1. The distance the bubble travelled for each environmental condition is given in Table 1 below.
   Convert the distance the bubble travelled (in mm) into water loss (mL). For every mm the bubble moved, 0.1 mL of water was lost by transpiration (e.g. 1 mm = 0.1 mL water loss).
   Determine the water loss for each environmental condition (the first has been completed for you).

Table 1. Water loss under different environmental conditions

Time (min) / Treatment	0	2	4	6	8	10	12	14	16	18	20
Ambient (mm)	0	0	0	1	1	2	3	3	4	5	5
Ambient (mL)	0	0	0	0.1	0.1	0.2	0.3	0.3	0.4	0.5	0.5
Fan (mm)	0	4	7	9	14	17	25	28	31	33	34
Fan (mL)											
High humidity (mm)	0	0	0	0	1	1	2	2	2	3	3
High humidity (mL)											
Bright light (mm)	0	1	3	5	6	7	8	9	10	11	13
Bright light (mL)											

2. Using an appropriate graph, plot the water loss (in mL) for each environmental condition on the grid below.

NEED HELP?
See Activity 17

3. (a) What is the control for this experiment? _____

   (b) Name the environmental conditions that increased water loss: _____
   _____

   (c) How do the environmental conditions in (b) cause water loss? _____
   _____
   _____
   _____

4. Why is it important that the potometer has no leaks in it? _____
   _____

© 2016 **BIOZONE** International
ISBN: 978-1-927309-46-9
Photocopying Prohibited

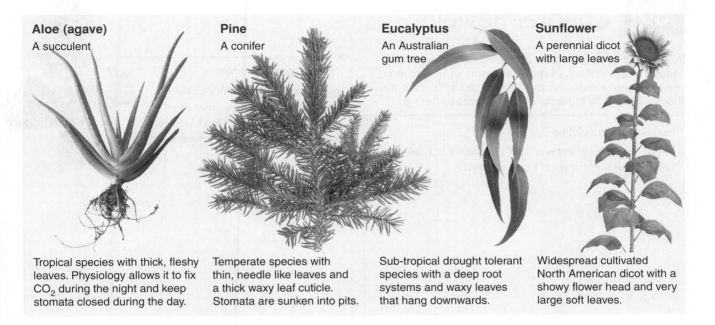

Aloe (agave)
A succulent

Tropical species with thick, fleshy leaves. Physiology allows it to fix CO_2 during the night and keep stomata closed during the day.

Pine
A conifer

Temperate species with thin, needle like leaves and a thick waxy leaf cuticle. Stomata are sunken into pits.

Eucalyptus
An Australian gum tree

Sub-tropical drought tolerant species with a deep root systems and waxy leaves that hang downwards.

Sunflower
A perennial dicot with large leaves

Widespread cultivated North American dicot with a showy flower head and very large soft leaves.

5. Read the information below and use it to design an experiment to measure transpiration rate in different plant species:

Do different plant species lose water at different rates?

▶ Some of the adaptations of each of the plants pictured above are described below the photographs. Adaptations that help to reduce water loss include reduced or sunken stomata, thick waxy leaves, and succulence. Deep root systems can help plants take up enough water despite losses by transpiration.

▶ Use the information about the use of the potometer to design an experiment to compare rates of water loss (as measured by water uptake) in the four different plants above. Do not include a full method, but think about the duration of your experiment, assumptions you are making, and how you will control the experimental conditions.

My experimental design to compare transpiration loss in different plants

6. (a) Write a hypothesis for the experiment and a prediction of its outcome: _____

_____

_____

(b) Explain the basis of your prediction: _____

_____

_____

80 Chapter Review

Summarize what you know about this topic under the headings given. You can draw diagrams or mind maps, or write short notes to organize your thoughts. Use the checklist in the introduction and the hints to help you.

Homeostatic regulation

HINT: Include reference to feedback mechanisms and the importance of homeostasis.

Homeostasis in humans

HINT: Examples could include control of blood glucose or thermoregulation. What happens when homeostatic mechanisms fail?

Homeostasis in plants

HINT: Include reference to transpiration and the interaction between plant organ systems.

81 KEY TERMS AND IDEAS: Did You Get It?

1. Test your vocabulary by matching each term to its definition, as identified by its preceding letter code.

diabetes mellitus _____

homeostasis _____

negative feedback _____

positive feedback _____

thermoregulation _____

transpiration _____

A A mechanism in which the output of a system acts to oppose changes to the input of the system. The effect is to stabilize the system and dampen fluctuations.

B The loss of water vapor by plants, mainly from leaves via the stomata.

C A destabilizing mechanism in which the output of the system causes an escalation in the initial response.

D Regulation of the internal environment to maintain a stable, constant condition.

E A condition in which the blood glucose level is elevated above normal levels, either because the body doesn't produce enough insulin, or because the cells do not respond to the insulin that is produced.

F The regulation of body temperature.

2. Test your knowledge about feedback mechanisms by studying the two graphs below, and answering the questions about them. In your answers, use biological terms appropriately to show your understanding.

A

B

Type of feedback mechanism:

_____

Mode of action:

_____

_____

_____

Biological examples of this mechanism:

_____

_____

_____

Type of feedback mechanism:

_____

Mode of action:

_____

_____

_____

Biological examples of this mechanism:

_____

_____

_____

3. (a) Why does a human expend a large amount of energy when exposed to cold temperatures for a long period of time?

_____

_____

(b) How is this different to a lizard exposed to the same conditions? _____

_____

_____

TEST

82 Summative Assessment

Maintaining homeostasis and the role of interacting systems

▶ The body is constantly interacting with the environment, with daily fluctuations in temperature, food and fluid intake, and physical exertion. In addition, the body must cope with potential pathogens, such as viruses and bacteria, and repair itself if injured. The maintenance of a steady state is the job of all the body's systems working together.

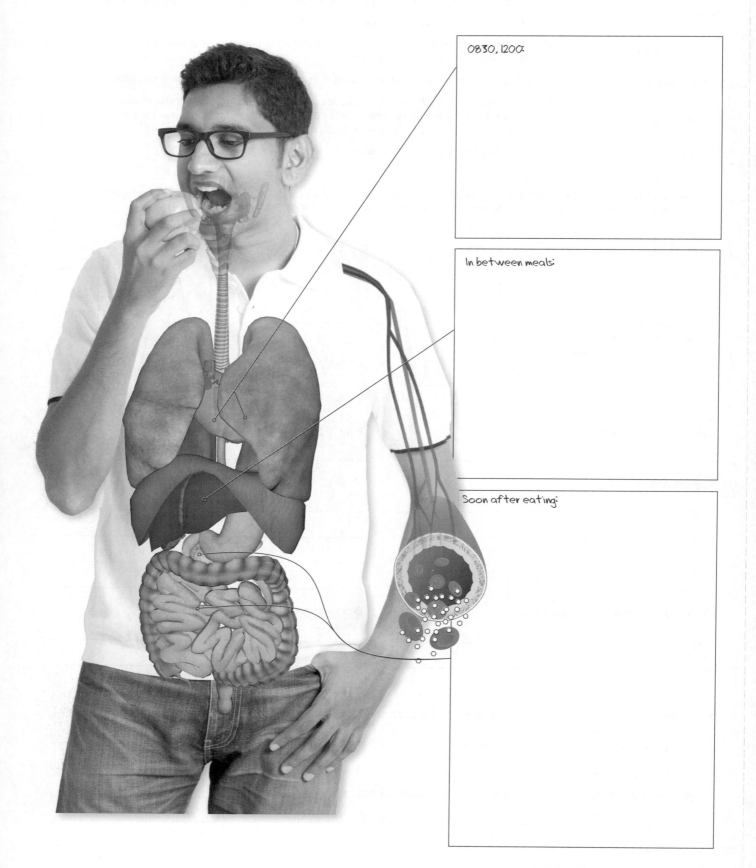

0830, 1200:

In between meals:

Soon after eating:

TEST

© 2016 **BIOZONE** International
ISBN: 978-1-927309-46-9
Photocopying Prohibited

A day in the life of Deepak, a 24 year old male of south Asian ethnicity:

Deepak is in good health, of normal weight, and likes to keep fit walking, jogging, and playing recreational basketball for a social team. He has a sedentary job.

▶ **0630**: Wake, drink a 200 mL glass of water, brush teeth.

▶ **0700**: 10 minutes stretching.

▶ **0730**: Breakfast 2000 kJ including 2 cups (400 mL) of green tea.

▶ **0800**: Toilet (urination (400 mL), defecation), shower, brushes teeth again.

▶ **0830**: Walks briskly to work as IT support at local school (20 min). The day is humid so he gets a little hot and sweaty.

▶ **0850**: Drinks a glass of water (200 mL). Quick shower. Begins work (sedentary job).

▶ **1000**: Snack of raw nuts, fruit, and 150 mL coffee (1200 kJ).

▶ **1200**: Toilet (urination 300 mL) then meets friends for a short lunchtime basketball practice.

▶ **1245**: Shower. Takeout lunch 2500 kJ, including juice (250 mL)

▶ **1300**: Resumes work.

▶ **1600**: Finishes work and walks home, stopping on the way to meet a friend for a coffee (200 mL) and cake at a local cafe (1000 kJ). Toilet (urination 200 mL).

▶ **1830**: Deepak visits his family and joins them for a traditional Indian meal involving several courses and 2 glasses of wine and 3 glasses of water (600 mL) (6000 kJ). Visits the bathroom once to urinate (200 mL).

▶ **2230**: Toilet (urination, 400 mL) and bed. He does not sleep well, feeling uncomfortable after such a large meal.

▶ **0630**: Wake.

Read the text outlining Deepak's day:

1. The opposite page shows some of the body systems involved in Deepak maintaining homeostasis during the day. In the boxes, summarize the role of each organ or organ system at the time indicated and describe how it interacts with other systems to maintain a steady state.

2. (a) Calculate Deepak's total energy intake for the day:

   _____

   (b) The recommended energy intake for an adult male is between 10,500 and 11,700 kJ per day. Is Deepak's energy intake within the recommended range? If not, how is it different?

   _____

   _____

   (c) Deepak's weight is stable. What does this tell us about his energy expenditure?

   _____

   _____

3. (a) Calculate total fluid intake for the day: _____

   (b) Calculate fluid losses via urination: _____

   (c) What is the difference between these values? Where has the extra fluid been lost from?

   _____

   _____

   (d) What organ system is primarily responsible for maintaining Deepak's fluid balance:

   _____

   (e) What other organ systems are involved and what do they do?

   _____

   _____

4. The graph below shows fluctuations in Deepak's blood glucose over the period described.
   (a) Use arrows to indicate meals and draw a second line to show what his insulin levels would be doing over this period.

   (b) Suggest why the rate of blood glucose decline after 2100 was less than at other times during the day:

   _____

   _____

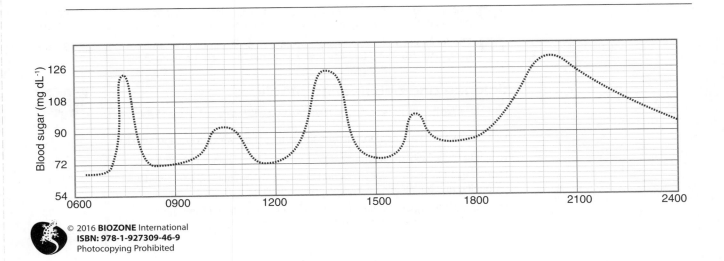

LS1.B   Growth and Development

Key terms

cell cycle

cell differentiation

cell division

DNA replication

interphase

mitosis

nucleotides

semi-conservative replication

stem cell

tissue

zygote

Disciplinary core ideas

Show understanding of these core ideas

Activity number

Organisms grow and develop through mitosis

☐ 1 Multicellular organisms develop from a single cell (a fertilized egg) called a zygote. — 83

☐ 2 The cell cycle describes the events in a cell leading to its division into two daughter cells. The cell cycle consists of two main phases, interphase and the mitotic or M phase (mitosis and cytokinesis). — 88

☐ 3 During mitosis (cell division) a cell divides to produce two genetically identical cells. — 88 89 90

☐ 4 Mitosis has three functions: growth of an organism, replacement of damaged cells, and asexual reproduction (in some organisms). — 87

☐ 5 DNA replication must take place before a cell can divide. DNA replication produces two identical copies of DNA. A copy goes to each new cell produced during mitosis. — 84 85

☐ 6 DNA replication is semi-conservative. Each replicated DNA molecule consists of one 'old' (parent) strand of DNA, and one 'new' (daughter) strand of DNA. — 84 85

Cells become differentiated to carry out specialized roles

☐ 7 A multicellular organism is made up of many different types of specialized cells. Specialized cells have specific roles in the organism. They arise through cellular differentiation, a process involving the activation of specific genes within a cell. — 92 93

☐ 8 Stem cells are unspecialized cells that can give rise to many different cell types. — 93

☐ 9 Differentiation and specialization of cells produces tissues and organs, which work together to meet the needs of the organism. — 94

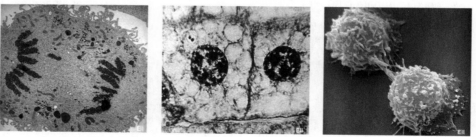

Crosscutting concepts

Understand how these fundamental concepts link different topics

Activity number

☐ 1 **SSM** Understand how models can be used to simulate DNA replication and mitosis to show how information flows within and between systems. — 84-86 88 91

☐ 2 **SF** The functions and properties of specialized cells can be inferred from their overall structure and the way their components are shaped and used. — 93 94

Science and engineering practices

Demonstrate competence in these science and engineering practices

Activity number

☐ 1 Use and evaluate a model to illustrate the semi-conservative replication of DNA. — 86

☐ 2 Evaluate the validity of experiments to determine how DNA is replicated. — 86

☐ 3 Use a model to show how mitosis and cellular differentiation are involved in producing and maintaining a multicellular organism. — 91 92

☐ 4 Use a model to show how the components of blood (a liquid tissue) are produced through cellular differentiation from stem cells. — 93

83 Growth and Development of Organisms

Key Idea: Multicellular organisms develop from a single cell. The cell divides by mitosis many times, producing genetically identical copies of the original cell.

Organisms grow and develop through mitosis

▶ Multicellular organisms begin as a single cell (a fertilized egg) and develop into complex organisms made up of many cells. This is achieved by the process of mitosis (mitotic cell division).

▶ In multicellular organisms, mitosis is responsible for growth and for the replacement of old and damaged cells. In some unicellular eukaryotic organisms (e.g. yeast cells), mitosis is also responsible for reproduction.

▶ Two important processes must occur in order for new cells to be produced. The first is the duplication of the genetic material (DNA). The second is the division (splitting) of the parent cell into two identical daughter cells. The two daughter cells have the same genetic material as the parent cell.

▶ In multicellular animals, mitosis only occurs in body cells (somatic cells). Sperm and egg cells (gametes) are produced by a different type of cell division (meiosis).

The role of mitosis in human development

Every cell in an adult's body has the exact same genetic material as the fertilized egg it has developed from.

An egg and sperm join together in fertilization and produce a zygote. The zygote has two sets of chromosomes, one set from the father and one from the mother. This condition is called **diploid** (or 2N).

The zygote begins to divide by mitosis. The size of the embryo begins to increase as more cells are formed. This early embryo contains about 100 cells.

Once born, mitosis continues to be important in the growth of an organism until they are fully grown.

Definitions
Zygote: A fertilized egg cell.
Embryo: The ball of cells that forms four to five days after fertilization.

In adults, cell division is involved in the replacement of old cells rather than growth. This continues through the adult's lifetime.

1. Briefly explain how multicellular organisms can develop from a single cell: _____

2. What two things must occur for a new cell to be produced? _____

3. Explain the role of mitosis in:

   (a) A developing embryo: _____

   (b) An adult: _____

KNOW

84 DNA Replication

Key Idea: Before a cell can divide, the DNA must be copied. DNA replication produces two identical copies of DNA. A copy goes to each new daughter cell.

▸ Before a cell can divide, its DNA must be copied (replicated). **DNA replication** ensures that the two daughter cells receive identical genetic information.

▸ In eukaryotes, DNA is organized into structures called chromosomes in the nucleus.

▸ After the DNA has replicated, each chromosome is made up of two chromatids, which are joined at the centromere. The two chromatids will become separated during cell division to form two separate chromosomes.

▸ DNA replication takes place in the time between cell divisions. The process is **semi-conservative**, meaning that each chromatid contains half original (parent) DNA and half new (daughter) DNA.

DNA replication duplicates chromosomes

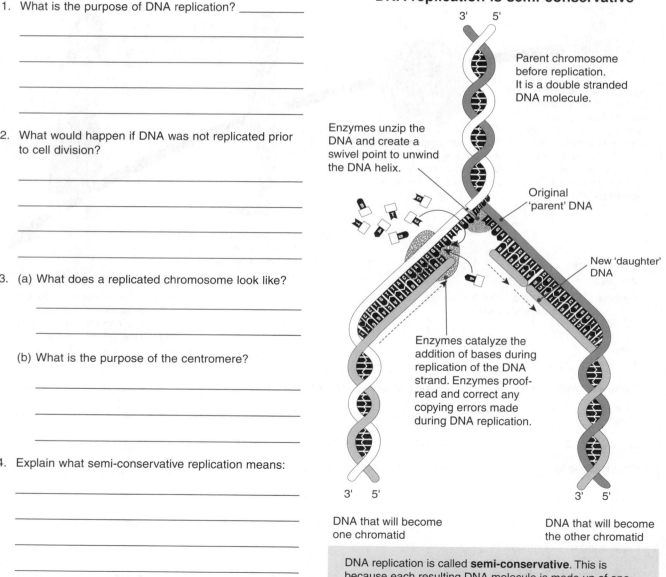

DNA replication creates a chromosome with two identical chromatids

Parent chromosome

Replicated chromosome consists of two chromatids joined at the centromere.

Centromere links sister chromatids

Chromatid

The centromere keeps sister chromatids together in an organized way until they are separated prior to nuclear division.

1. What is the purpose of DNA replication? _____

_____

_____

_____

2. What would happen if DNA was not replicated prior to cell division?

_____

_____

_____

_____

3. (a) What does a replicated chromosome look like?

_____

_____

(b) What is the purpose of the centromere?

_____

_____

_____

4. Explain what semi-conservative replication means:

_____

_____

_____

_____

_____

DNA replication is semi-conservative

3'   5'

Parent chromosome before replication. It is a double stranded DNA molecule.

Enzymes unzip the DNA and create a swivel point to unwind the DNA helix.

Original 'parent' DNA

New 'daughter' DNA

Enzymes catalyze the addition of bases during replication of the DNA strand. Enzymes proof-read and correct any copying errors made during DNA replication.

3'   5'

3'   5'

DNA that will become one chromatid

DNA that will become the other chromatid

DNA replication is called **semi-conservative**. This is because each resulting DNA molecule is made up of one parent strand and one daughter strand of DNA.

 © 2016 **BIOZONE** International
ISBN: 978-1-927309-46-9
Photocopying Prohibited

85 Details of DNA Replication

Key Idea: DNA replication is achieved by enzymes attaching new nucleotides to the growing DNA strand at the replication fork.

▶ During DNA replication, new nucleotides (the units that make up the DNA molecule) are added at a region called the **replication fork**. The replication fork moves along the chromosome as replication progresses.

▶ Nucleotides are added in by complementary base-pairing: Nucleotide A is always paired with nucleotide T. Nucleotide C is always paired with nucleotide G.

▶ The DNA strands can only be replicated in one direction, so one strand has to be copied in short segments, which are joined together later.

▶ This whole process occurs simultaneously for each chromosome of a cell and the entire process is tightly controlled by enzymes.

1. How are the new strands of DNA lengthened?

_____

_____

_____

_____

2. What rule ensures that the two new DNA strands are identical to the original strand?

_____

_____

3. Why does one strand of DNA need to be copied in segments?

_____

_____

_____

_____

_____

4. Describe three activities carried out by enzymes during DNA replication:

(a) _____

_____

(b) _____

_____

(c) _____

_____

Stages in DNA replication

Parent DNA is made up of two **anti-parallel** strands coiled into a double helix.

The two strands are joined by base pairing

1 Enzymes unwind parent DNA double helix

5'   3'

Parent strand of DNA acts as a template to match nucleotides for the new DNA strand.

2 Enzymes unzip parent DNA at the replication fork.

Original 'parent' DNA

3'

New 'daughter' DNA

5'

3 Enzymes add free nucleotides to the exposed bases on the template.

The enzymes can work in only one direction and the strands are anti-parallel, so one strand is made in fragments that are later joined by other enzymes.

Nucleotide symbols

G   C   A   T

DNA base pairing rule

G pairs with C
A pairs with T

4 Two new double-stranded DNA molecules

Original 'parent' DNA

New 'daughter' DNA

Enzymes are involved at every step of DNA replication. They unzip the parent DNA, add the free nucleotides to the 3' end of each single strand, join DNA fragments, and check and correct the new DNA strands.

CCC   WEB

SSM   85   KNOW

86 Modeling DNA Replication

Key Idea: Meselson and Stahl devised an experiment that showed DNA replication is semi-conservative. This practical activity models DNA replication.

Models for DNA replication

▶ Several models were proposed for DNA replication (right).

▶ The **conservative model** proposed that when DNA is replicated one of the new DNA molecules receives two newly replicated DNA strands and the other receives the two original DNA strands.

▶ The **semi-conservative model** proposed that each DNA strand in the original molecule served as a template for the new DNA molecules. Therefore the two new DNA molecules each contained one strand of original DNA and one strand of new DNA.

▶ On the next page you will model DNA replication.

Conservative model

Semi-conservative model

—— Original DNA strand   —— New DNA strand

Meselson and Stahl's experiment

▶ Two scientists, called Meselson and Stahl, determined that DNA replication is semi-conservative.

▶ They grew bacteria in a solution containing heavy nitrogen (^{15}N) until their DNA contained only ^{15}N. The bacteria were placed into a growth solution containing normal nitrogen (^{14}N), which is lighter than ^{15}N. After set generation times the DNA was extracted and centrifuged in a cesium chloride (CsCl) solution. The CsCl provides a density gradient to separate the DNA. Heavy DNA sinks to the bottom, light DNA rises to the top, and intermediate DNA (one light and one heavy strand) settles in between (right).

▶ If DNA replicated conservatively, after one generation there would be two kinds of DNA. Heavy DNA consisting of two stands of ^{15}N DNA, and light DNA consisting of two ^{14}N stands of DNA.

▶ If DNA replicated semi-conservatively, after one generation there would be one type of DNA, an intermediate density DNA consisting of one strand of heavy ^{15}N and one strand of light ^{14}N DNA. This is the result that Meselson and Stahl obtained.

Separating DNA by density

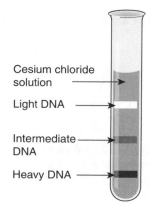

Cesium chloride solution

Light DNA

Intermediate DNA

Heavy DNA

1. Why did Meselson and Stahl's experiment support the semi-conservative replication model?

_____

_____

_____

_____

_____

2. (a) Complete the diagram (right) for the next DNA replication.

(b) Predict what would happen to the proportion of intermediate DNA under the semi-conservative method in future generations:

_____

_____

_____

Semi-conservative model

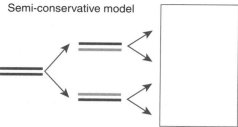

Conservative model

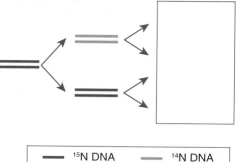

—— ^{15}N DNA   —— ^{14}N DNA

Instructions

1. Cut out the DNA shapes provided on this page.

2. Intertwine the first pair (labelled 0) of heavy ^{15}N (black) DNA. This forms Generation 0 (parental DNA).

3. Use the descriptions of the two possible models for DNA replication on the previous page to model semi-conservative and conservative DNA replication.

4. For each replication method, record on a separate sheet of paper the percentage of heavy ^{15}N-^{15}N (black-black), intermediate ^{15}N-^{14}N (black-gray), and light ^{14}N-^{14}N (gray-gray).

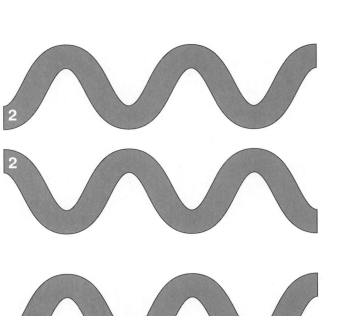

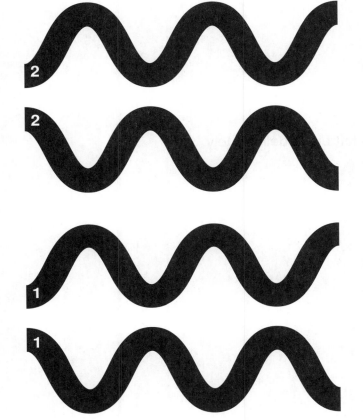

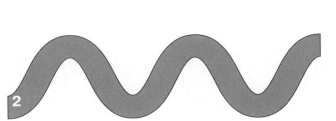

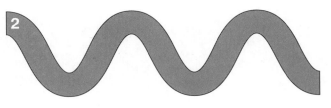

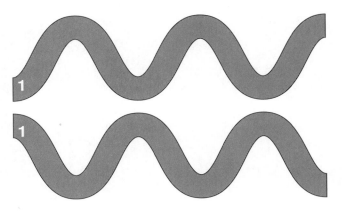

This page is left blank deliberately

87 The Functions of Mitosis

Key Idea: Mitosis has three primary functions: growth of the organism, replacement of damaged or old cells, and asexual reproduction (in some organisms).

Mitotic cell division has three purposes

▶ **Growth**: Multicellular organisms grow from a single fertilized cell into a mature organism. Depending on the organism, the mature form may consist of several thousand to several trillion cells. These cells that form the building blocks of the body are called somatic cells.

▶ **Repair**: Damaged and old cells are replaced with new cells.

▶ **Asexual reproduction**: Some unicellular eukaryotes (such as yeasts) and some multicellular organisms (e.g. *Hydra*) reproduce asexually by mitotic division.

Egg cell   Embryo   Adult

Matthias Zepper

Asexual reproduction

Some simple eukaryotic organisms reproduce asexually by mitosis. Yeasts (such as baker's yeast, used in baking) can reproduce by budding. The parent cell buds to form a daughter cell (right). The daughter cell continues to grow, and eventually separates from the parent cell.

Parent cell

Daughter cell

Growth

Multicellular organisms develop from a single fertilized egg cell and grow by increasing in cell numbers. Cells complete a cell cycle, in which the cell copies its DNA and then divides to produce two identical cells. During the period of growth, the production of new cells is faster than the death of old ones. Organisms, such as the 12 day old mouse embryo (above, middle), grow by increasing their total cell number and the cell become specialized as part of development. Cell growth is highly regulated and once the mouse reaches its adult size (above, right), physical growth stops and the number of cell deaths equals the number of new cells produced.

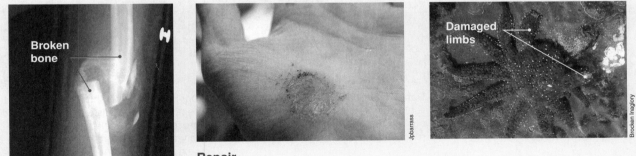

Broken bone

Damaged limbs

Jpbarrass

Brocken Inaglory

Repair

Mitosis is vital in the repair and replacement of damaged cells. When you break a bone or graze your skin, new cells are generated to repair the damage. Some organisms, like this sea star (above right) are able to generate new limbs if they are broken off.

1. Use examples to explain the role of mitosis in:

   (a) Growth of an organism: _____

   _____

   (b) Replacement of damaged cells: _____

   _____

   (c) Asexual reproduction: _____

   _____

© 2016 **BIOZONE** International
ISBN: 978-1-927309-46-9
Photocopying Prohibited

KNOW

88 The Eukaryotic Cell Cycle

Key Idea: The eukaryotic cell cycle can be divided into phases, although the process is continuous. Specific cellular events occur in each phase.

The life cycle of a eukaryotic cell is called the cell cycle. The cell cycle can be divided into two broad phases; interphase and M phase. Specific activities occur in each phase.

Interphase

Cells spend most of their time in interphase. Interphase is divided into three stages:

▶ The first gap phase (G1).

▶ The S-phase (S).

▶ The second gap phase (G2).

During interphase the cell increases in size, carries out its normal activities, and replicates its DNA in preparation for cell division. Interphase is not a stage in mitosis.

Mitosis and cytokinesis (M-phase)

Mitosis and cytokinesis occur during M-phase. During mitosis, the cell nucleus (containing the replicated DNA) divides in two equal parts. Cytokinesis occurs at the end of M-phase. During cytokinesis the cell cytoplasm divides, and two new daughter cells are produced.

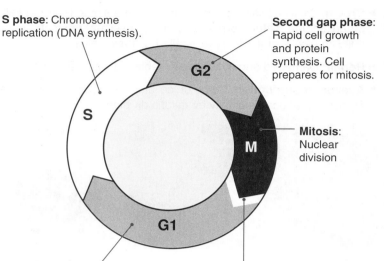

S phase: Chromosome replication (DNA synthesis).

Second gap phase: Rapid cell growth and protein synthesis. Cell prepares for mitosis.

Mitosis: Nuclear division

First gap phase: Cell increases in size and makes the mRNA and proteins needed for DNA synthesis.

Cytokinesis: The cytoplasm divides and the two cells separate. Cytokinesis is part of M phase but distinct from nuclear division.

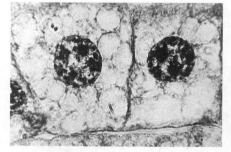

During interphase, the cell grows and acquires the materials needed to undergo mitosis. It also prepares the nuclear material for separation by replicating it.

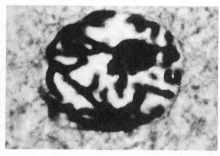

During interphase the nuclear material is unwound. As mitosis approaches, the nuclear material begins to reorganize in readiness for nuclear division.

During mitosis the chromosomes are separated. Mitosis is a highly organized process and the cell must pass "checkpoints" before it proceeds to the next phase.

1. Briefly outline what occurs during the following phases of the cell cycle:

(a) Interphase: _____

_____

_____

(b) Mitosis: _____

_____

_____

(c) Cytokinesis: _____

_____

_____

© 2016 **BIOZONE** International
ISBN: 978-1-927309-46-9
Photocopying Prohibited

89 Mitosis

Key Idea: Mitosis is an important part of the eukaryotic cell cycle in which the replicated chromosomes are separated and the cell divides, producing two new cells.

Mitosis is a stage in the cell cycle

▶ M-phase (**mitosis** and cytokinesis) is the part of the cell cycle in which the parent cell divides in two to produce two genetically identical daughter cells (right).

▶ Mitosis results in the separation of the nuclear material and division of the cell. It does not result in a change of chromosome number.

▶ Mitosis is one of the shortest stages of the cell cycle. When a cell is not undergoing mitosis, it is said to be in interphase.

▶ In animals, mitosis takes place in the somatic (body) cells. Somatic cells are any cell of the body except sperm and egg cells.

▶ In plants, mitosis takes place in the meristems. The meristems are regions of growth (where new cells are produced), such as the tips of roots and shoots.

Mitosis produces identical daughter cells

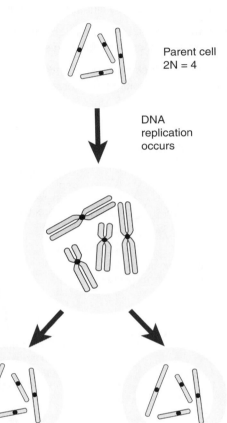

Parent cell 2N = 4

DNA replication occurs

Daughter cell, 2N = 4   Daughter cell, 2N = 4

The cell divides forming two identical daughter cells. The chromosome number remains the same as the parent cell.

Onion cells

At any one time, only a small proportion of the cells in an organism will be undergoing mitosis. The majority of the cells will be in interphase.

The meristematic tissue (M) at the growing tip is the site of mitosis in this plant root. The root cap below the meristem protects the dividing cells.

1. Briefly outline the events in mitosis: _____

    _____

    _____

2. Where does mitosis take place in:

    (a) Animals: _____

    (b) Plants: _____

3. A cell with 10 chromosomes undergoes mitosis.

    (a) How many daughter cells are created: _____

    (b) How many chromosomes does each daughter cell have? _____

    (c) The genetic material of the daughter cells is the same as / different to the parent cell (delete one).

 © 2016 **BIOZONE** International
ISBN: 978-1-927309-46-9
Photocopying Prohibited

WEB
 89   **KNOW**

90 Mitosis and Cytokinesis

Key Idea: Mitosis is a continuous process, but it is divided into stages to help identify and describe what is occurring.

The cell cycle and stages of mitosis

▶ Mitosis is continuous, but is divided into stages for easier reference (1-6 below). Enzymes are critical at key stages. The example below illustrates the cell cycle in an animal cell.

▶ In animal cells, centrioles (located in the centrosome), form the spindle. During **cytokinesis** (division of the cytoplasm) a constriction forms dividing the cell in two. Cytokinesis is part of M-phase, but it is distinct from mitosis.

▶ Plant cells lack centrioles, and the spindle is organized by structures associated with the plasma membrane. In plant cells, cytokinesis involves formation of a cell plate in the middle of the cell. This will form a new cell wall.

The animal cell cycle and stages of mitosis

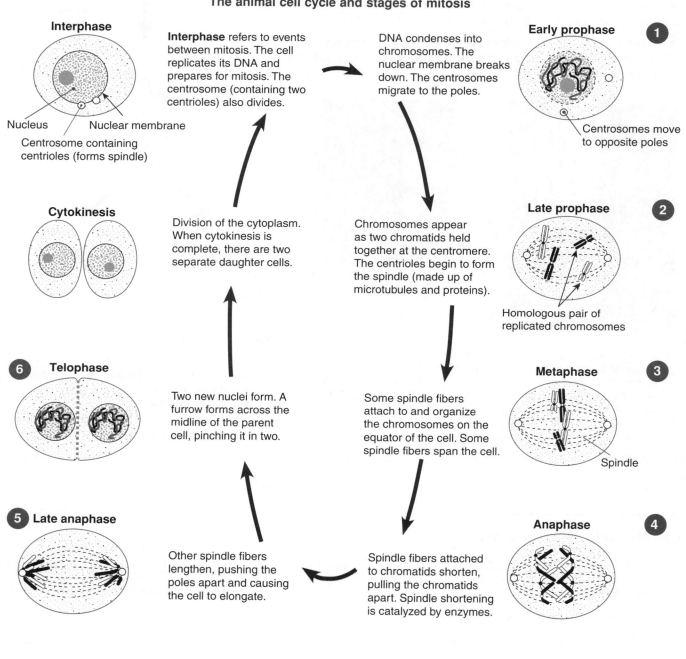

Interphase

Nucleus
Nuclear membrane
Centrosome containing centrioles (forms spindle)

Interphase refers to events between mitosis. The cell replicates its DNA and prepares for mitosis. The centrosome (containing two centrioles) also divides.

DNA condenses into chromosomes. The nuclear membrane breaks down. The centrosomes migrate to the poles.

Early prophase ①

Centrosomes move to opposite poles

Cytokinesis

Division of the cytoplasm. When cytokinesis is complete, there are two separate daughter cells.

Chromosomes appear as two chromatids held together at the centromere. The centrioles begin to form the spindle (made up of microtubules and proteins).

Late prophase ②

Homologous pair of replicated chromosomes

⑥ **Telophase**

Two new nuclei form. A furrow forms across the midline of the parent cell, pinching it in two.

Some spindle fibers attach to and organize the chromosomes on the equator of the cell. Some spindle fibers span the cell.

Metaphase ③

Spindle

⑤ **Late anaphase**

Other spindle fibers lengthen, pushing the poles apart and causing the cell to elongate.

Spindle fibers attached to chromatids shorten, pulling the chromatids apart. Spindle shortening is catalyzed by enzymes.

Anaphase ④

1. What must occur before mitosis takes place? _____

_____

© 2016 **BIOZONE** International
ISBN: 978-1-927309-46-9
Photocopying Prohibited

Cytokinesis

In plant cells (below right), cytokinesis (division of the cytoplasm) involves construction of a cell plate (a precursor of the new cell wall) in the middle of the cell. The cell wall materials are delivered by vesicles derived from the Golgi. The vesicles join together to become the plasma membranes of the new cell surfaces. Animal cell cytokinesis (below left) begins shortly after the sister chromatids have separated in anaphase of mitosis. A ring of microtubules assembles in the middle of the cell, next to the plasma membrane, constricting it to form a cleavage furrow. In an energy-using process, the cleavage furrow moves inwards, forming a region of separation where the two cells will separate.

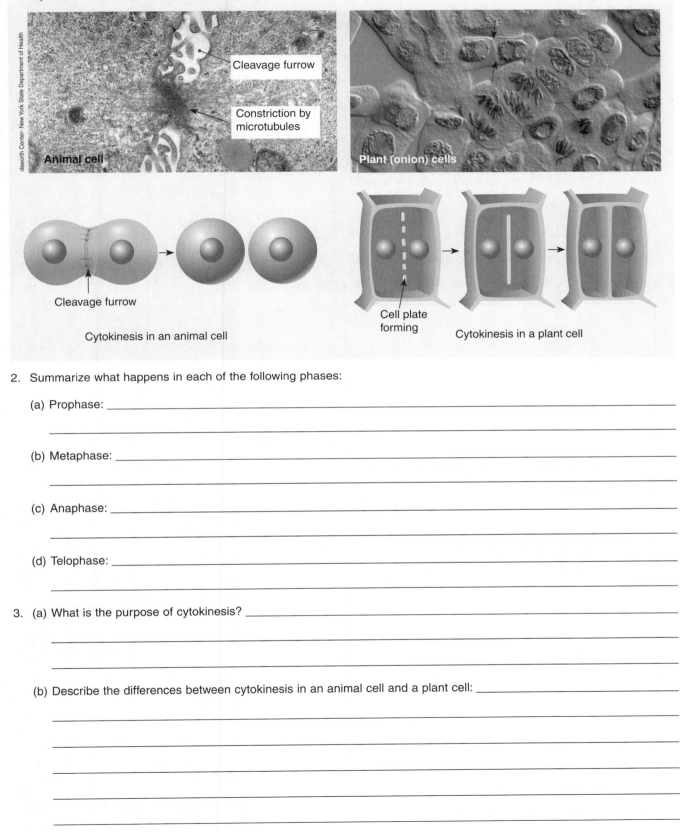

Animal cell

Cleavage furrow

Constriction by microtubules

Plant (onion) cells

Cleavage furrow

Cytokinesis in an animal cell

Cell plate forming

Cytokinesis in a plant cell

2. Summarize what happens in each of the following phases:

(a) Prophase: _____

_____

(b) Metaphase: _____

_____

(c) Anaphase: _____

_____

(d) Telophase: _____

_____

3. (a) What is the purpose of cytokinesis? _____

_____

_____

(b) Describe the differences between cytokinesis in an animal cell and a plant cell: _____

_____

_____

_____

_____

_____

91 Modeling Mitosis

Key Idea: Using pipe cleaners to model the stages of mitosis will help you to visualize and understand the process.

▶ Students used pipe cleaners and yarn to model mitosis in an animal cell. Four chromosomes were used for simplicity (2N = 4). Images of their work are displayed below.

1. (a) Photo 1 represents a cell in interphase before mitosis begins. The circular structures are the centrosomes. Name the labeled structures:

   A: _____

   B: _____

   C: _____

   (b) Why are there two copies of the centrosomes?_____

   _____

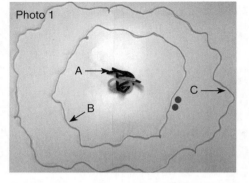

Photo 1

2. (a) What is happening to the structure labelled A in photo 2?

   _____

   _____

   (b) On photo 2, draw the new position of the centrosomes:

3. In the space below, draw what happens next, and give an explanation:

   _____
   _____
   _____
   _____

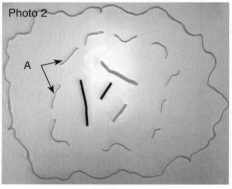

Photo 2

4. (a) What is happening in photo 3: _____

   _____

   (b) On the photo, draw the centrosomes in their correct positions.

5. In the space below, draw what happens next:

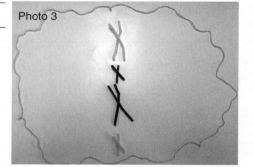

Photo 3

6. Photo 4 shows the completion of mitosis.

   (a) How many cells are formed? _____

   (b) How many chromosomes are in each cell? _____

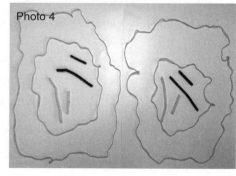

Photo 4

© 2016 **BIOZONE** International
ISBN: 978-1-927309-46-9
Photocopying Prohibited

92 Differentiation of Cells

Key Idea: Many different cell types arise during development of the embryo. Activation of specific genes determines what type of cell will develop.

▶ When a cell divides by mitosis, it produces genetically identical cells. However, a multicellular organism is made up of many different types of cells, each specialized to carry out a particular role. How can it be that all of an organism's cells have the same genetic material, but the cells have a wide variety of shapes and functions? The answer is through **cellular differentiation** (transformation) of unspecialized cells called **stem cells**.

▶ Although each cell has the same genetic material (genes), different genes are turned on (activated) or off in different patterns during development in particular types of cells. The differences in gene activation controls what type of cell forms (below). Once the developmental pathway of a cell is determined, it cannot alter its path and change into another cell type.

How stem cells give rise to different cell types

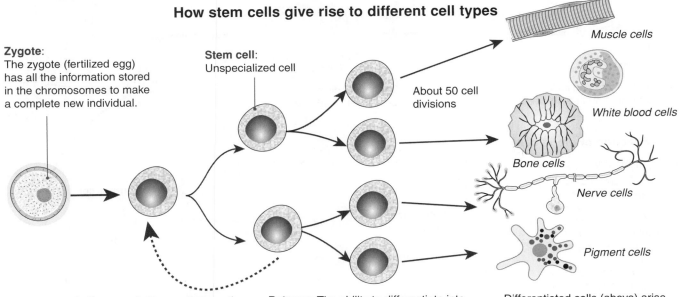

Zygote:
The zygote (fertilized egg) has all the information stored in the chromosomes to make a complete new individual.

Stem cell:
Unspecialized cell

About 50 cell divisions

Muscle cells

White blood cells

Bone cells

Nerve cells

Pigment cells

Self renewal: Stem cells have the ability to divide many times while maintaining an unspecialized state.

Potency: The ability to differentiate into specialized cells. Different types of stem cell have different levels of potency.

Differentiated cells (above) arise because genes are turned on or off in a particular sequence.

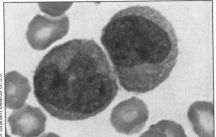

Dr Graham Beards cc 3.0

White blood cells function in immunity

Neurons transmit impulses

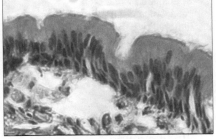

Ciliated epithelium lines the trachea

1. Name the cell from which all other cells are derived: _____

2. Explain how so many different types of cells can be formed, even though all cells have the same DNA: _____

   _____

   _____

3. (a) What are stem cells? _____

   _____

   (b) What are the two defining properties of stem cells? _____

      i _____

      ii _____

© 2016 **BIOZONE** International
ISBN: 978-1-927309-46-9
Photocopying Prohibited

PRACTICES   WEB

92

KNOW

124

93 Stem Cells Give Rise to Other Cells

Key Idea: Stem cells are undifferentiated cells, which can develop into many different cell types. Related cell types come together to form tissues, such as blood.

Totipotent stem cells

These stem cells can differentiate into all the cells in an organism. Example: In humans, the zygote and its first few divisions. The tissue at the root and shoot tips of plants is also totipotent.

Pluripotent stem cells

These stem cells can give rise to any cells of the body, except extra-embryonic cells (e.g. placenta and chorion). Example: Embryonic stem cells.

Multipotent stem cells

These adult stem cells can give rise to a limited number of cell types, related to their tissue of origin. Example: Bone marrow stem cells (below), skin stem cells, bone stem cells, umbilical cord blood.

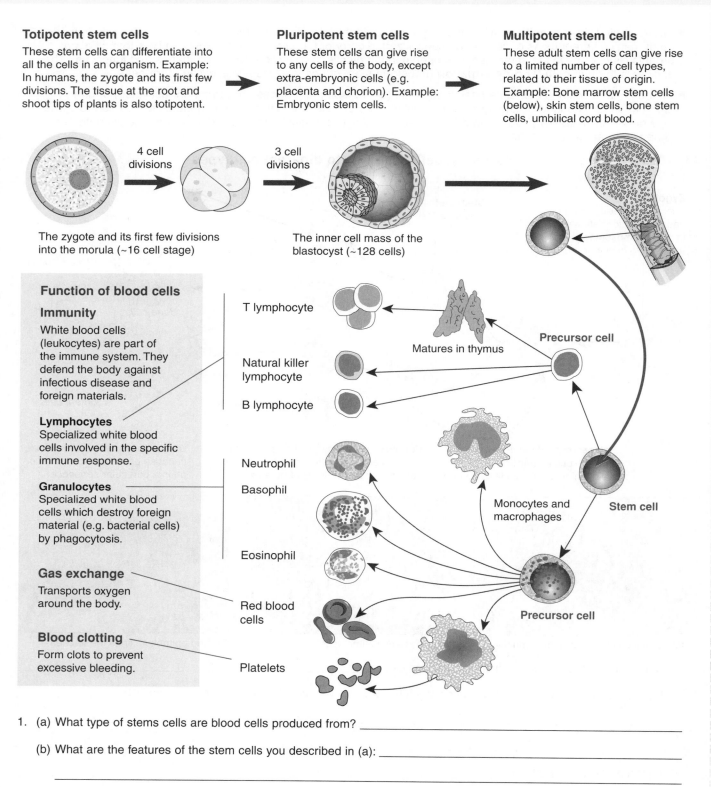

4 cell divisions

3 cell divisions

The zygote and its first few divisions into the morula (~16 cell stage)

The inner cell mass of the blastocyst (~128 cells)

Function of blood cells

Immunity

White blood cells (leukocytes) are part of the immune system. They defend the body against infectious disease and foreign materials.

Lymphocytes

Specialized white blood cells involved in the specific immune response.

Granulocytes

Specialized white blood cells which destroy foreign material (e.g. bacterial cells) by phagocytosis.

Gas exchange

Transports oxygen around the body.

Blood clotting

Form clots to prevent excessive bleeding.

T lymphocyte

Natural killer lymphocyte

B lymphocyte

Neutrophil

Basophil

Eosinophil

Red blood cells

Platelets

Matures in thymus

Precursor cell

Monocytes and macrophages

Stem cell

Precursor cell

1. (a) What type of stems cells are blood cells produced from? _____

   (b) What are the features of the stem cells you described in (a): _____

   _____

2. Describe the functional roles of blood cells: _____

   _____

   _____

   _____

WEB CCC PRACTICES

KNOW **93** SF

© 2016 **BIOZONE** International
ISBN: 978-1-927309-46-9
Photocopying Prohibited

94 Tissues Work Together

Key Idea: Tissues are specialized to perform particular tasks. Different tissue types work together to meet the body's needs efficiently.

A tissue is a collection of related cell types that work together to carry out a specific function. Different tissues come together to form organs. The cells, tissues, and organs of the body interact to meet the needs of the entire organism. This activity explains the role of the four tissue types (below) in humans.

Muscle tissue	Epithelial tissue	Nervous tissue	Connective tissue
▶ Contractile tissue	▶ Lining tissue	▶ Receives and responds to stimuli	▶ Supports, protects, and binds other tissues
▶ Produces movement of the body or its parts	▶ Covers the body and lines internal surfaces	▶ Makes up the structures of the nervous system	▶ Contains cells in an extracellular matrix
▶ Includes smooth, skeletal, and cardiac muscle	▶ Can be modified to perform specific roles	▶ Regulates function of other tissues	▶ Can be hard or fluid

Cilia

Mucus-producing cell

The upper respiratory tract is lined with ciliated epithelium to move irritants before they reach the lungs. The lungs and cardiovascular system work together to respond to changes in oxygen demand.

Processes

Neuron

Nervous tissue is made up of nerve cells (neurons) and supporting cells. The long processes of neurons control the activity of muscles and glands.

E   CT   SM

The digestive tract is lined with epithelial tissue (E) and held in place by connective tissue (CT). It is moved by smooth muscle (SM) in response to messages from neurons.

CT
E.
SM

Epithelial tissue (E) lines organs such as the bladder. Connective tissue (CT) supports the organ. This epithelium is layered so that it can stretch. The bladder's activity is controlled by smooth muscle (SM) and neurons.

Bone is a type of connective tissue. It provides shape to the body and works with muscle to produce movement. Ligaments are also connective tissue structures. They hold bones together.

Skeletal muscle tissue contracts to pull on the rigid bones of the skeleton to bring about movement of the body. Tendons are connective tissue structures that attach muscles to bones.

CCC   WEB

SF   **94**   **KNOW**

Tissues work together and make up organs, which perform specific functions

The body's tissues work together in order for the body to function. Tissues also group together to form organs. For example, epithelial tissues are found associated with other tissues in every organ of the body. Other examples include:

Nerves, muscles, and movement

▶ Nerves stimulate muscles to move.

▶ Connective tissue binds other tissues together and holds them in place (e.g. skeletal muscle tissue is held together by connective tissue sheaths to form discrete muscle, neurons are bundled together by connective tissue to form nerves).

▶ Bones are held together by connective tissue ligaments at joints, allowing the skeleton to move. Skeletal muscles are attached to bone by connective tissue tendons. Muscle contraction causes the tendon to pull on the bone, moving it.

Heart, lungs, blood vessels, and blood

▶ Cardiac muscle pumps blood (a specialized connective tissue) around the body within blood vessels.

▶ In the lungs, blood vessels surround the epithelium of the tiny air sacs to enable the exchange of gases between the blood and the air in the lungs.

▶ Neurons regulate the activity of heart and lungs to respond to changes in oxygen demand, as when a person is exercising.

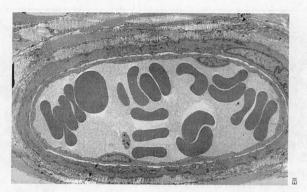

This image shows a neuron branching to supply muscle fibers. Impulses from the neuron will cause the muscle to contract. Neurons are bundled together by connective tissue wrappings to form nerves.

This image shows red blood cells within a vein. Veins return blood to the heart. When oxygen demand increases, heart and breathing rates increase and blood is delivered more quickly to working tissues.

1. Describe the main function of each of the following types of tissues:

(a) Epithelial tissue: _____

_____

(b) Connective tissue: _____

_____

(c) Nervous tissue: _____

_____

(d) Muscle tissue: _____

_____

2. Describe how different tissues interact to bring about movement of a body part: _____

_____

_____

_____

_____

_____

_____

© 2016 **BIOZONE** International
ISBN: 978-1-927309-46-9

95 Chapter Review

Summarize what you know about this topic so far under the headings provided. You can draw diagrams or mind maps, or write notes to organize your thoughts. Use the checklist in the introduction and the hints to help you:

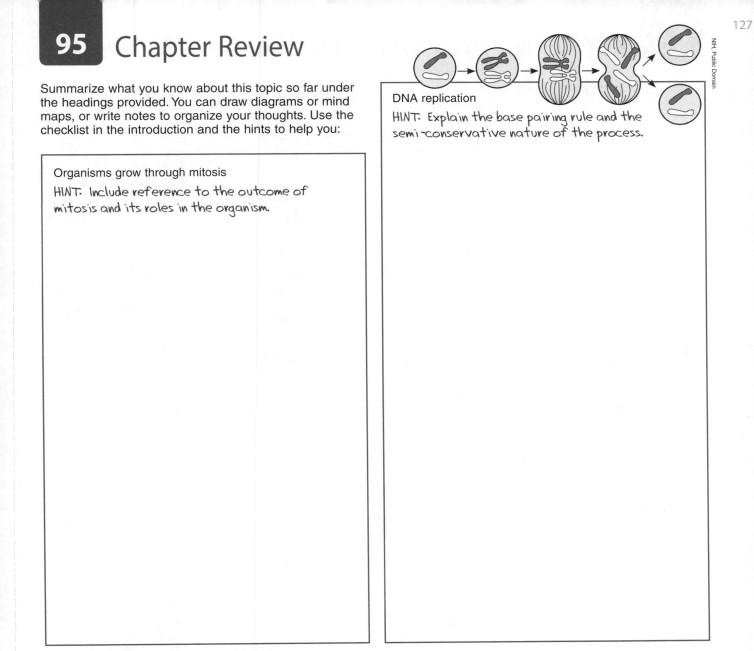

DNA replication

HINT: Explain the base pairing rule and the semi-conservative nature of the process.

Organisms grow through mitosis

HINT: Include reference to the outcome of mitosis and its roles in the organism.

Cell differentiation and the formation of tissues

HINT: Identify types of tissues and give examples of how they interact in the functioning organism.

REVISE

96 KEY TERMS AND IDEAS: Did You Get It?

1. The light micrograph (right) shows a section of cells in an onion root tip. These cells have a cell cycle of approximately 24 hours. The cells can be seen to be in various stages of the cell cycle. By counting the number of cells in the various stages it is possible to calculate how long the cell spends in each stage of the cycle.

   Count and record the number of cells in the image that are undergoing mitosis and those that are in interphase. Estimate the amount of time a cell spends in each phase.

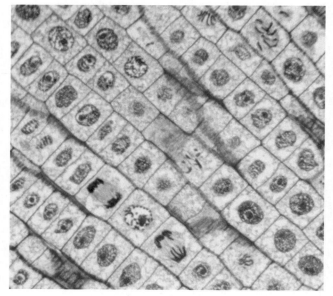

Stage	No. of cells	% of total cells	Estimated time in stage
Interphase			
Mitosis			
Total		100	

2. For each of the following statements, circle the correct answer:

   (a) The stage in the cell cycle where each chromosome consists of two, not yet visible, chromatids in preparation for mitosis:

   A G1   B S   C M   D G2

   (b) The stage of the cell cycle where the cell is preparing to begin DNA replication:

   A G1   B S   C M   D G2

   (c) Which sequence of the cell cycle is common to eukaryotes:

   A G1 to G2 to S to M to cytokinesis   C G1 to S to M to G2 to cytokinesis

   B G1 to M to G2 to S to cytokinesis   D G1 to S to G2 to M to cytokinesis

3. (a) What are the structures pictured right and what is their role?

   _____

   _____

   (b) Where are they found? _____

   _____

Microtubules

4. Use the numbers 1-4 to identify the order in which the following stages of mitosis occur:

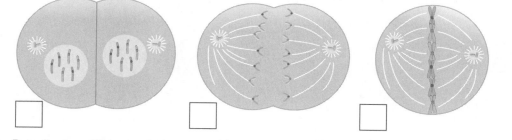

5. Describe two differences between cytokinesis (division of the cytoplasm) in plant and animal cells:

   (a) _____

   (b) _____

6. Match the following stem cell types with their potency: Zygote, Adult stem cell, Embryonic stem cell

   Pluripotent _____   Totipotent _____   Multipotent _____

 © 2016 **BIOZONE** International
ISBN: 978-1-927309-46-9
Photocopying Prohibited

97 Summative Assessment

Multicellular organisms begin as a single cell and develop into complex organisms made up of many cells.

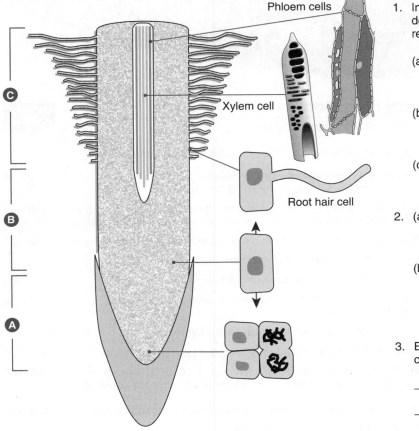

Phloem cells

Xylem cell

Root hair cell

C

B

A

1. In the diagram of a growing root tip (left) describe the cellular events happening at the regions labeled:

(a) _____

_____

(b) _____

_____

(c) _____

_____

2. (a) What type of cells occupy region A?

_____

(b) What general properties do they show?

_____

_____

3. Explain how the events happening in region C come about:

_____

_____

_____

4. (a) What process must take place before the event pictured in region A can occur? _____

(b) What does this involve?_____

_____

5. The annotated photograph right shows a tissue from the small intestine of the digestive tract. Explain how each of the tissue types identified contributes to the functioning of the organ as a whole:

(a) Muscle tissue: _____

_____

_____

(b) Connective tissue: _____

_____

_____

(c) Epithelial tissue: _____

_____

_____

(d) What type of tissue is not labeled (or visible). What does it do?

_____

_____

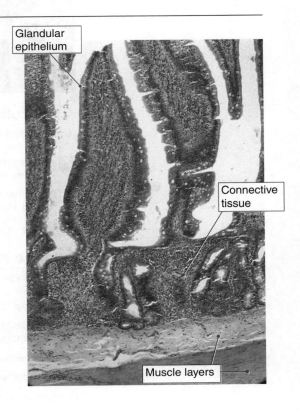

Glandular epithelium

Connective tissue

Muscle layers

TEST

Key terms
aerobic
anaerobic
cellular respiration
chloroplast
fermentation
glucose
heat energy
mitochondria
photosynthesis

Disciplinary core ideas

Show understanding of these core ideas

Activity number

Photosynthesis converts light energy into stored chemical energy

☐ 1 ATP is an energy transfer molecule. It provides the energy to drive cellular reactions. Energy is released during the hydrolysis of ATP.   98 99

☐ 2 Photosynthesis is the process that captures light energy and converts it into stored chemical energy. In plants, photosynthesis occurs in organelles called chloroplasts.   100 102 103

☐ 3 In photosynthesis, carbon dioxide and water are converted into glucose and oxygen.   100 103

Glucose can be used to make other macromolecules

☐ 4 Glucose consists of carbon, oxygen, and hydrogen atoms.   104

☐ 5 Glucose is used as a precursor to make many other biologically important molecules (e.g. DNA and proteins).   104

Matter and energy flow between different living systems

☐ 6 The glucose produced during photosynthesis is used in cellular respiration.   105

☐ 7 In cellular respiration, glucose and oxygen are used to produce ATP, which provides the energy needed to perform cellular work, such as muscle contraction.   106 107

☐ 8 Cellular respiration takes place in the cell cytoplasm and in the mitochondrion.   106 107

Heat is released during chemical reactions

☐ 9 The heat energy released from chemical reactions is lost to the surrounding environment and can be used to maintain body temperature   107

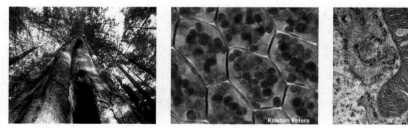

Kristian Peters

Crosscutting concepts

Understand how these fundamental concepts link different topics

Activity number

☐ 1 **EM** ▶ Changes of energy and matter in a system (e.g. cell) can be described in terms of energy and matter movements into, out of, and within that system.   98 100 102 - 105

☐ 2 **EM** ▶ Energy cannot be created or destroyed; it only moves between one place and another, e.g. between cells or between organisms and their environment.   98 106 - 107

☐ 3 **SF** ▶ The structure of organelles tells us something about their role in the cell.   98 102 107

Science and engineering practices

Demonstrate competence in these science and engineering practices

Activity number

☐ 1 Use a model to show how ATP provides energy to carry out life's functions.   99

☐ 2 Use a model to show how photosynthesis transforms light energy into stored chemical energy.   100 105 109

☐ 3 Based on evidence, explain the fate of glucose in living systems.   104

☐ 4 Use a model based on evidence to show understanding of cellular respiration.   106-07 109

☐ 5 Conduct an investigation to demonstrate that light drives photosynthesis.   101

☐ 6 Conduct an investigation to show that respiration uses oxygen and produces CO_2.   108

98 Energy in Cells

Key Idea: Cells need energy to perform the functions essential to life. This energy is provided by cellular respiration and stored in the molecule ATP.

Energy for metabolism

▶ All organisms require energy to be able to perform the metabolic processes required for them to function and reproduce.

▶ This energy is obtained by **cellular respiration**, a set of metabolic reactions which ultimately convert biochemical energy from 'food' into the energy-carrying molecule **adenosine triphosphate (ATP)**.

▶ The steps of cellular respiration take place in the cell cytoplasm and in the mitochondria.

▶ ATP is considered to be a universal energy carrier, transporting chemical energy within the cell for use in metabolic processes such as biosynthesis, cell division, cell signaling, thermoregulation, cell movement, and active transport of substances across membranes.

The mitochondrion

A mitochondrion is bounded by a double membrane. The inner and outer membranes are separated by an inter-membrane space, compartmentalizing the regions of the mitochondrion in which the different reactions of cellular respiration occur.

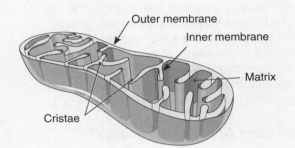

Outer membrane
Inner membrane
Matrix
Cristae

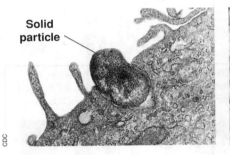

Solid particle

Energy is needed to actively transport molecules and substances across the cellular membrane such as engulfing solid particles, phagocytosis (above).

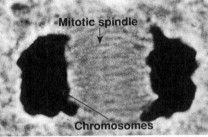

Mitotic spindle
Chromosomes

Cell division (mitosis) (above), requires energy to proceed. ATP provides energy for the mitotic spindle formation and chromosome separation.

The maintenance of body temperature requires energy. Both heating and cooling the body require energy by shivering and secretion of sweat.

1. What process produces usable energy in cells? _____

2. How is energy carried around the cell? _____

3. (a) Describe the general role of mitochondria in the cell: _____

(b) What is the purpose of the folded inner membrane in mitochondria? _____

4. (a) What energy-using process helps warm the body? _____

(b) What energy-using process helps cool the body? _____

© 2016 **BIOZONE** International
ISBN: 978-1-927309-46-9
Photocopying Prohibited

CCC CCC WEB
EM SF 98 KNOW

99 ATP

Key Idea: ATP transfers energy to where it is needed in the cell. Hydrolysis of the phosphate group releases energy, which can be used to do work.

Adenosine triphosphate (ATP)

▶ The ATP molecule (right) is a nucleotide derivative. It consists of three components;

- A purine base (**adenine**)
- A pentose sugar (**ribose**)
- Three **phosphate groups**.

▶ ATP acts as a store of energy within the cell. The bonds between the phosphate groups contain electrons in a high energy state, which store a large amount of energy that is released during a chemical reaction. The removal of one phosphate group from ATP results in the formation of adenosine diphosphate (ADP).

How does ATP provide energy?

▶ The bonds between the phosphate groups of ATP are unstable, very little energy is needed to break them. The energy in the ATP molecule is transferred to a target molecule (e.g. a protein) by a hydrolysis reaction. Water is split during the reaction and added to the terminal phosphate on ATP, forming ADP and an inorganic phosphate molecule (Pi).

▶ When the Pi molecule combines with a target molecule, energy is released. Most of the energy (about 60%) is lost as heat (this helps keep you warm). The rest of the energy is transferred to the target molecule, allowing it to do work, e.g. joining with another molecule (right).

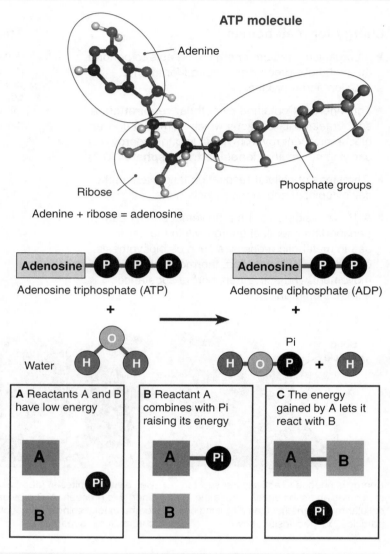

ATP molecule

Adenine

Ribose

Phosphate groups

Adenine + ribose = adenosine

| Adenosine P P P | Adenosine P P |
| Adenosine triphosphate (ATP) | Adenosine diphosphate (ADP) |

A Reactants A and B have low energy

B Reactant A combines with Pi raising its energy

C The energy gained by A lets it react with B

Note! The phosphate bonds in ATP are often referred to as being high energy bonds. This can be misleading. The bonds contain *electrons* in a high energy state (making the bonds themselves relatively weak). A small amount of energy is required to break the bonds, but when the intermediaries recombine and form new chemical bonds a large amount of energy is released. The final product is less reactive than the original reactants.

1. What are the three components of ATP? _____

2. (a) What is the biological role of ATP? _____

_____

   (b) Where is the energy stored in ATP? _____

_____

   (c) What products are formed during hydrolysis of ATP? _____

_____

3. Why does the conversion of ATP to ADP help keep us warm? _____

_____

_____

WEB   PRACTICES

KNOW **99**

© 2016 **BIOZONE** International
ISBN: 978-1-927309-46-9
Photocopying Prohibited

100 Introduction to Photosynthesis

Key Idea: Photosynthesis is the process of converting sunlight, carbon dioxide, and water into glucose and oxygen.

▶ Plants, algae, and some bacteria are photoautotrophs. They use pigments called chlorophylls to absorb light of specific wavelengths and capture light energy. The light energy is used in a process called photosynthesis.

▶ During photosynthesis carbon dioxide and water are converted into glucose and oxygen. The reaction requires sunlight energy which is transformed into chemical energy within the bonds of the glucose molecule. This chemical energy fuels life's essential processes.

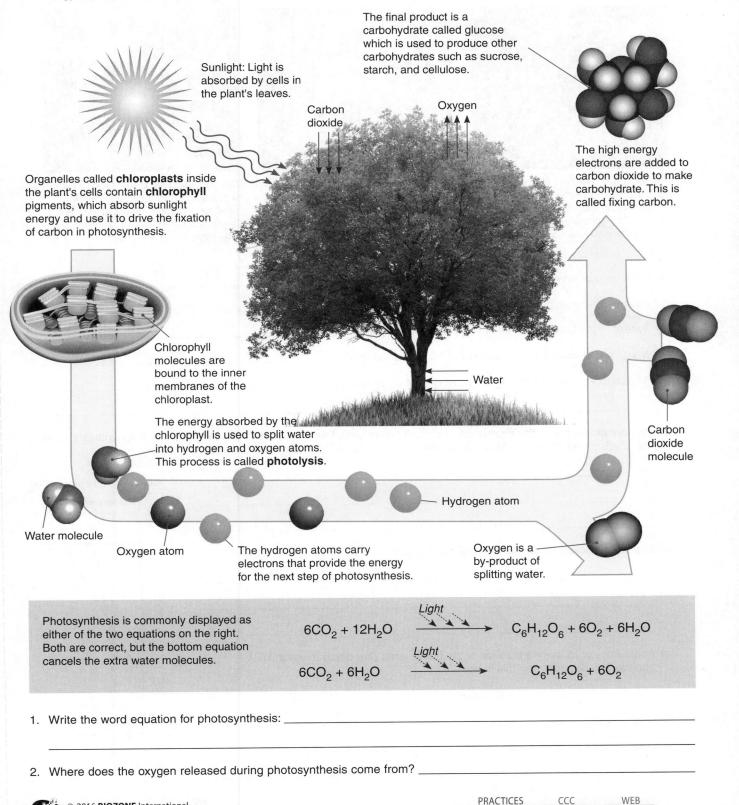

The final product is a carbohydrate called glucose which is used to produce other carbohydrates such as sucrose, starch, and cellulose.

Sunlight: Light is absorbed by cells in the plant's leaves.

Carbon dioxide

Oxygen

The high energy electrons are added to carbon dioxide to make carbohydrate. This is called fixing carbon.

Organelles called **chloroplasts** inside the plant's cells contain **chlorophyll** pigments, which absorb sunlight energy and use it to drive the fixation of carbon in photosynthesis.

Chlorophyll molecules are bound to the inner membranes of the chloroplast.

Water

Carbon dioxide molecule

The energy absorbed by the chlorophyll is used to split water into hydrogen and oxygen atoms. This process is called **photolysis**.

Hydrogen atom

Water molecule

Oxygen atom

The hydrogen atoms carry electrons that provide the energy for the next step of photosynthesis.

Oxygen is a by-product of splitting water.

Photosynthesis is commonly displayed as either of the two equations on the right. Both are correct, but the bottom equation cancels the extra water molecules.

$$6CO_2 + 12H_2O \xrightarrow{\text{Light}} C_6H_{12}O_6 + 6O_2 + 6H_2O$$

$$6CO_2 + 6H_2O \xrightarrow{\text{Light}} C_6H_{12}O_6 + 6O_2$$

1. Write the word equation for photosynthesis: _____

_____

2. Where does the oxygen released during photosynthesis come from? _____

 © 2016 **BIOZONE** International
ISBN: 978-1-927309-46-9
Photocopying Prohibited

PRACTICES   CCC   WEB

EM   100   **KNOW**

Requirements for photosynthesis

Plants need only a few raw materials to make their own food:

▶ Light energy from the sun

▶ Chlorophyll absorbs light energy

▶ CO_2 gas is reduced to carbohydrate

▶ Water is split to provide the electrons for the fixation of carbon as carbohydrate

Photosynthesis is not a single process but two complex processes (the light dependent and light independent reactions) each with multiple steps.

Production of carbohydrate (light independent reactions of Calvin cycle) occur in the fluid stroma of chloroplast. This is commonly called carbon 'fixation'.

Energy capture (light dependent reactions) occurs in the inner membranes (thylakoids) of the chloroplast.

The photosynthesis of marine algae, such as these diatoms, supplies a substantial portion of the world's oxygen. The oceans also act as sinks for absorbing large amounts of CO_2.

Macroalgae, like this giant kelp, are important marine producers. Algae living near the ocean surface get access to light used in photosynthesis (the red wavelength).

On land, vascular plants (such as trees with transport vessels) are the main producers of food. Plants at different levels in a forest receive different intensity and quality of light.

3. (a) What form of energy is used to drive photosynthesis? _____

(b) What is the name of the molecule that captures this energy? _____

(c) Where in the plant cell does photosynthesis take place? _____

(d) What form of energy is your answer to (a) converted into? _____

(e) What happens in each of the two phases of photosynthesis? _____

_____

4. (a) Primary production is the production of carbon compounds from carbon dioxide, generally by photosynthesis. Study the graph (right) showing primary production in the oceans. Describe what the graph is showing:

_____

_____

_____

(b) Explain the shape of the curves described in (a):

_____

_____

_____

(c) About 90% of all marine life lives in the photic zone (the depth to which light penetrates). Suggest why this is so:

_____

_____

_____

Ocean primary production

Primary production ($mgC\ m^{-3}\ d^{-1}$)

Depth (m)

2011 ——
2012 ········
2013 ——

© 2016 **BIOZONE** International
ISBN: 978-1-927309-46-9
Photocopying Prohibited

101 Investigating Photosynthetic Rate

Key Idea: Measuring the production of oxygen provides a simple means of measuring the rate of photosynthesis.

Background

Photosynthetic rate can be investigated by measuring the uptake of carbon dioxide (CO_2) and production of oxygen (O_2) over time. Measuring the rate of oxygen production provides an approximation of photosynthetic rate.

Aim

To investigate the effect of light intensity on the rate of photosynthesis in an aquatic plant, *Cabomba aquatica*.

Hypothesis

If photosynthetic rate is dependent on light intensity, more oxygen bubbles will be produced by *Cabomba* per unit time at higher light intensities.

Method

- 0.8-1.0 g of *Cabomba* stem were weighed. The stem was cut and inverted to ensure a free flow of oxygen.
- The stem was placed into a beaker filled with a 20°C solution of 0.2 mol L^{-1} sodium bicarbonate ($NaHCO_3$) to supply CO_2. An inverted funnel and a test tube filled with the $NaHCO_3$ solution collected the gas produced.
- The beaker was placed at distances (20, 25, 30, 35, 40, 45 cm) from a 60W light source and the light intensity measured with a lux (lx) meter at each interval. One beaker was not exposed to the light source (5 lx).
- Before recording data, the stem was left to acclimatize to the new light level for 5 minutes. Bubbles were counted for a period of three minutes at each distance.

Experimental set up

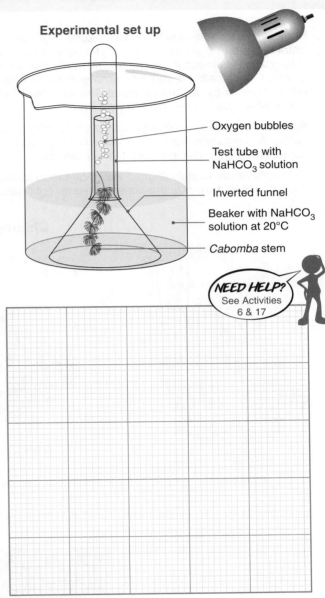

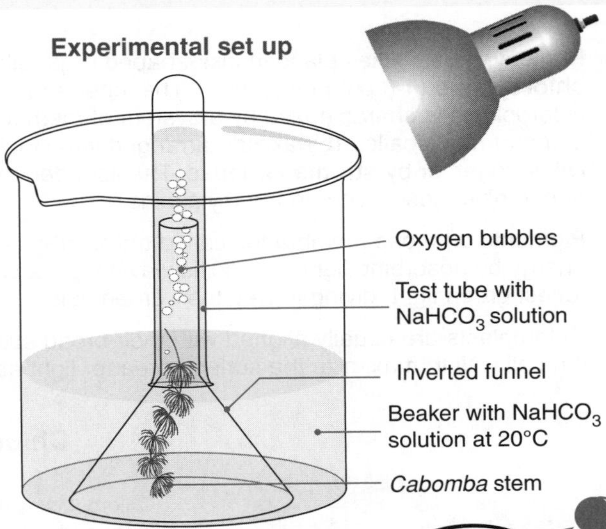
Oxygen bubbles
Test tube with NaHCO3 solution
Inverted funnel
Beaker with NaHCO3 solution at 20°C
Cabomba stem

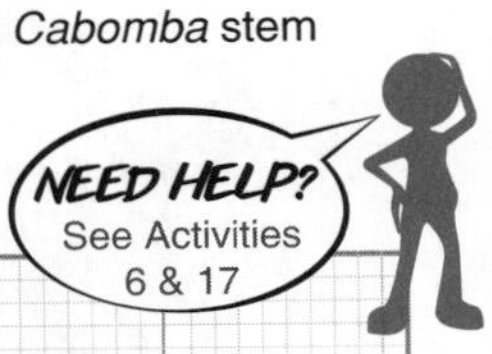
NEED HELP?
See Activities 6 & 17

The results

Light intensity in lx (distance)	Bubbles counted in three minutes	Bubbles per minute
5	0	
13 (45 cm)	6	
30 (40 cm)	9	
60 (35 cm)	12	
95 (30 cm)	18	
150 (25 cm)	33	
190 (20 cm)	35	

1. Complete the table (left) by calculating the rate of oxygen production (bubbles of oxygen gas per minute):

2. Use the data to draw a graph on the grid above of the bubbles produced per minute vs light intensity:

3. Although the light source was placed set distances from the *Cabomba* stem, light intensity in lux was recorded at each distance rather than distance *per se*. Explain why this would be more accurate:

4. The sample of gas collected during the experiment was tested with a glowing splint. The splint reignited when placed in the gas. What does this confirm about the gas produced?

© 2016 **BIOZONE** International
ISBN: 978-1-927309-46-9
Photocopying Prohibited

102 Chloroplasts

Key Idea: Photosynthesis occurs in organelles called chloroplasts. Chloroplasts contain the pigment chlorophyll, which captures light energy.

Photosynthesis takes place in disk-shaped organelles called **chloroplasts** (4-6 µm in diameter). The inner structure of chloroplasts is characterized by a system of membrane-bound compartments called **thylakoids** arranged into stacks called **grana** linked together by **stroma lamellae**. The light dependent reactions of photosynthesis occur in the thylakoids.

Pigments on these membranes called **chlorophylls** capture light energy by absorbing light of specific wavelengths. Chlorophylls reflect green light, giving leaves their green color.

Chloroplasts are usually aligned with their broad surface parallel to the cell wall to maximize the surface area for light absorption.

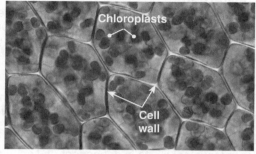

A mesophyll leaf cell contains 50-100 chloroplasts.

Chloroplast structure

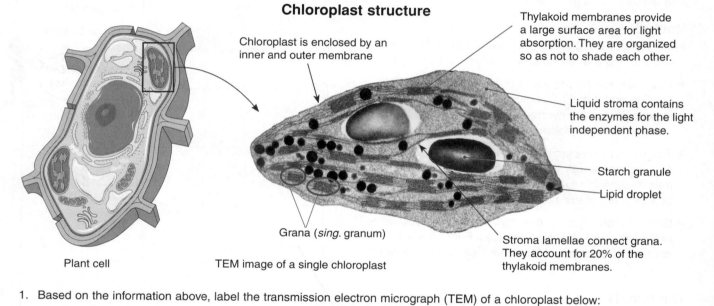

Chloroplast is enclosed by an inner and outer membrane

Thylakoid membranes provide a large surface area for light absorption. They are organized so as not to shade each other.

Liquid stroma contains the enzymes for the light independent phase.

Starch granule

Lipid droplet

Grana (*sing*. granum)

Stroma lamellae connect grana. They account for 20% of the thylakoid membranes.

Plant cell

TEM image of a single chloroplast

1. Based on the information above, label the transmission electron micrograph (TEM) of a chloroplast below:

(a)

(b)

(c)

(d)

(e)

(f)

2. What does chlorophyll do? _____
_____

3. What features of chloroplasts help maximize the amount of light that can be absorbed? _____
_____
_____
_____

 © 2016 **BIOZONE** International
ISBN: 978-1-927309-46-9
Photocopying Prohibited

103 Stages in Photosynthesis

Key Idea: Photosynthesis consists of two phases, the light dependent phase (reactions) and the light independent phase (reactions).

▶ Photosynthesis has two phases, the light dependent phase and the light independent phase.

▶ In the reactions of the **light dependent phase**, light energy is converted to chemical energy (ATP and NADPH). This phase occurs in the thylakoid membranes of the chloroplasts.

▶ In the reactions of the **light independent phase**, the chemical energy is used to synthesize carbohydrate. This phase occurs in the stroma of chloroplasts.

Light dependent phase (LDP):

In the first phase of photosynthesis, chlorophyll captures light energy, which is used to split water, producing O_2 gas (waste), electrons and H^+ ions, which are transferred to the molecule NADPH. ATP is also produced. The light dependent phase occurs in the thylakoid membranes of the grana.

Light independent phase (LIP):

The second phase of photosynthesis occurs in the stroma and uses the NADPH and the ATP to drive a series of enzyme-controlled reactions (the **Calvin cycle**) that fix carbon dioxide to produce triose phosphate. This phase does not need light to proceed.

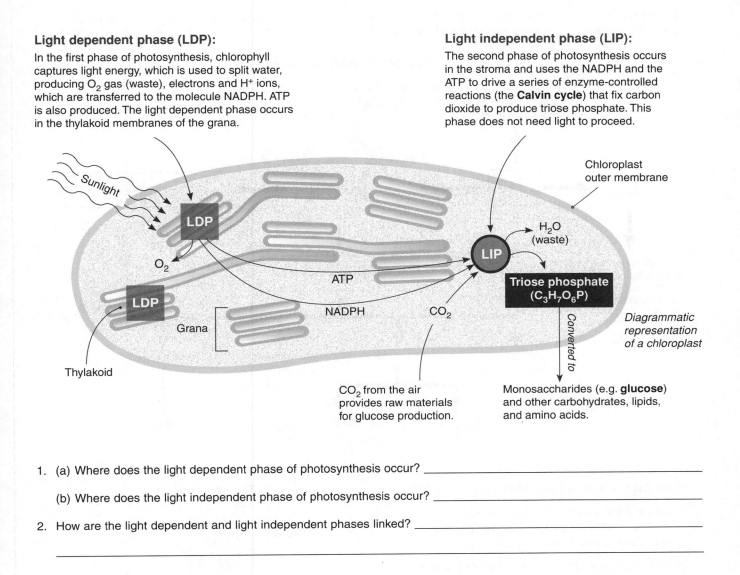

Diagrammatic representation of a chloroplast

CO_2 from the air provides raw materials for glucose production.

Monosaccharides (e.g. **glucose**) and other carbohydrates, lipids, and amino acids.

1. (a) Where does the light dependent phase of photosynthesis occur? _____

   (b) Where does the light independent phase of photosynthesis occur? _____

2. How are the light dependent and light independent phases linked? _____

   _____

   _____

3. In two experiments, radioactively-labeled oxygen (shown in blue) was used to follow oxygen through the photosynthetic process. The results of the experiment are shown below:

   Experiment A: $6CO_2 + 12H_2O$ + sunlight energy → $C_6H_{12}O_6 + 6O_2 + 6H_2O$

   Experiment B: $6CO_2 + 12H_2O$ + sunlight energy → $C_6H_{12}O_6 + 6O_2 + 6H_2O$

   From these results, what would you conclude about the source of the oxygen in:

   (a) The carbohydrate produced? _____

   (b) The oxygen released? _____

104 The Fate of Glucose

Key Idea: Glucose is an important precursor molecule used to produce a wide range of other molecules.

Glucose is a multipurpose molecule

▶ Glucose is a versatile biological molecule. It contains the elements carbon, oxygen, and hydrogen, which are used to build many other molecules produced by plants, animals, and other living organisms.

▶ Plants make their glucose directly through the process of photosynthesis and use it to build all the molecules they require. Animals obtain their glucose (as carbohydrates) by consuming plants or other animals. Other molecules (e.g. amino acids and fatty acids) are also obtained by animals this way.

▶ Glucose has three main fates: immediate use to produce ATP molecules (available energy for work), storage for later ATP production, or for use in building other molecules.

The fate of glucose

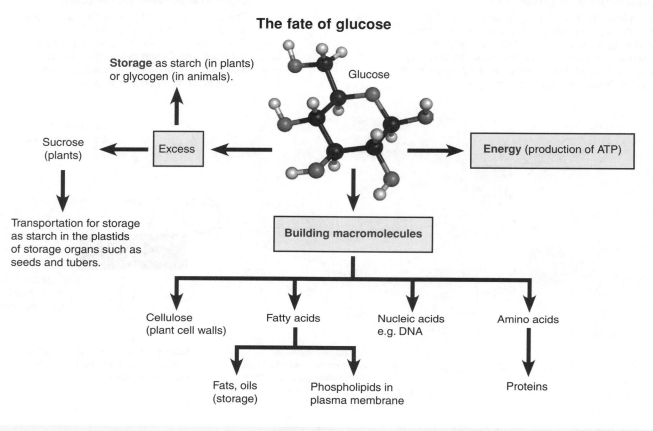

Storage as starch (in plants) or glycogen (in animals).

Glucose

Sucrose (plants)

Excess

Energy (production of ATP)

Transportation for storage as starch in the plastids of storage organs such as seeds and tubers.

Building macromolecules

Cellulose (plant cell walls)

Fatty acids

Nucleic acids e.g. DNA

Amino acids

Fats, oils (storage)

Phospholipids in plasma membrane

Proteins

How do we know how glucose is used?

▶ Labeling the carbon atoms in a glucose molecule with isotopes shows how glucose is incorporated into other molecules.

▶ An isotope is an element (e.g. carbon) whose atoms have a particular number of neutrons in their nucleus. The different number of neutrons allows the isotopes to be identified by their density (e.g. a carbon atom with 13 neutrons is denser than a carbon atom with 12 neutrons).

▶ Some isotopes are radioactive. These radioactive isotopes can be traced using X-ray film or devices that detect the disintegration of the isotopes, such as Geiger counters.

The carbon atom

Nucleus

Proton

Neutron

The nucleus of an atom is made up of neutrons and protons. For any element, the number of protons remains the same, but the number of neutrons can vary. Electrons (not shown) are found outside the nucleus.

Naturally occurring C isotopes

^{12}C	^{13}C	^{14}C
6 protons	6 protons	6 protons
6 neutrons	7 neutrons	8 neutrons
Stable. 99.9% of all C isotopes.	Stable	Radioactive

Isotope experiments with animals

Experiments using ^{13}C isotopes to identify the fate of glucose in guinea pigs showed that 25% of the glucose intake was used as fuel for cellular respiration. The rest of the glucose was incorporated into proteins, fats, and glycogen.

Corals are small sea anemone-like organisms that live in a symbiotic relationship with algae. The algae transfer sugars to the coral in return for the safe environment provided by the coral. Experiments with ^{13}C showed that the major molecule being transferred to the coral was glucose.

Isotope experiments with plants

^{13}C isotopes were used to trace the movement of glucose in plant leaves. It was found that some glucose is converted to fructose (a sugar molecule similar to glucose). Fructose molecules can be joined together to be stored as fructan in plant vacuoles. Fructose is also added to glucose to produce sucrose. Sucrose is transported out of the leaf.

Four molecule model of glucose use in a plant

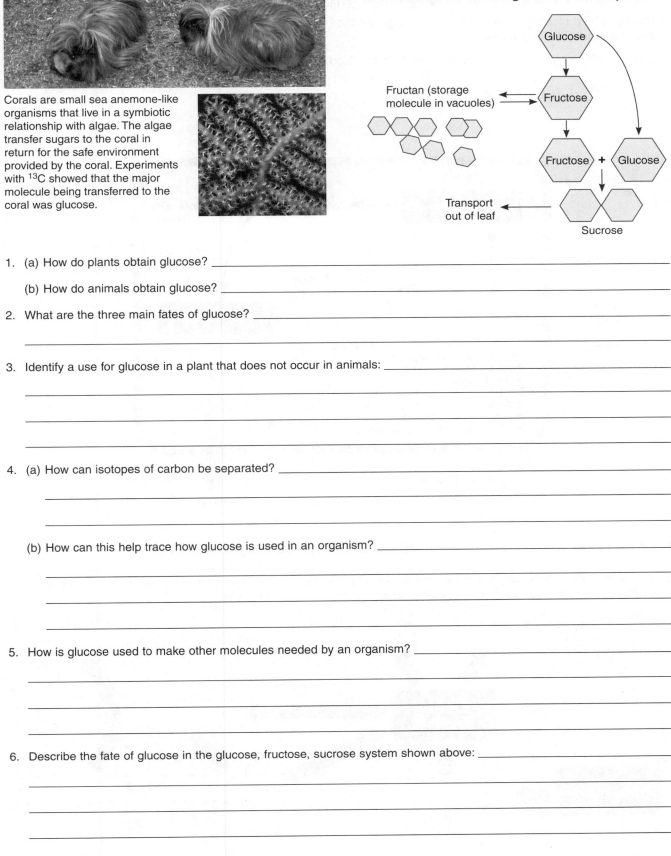

1. (a) How do plants obtain glucose? _____

   (b) How do animals obtain glucose? _____

2. What are the three main fates of glucose? _____
   _____

3. Identify a use for glucose in a plant that does not occur in animals: _____
   _____
   _____
   _____

4. (a) How can isotopes of carbon be separated? _____
   _____
   _____

   (b) How can this help trace how glucose is used in an organism? _____
   _____
   _____
   _____

5. How is glucose used to make other molecules needed by an organism? _____
   _____
   _____
   _____

6. Describe the fate of glucose in the glucose, fructose, sucrose system shown above: _____
   _____
   _____
   _____

105 Energy Transfer Between Systems

Key Idea: The energy from sunlight is captured and stored as glucose, which powers the production of ATP. ATP provides the energy for the various chemical reactions in living systems.

▶ During photosynthesis light energy is converted into chemical energy in the form of glucose. Glucose is used by plants and animals to provide the energy for cellular respiration.

▶ During cellular respiration ATP is formed through a series of chemical reactions. The ATP provides the energy to drive life's essential processes.

▶ Heterotrophs (organisms that cannot make their own food) obtain their glucose by eating plants or other organisms.

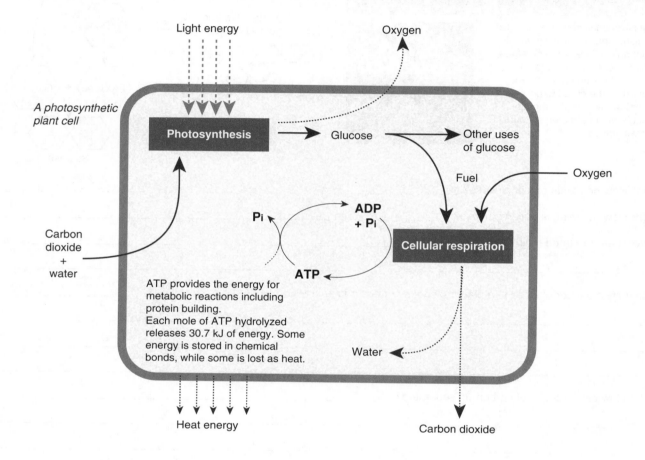

1. On the diagram above, mark with an arrow where heterotrophs join the energy transfer system.

2. Complete the schematic diagram of the transfer of energy and the production of macromolecules below using the following word list: water, ADP, protein, carbon dioxide, amino acid, glucose, ATP.

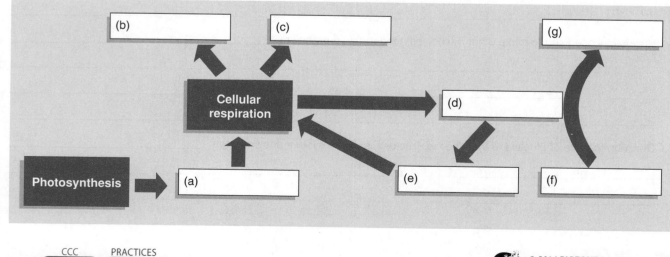

CCC   PRACTICES

KNOW   EM

© 2016 **BIOZONE** International
ISBN: 978-1-927309-46-9
Photocopying Prohibited

106 Energy From Glucose

Key Idea: Cellular respiration is the process by which energy is released from molecules via a series of chemical reactions.

▶ Energy is released in cells by the breakdown of sugars and other substances in cellular respiration. During aerobic respiration oxygen is consumed and carbon dioxide is released. These gases need to be exchanged with the environment by diffusion. Diffusion gradients are maintained by transport of gases away from the gas exchange surface (e.g. the cell surface). In anaerobic pathways, ATP is generated but oxygen is not used.

The overall equation for cellular respiration is:

Glucose + Oxygen ⟶ Carbon dioxide + Water + Energy

$$C_6H_{12}O_6 + 6O_2 \longrightarrow 6CO_2 + 6H_2O + Energy$$

To carry out aerobic respiration, the body needs oxygen and must remove carbon dioxide. Gas exchange surfaces provide a way for these respiratory gases to enter and leave the body by diffusion. Some organisms use the body surface as the gas exchange surface, but many have specialized gas exchange structures (e.g. lungs, gills, or stomata).

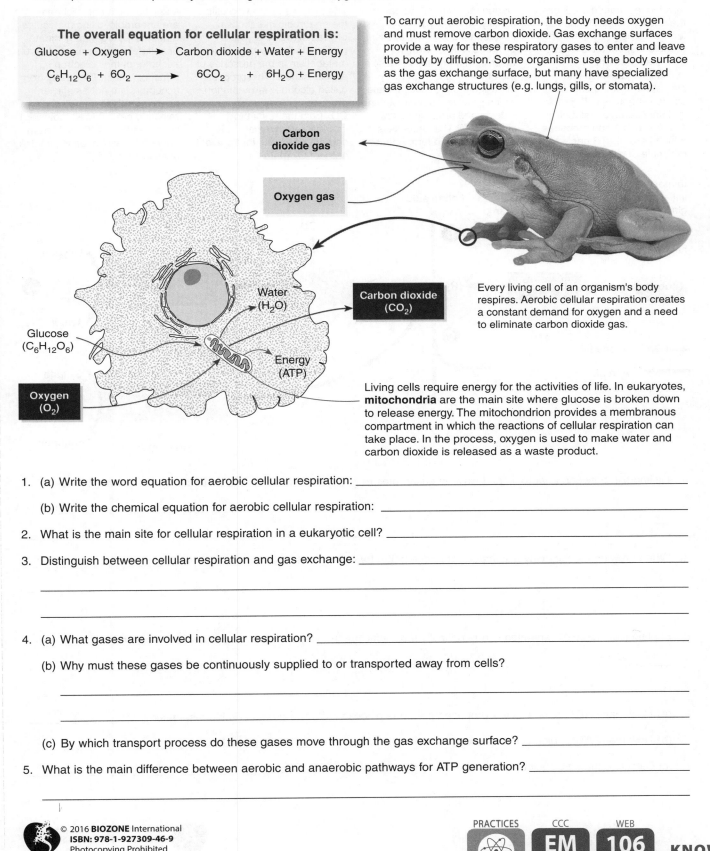

Every living cell of an organism's body respires. Aerobic cellular respiration creates a constant demand for oxygen and a need to eliminate carbon dioxide gas.

Living cells require energy for the activities of life. In eukaryotes, **mitochondria** are the main site where glucose is broken down to release energy. The mitochondrion provides a membranous compartment in which the reactions of cellular respiration can take place. In the process, oxygen is used to make water and carbon dioxide is released as a waste product.

1. (a) Write the word equation for aerobic cellular respiration: _____

(b) Write the chemical equation for aerobic cellular respiration: _____

2. What is the main site for cellular respiration in a eukaryotic cell? _____

3. Distinguish between cellular respiration and gas exchange: _____

4. (a) What gases are involved in cellular respiration? _____

(b) Why must these gases be continuously supplied to or transported away from cells?

(c) By which transport process do these gases move through the gas exchange surface? _____

5. What is the main difference between aerobic and anaerobic pathways for ATP generation? _____

PRACTICES CCC WEB EM 106 **KNOW**

Aerobic and anaerobic pathways for ATP production

A Aerobic respiration

B Lactic acid fermentation

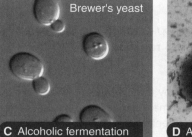

Brewer's yeast

C Alcoholic fermentation

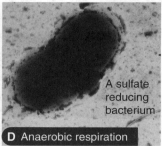

A sulfate reducing bacterium

D Anaerobic respiration

Aerobic respiration produces the energy (as ATP) needed for metabolism. The rate of aerobic respiration is limited by the amount of oxygen available. In animals and plants, most of the time the oxygen supply is sufficient to maintain aerobic metabolism. Aerobic respiration produces a high yield of ATP per molecule of glucose (path A).

During maximum physical activity, when oxygen is limited, anaerobic metabolism provides ATP for working muscle. In mammalian muscle, metabolism of a respiratory intermediate produces lactate, which provides fuel for working muscle and produces a low yield of ATP. This process is called lactic acid fermentation (path B).

The process of brewing utilizes the anaerobic metabolism of yeasts. Brewers yeasts preferentially use anaerobic metabolism in the presence of excess sugars. This process, called alcoholic fermentation, produces ethanol and CO_2 from the respiratory intermediate pyruvate. It is carried out in vats that prevent entry of O_2 (path C)

Many bacteria and archaea are anaerobic, using molecules other than oxygen (e.g. nitrate or sulfate) as a terminal electron acceptor of their electron transport chain. These electron acceptors are not as efficient as oxygen (less energy is released per oxidized molecule) so the energy (ATP) yield from anaerobic respiration is generally quite low (path D).

In most energy-yielding pathways the initial source of chemical energy is glucose. The first step, glycolysis, is an almost universal pathway. The paths differ in what happens after glucose has been converted to pyruvate.

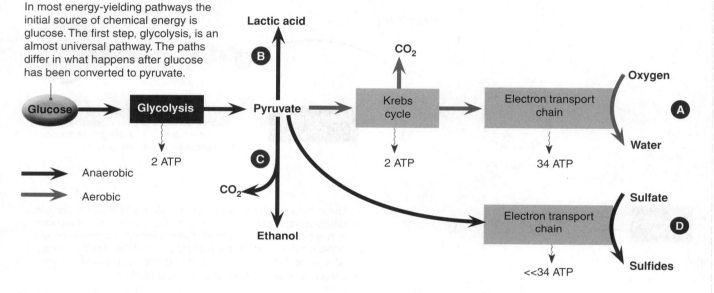

6. Distinguish between anaerobic pathways in eukaryotes (e.g. yeasts) and anaerobic respiration in anaerobic microbes:

_____

_____

7. When brewing alcohol, why is it important to prevent entry of oxygen to the fermentation vats? _____

_____

_____

8. Explain why aerobic respiration is energetically more efficient than fermentation and anaerobic respiration: _____

_____

_____

9. (a) How many ATP molecules are produced from one glucose molecule during aerobic respiration? _____

    (b) How many ATP molecules are produced during lactic acid or alcoholic fermentation? _____

    (c) Calculate the efficiency of fermentation compared to aerobic respiration: _____

_____

107 Aerobic Cellular Respiration

Key Idea: Cellular respiration is an aerobic process that converts the chemical energy in glucose into usable energy (in the form of ATP), carbon dioxide, and water.

▶ **Cellular respiration** is the process of extracting the energy stored in the chemical bonds in glucose and storing it in ATP molecules. The process includes many chemical reactions, some of which produce ATP molecules and some that prepare molecules for further chemical reactions.

▶ Cellular respiration can be divided into four major steps, each with its own set of chemical reactions. Every step, except the link reaction, produces ATP. The four steps are: glycolysis, the link reaction, the Krebs cycle, and the electron transport chain (ETC).

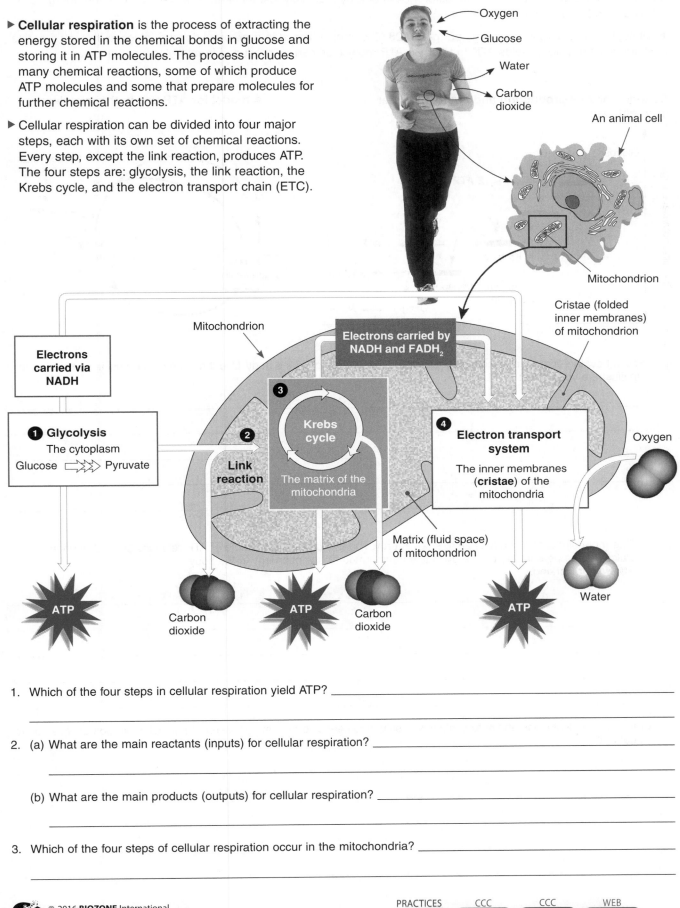

1. Which of the four steps in cellular respiration yield ATP? _____

_____

2. (a) What are the main reactants (inputs) for cellular respiration? _____

_____

(b) What are the main products (outputs) for cellular respiration? _____

_____

3. Which of the four steps of cellular respiration occur in the mitochondria? _____

_____

PRACTICES   CCC   CCC   WEB

EM   SF   107   **KNOW**

How does cellular respiration provide energy?

▶ A molecule's energy is contained in the electrons within the molecule's chemical bonds. During a chemical reaction, energy (e.g. heat) can break the bonds of the reactants.

▶ When the reactants form products, the new bonds within the product will contain electrons with less energy, making the bonds more stable. The difference in energy is usually lost as heat. However, some of the energy can be captured to do work.

▶ Glucose contains 16 kJ of energy per gram (2870 kJ mol^{-1}). The step-wise breakdown of glucose through a series of chemical reactions yields ATP. In total, 38 ATP molecules can be produced from 1 glucose molecule.

A model for ATP production and energy transfer from glucose

A model for ATP use in the muscles

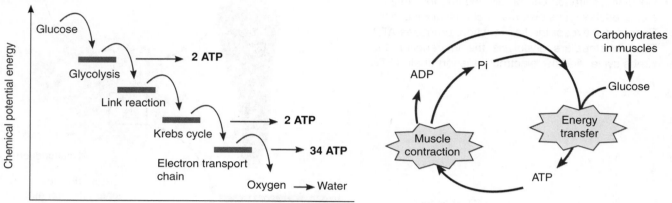

4. Explain how the energy in glucose is converted to useful energy in the body. Use the example of muscle contraction to help illustrate your ideas:

_____

_____

_____

_____

_____

5. (a) One mole of glucose contains 2870 kJ of energy. The hydrolysis of one mole of ATP releases 30.7 kJ of energy. Calculate the percentage of energy that is transformed to useful energy in the body. Show your working.

_____

_____

_____

_____

_____

(b) Use your calculations above to explain why shivering keeps you warm and extreme muscular exertion causes you to get hot:

_____

_____

_____

_____

 © 2016 **BIOZONE** International
ISBN: 978-1-927309-46-9
Photocopying Prohibited

108 Measuring Respiration

Key Idea: Oxygen consumption and carbon dioxide production in respiring organisms can be measured with a respirometer.

A respirometer measures the amount of oxygen consumed and the amount of carbon dioxide produced during cellular respiration. The diagram below shows a **simple respirometer**.

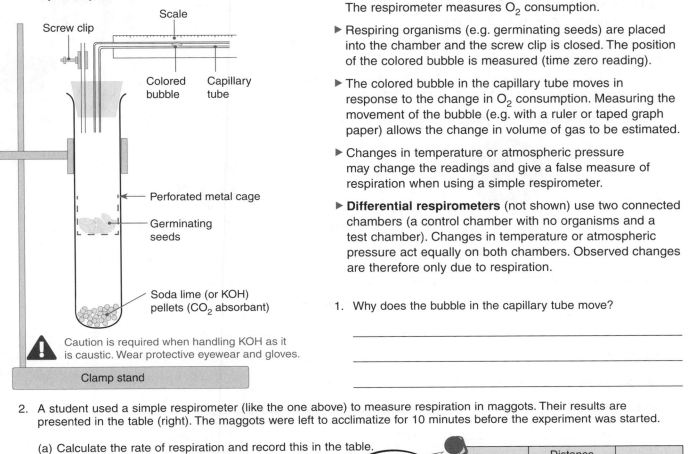

Scale

Screw clip

Colored bubble   Capillary tube

Perforated metal cage

Germinating seeds

Soda lime (or KOH) pellets (CO_2 absorbant)

⚠ Caution is required when handling KOH as it is caustic. Wear protective eyewear and gloves.

Clamp stand

Measuring respiration with a simple respirometer

▶ Soda lime or potassium hydroxide is placed into the chamber to absorb any CO_2 produced during respiration. The respirometer measures O_2 consumption.

▶ Respiring organisms (e.g. germinating seeds) are placed into the chamber and the screw clip is closed. The position of the colored bubble is measured (time zero reading).

▶ The colored bubble in the capillary tube moves in response to the change in O_2 consumption. Measuring the movement of the bubble (e.g. with a ruler or taped graph paper) allows the change in volume of gas to be estimated.

▶ Changes in temperature or atmospheric pressure may change the readings and give a false measure of respiration when using a simple respirometer.

▶ **Differential respirometers** (not shown) use two connected chambers (a control chamber with no organisms and a test chamber). Changes in temperature or atmospheric pressure act equally on both chambers. Observed changes are therefore only due to respiration.

1. Why does the bubble in the capillary tube move?

_____

_____

_____

2. A student used a simple respirometer (like the one above) to measure respiration in maggots. Their results are presented in the table (right). The maggots were left to acclimatize for 10 minutes before the experiment was started.

   (a) Calculate the rate of respiration and record this in the table. The first two calculations have been done for you.

   (b) Plot the rate of respiration on the grid, below right.

NEED HELP? See Activities 6 and 17

   (c) Describe the results in your plot: _____

_____

_____

_____

Time (minutes)	Distance bubble moved (mm)	Rate (mm min^{-1})
0	0	–
5	25	5
10	65	
15	95	
20	130	
25	160	

   (d) Why was there an acclimatization period before the experiment began?

_____

_____

3. Why would it have been better to use a differential respirometer? _____

_____

_____

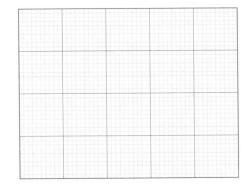

PRACTICES   WEB

108   DATA

109 Modeling Photosynthesis and Cell Respiration

Key Idea: Modeling photosynthesis and cellular respiration using paper cut outs will help you better understand the chemical processes going on.

▶ During photosynthesis and cellular respiration, molecules are broken down and recombined to form new molecules.

▶ In this activity you will model the inputs and outputs of each of these processes using the atoms (carbon, hydrogen, and oxygen) on the next page. We have placed the atoms in boxes to make it easier to cut them out.

▶ At the end of this activity you will be able to see how the reactants (starting molecules) are recombined to form the final products.

Glucose

Carbon dioxide

Water

Note: You can either work by yourself or team up with a partner. If you have beads or molecular models you could use these instead of the shapes on the next page.

1. Cut out the atoms and shapes on the following page. They are color coded as follows:

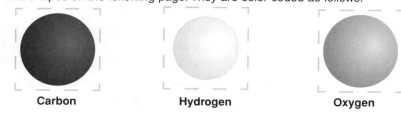

Carbon   **Hydrogen**   **Oxygen**

2. Write the equation for **photosynthesis** here: _____

   (a) State the starting reactants in photosynthesis: _____

   (b) State the total number of atoms of each type needed to make the starting reactants:

   Carbon: _____   Hydrogen: _____   Oxygen: _____

   (c) Use the atoms you have cut out to make the starting reactants in photosynthesis.

   (d) State the end products of photosynthesis: _____

   (e) State the total number atoms of each type needed to make the end products of photosynthesis:

   Carbon: _____   Hydrogen: _____   Oxygen: _____

   (f) Use the atoms you have cut out to make the end products of photosynthesis.

   (g) What do you notice about the number of C, H, and O atoms on each side of the photosynthesis equation? _____

   _____

   (h) Name the energy source for this process and add it to the model you have made: _____

3. Write the equation for **cellular respiration** here: _____

   (a) State the starting reactants in cellular respiration: _____

   (b) State the total number of atoms of each type needed to make the starting reactants:

   Carbon: _____   Hydrogen: _____   Oxygen: _____

   (c) Use the atoms you have cut out to make the starting reactants in cellular respiration.

   (d) State the end products of cellular respiration: _____

   (e) State the total number of atoms of each type needed to make the end products of cellular respiration:

   Carbon: _____   Hydrogen: _____   Oxygen: _____

   (f) Use the atoms you have cut out to make the end products of cellular respiration.

   (g) Name the end products of cellular respiration that are utilized in photosynthesis: _____

© 2016 **BIOZONE** International
ISBN: 978-1-927309-46-9
Photocopying Prohibited

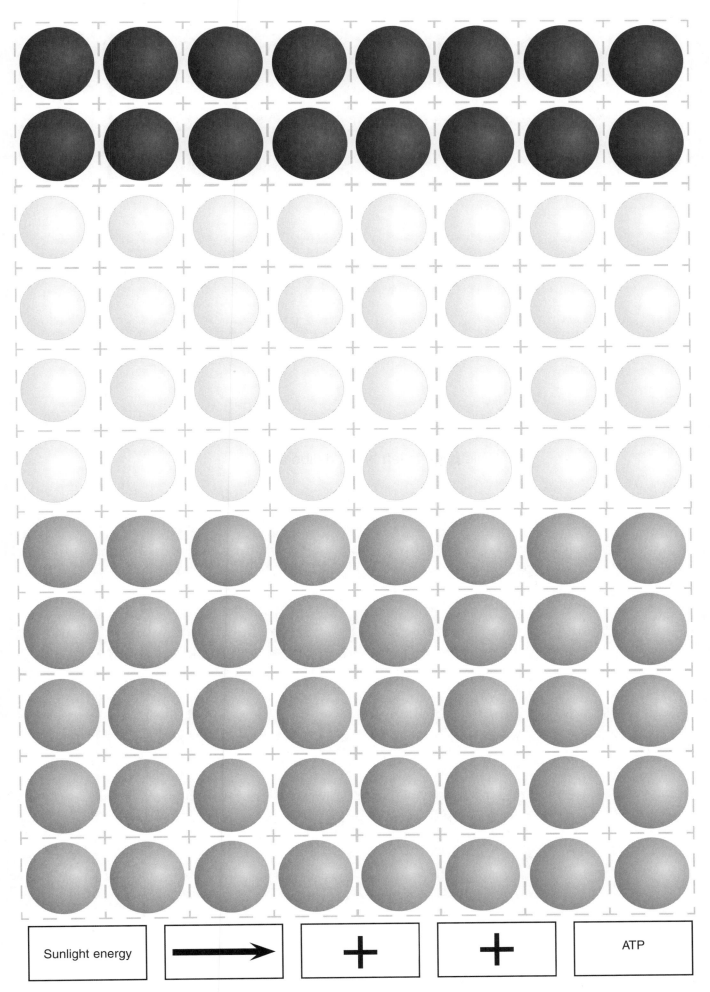

| Sunlight energy | → | + | + | ATP |

This page is left blank deliberately

110 Chapter Review

Summarize what you know about this topic so far under the headings provided. You can draw diagrams or mind maps, or write notes to organize your thoughts. Use the checklist in the introduction and the hints to help you.

The structure and role of ATP

HINT: How does the structure of ATP contribute to its function?

ATP

The fate of glucose

HINT: Describe the role of glucose in living systems (e.g. as a component of macromolecules with different properties).

Photosynthesis

HINT: Summarize the process of photosynthesis and explain its central role in living systems.

Cellular respiration

HINT: Summarize the process of cellular respiration and its role in living systems.

© 2016 BIOZONE International
ISBN: 978-1-927309-46-9
Photocopying Prohibited

REVISE

111 KEY TERMS AND IDEAS: Did You Get It?

1. Test your vocabulary by matching each term to its definition, as identified by its preceding letter code.

ATP ....

cellular respiration ....

chloroplast ....

glucose ....

heat energy ....

mitochondria ....

photosynthesis ....

A These organelles are the cell's energy transformers, in which chemical energy is converted into ATP.

B The biochemical process that uses light energy to convert carbon dioxide and water into glucose molecules and oxygen.

C One of the main products of photosynthesis, this molecule consists of carbon, oxygen, and hydrogen.

D A set of biochemical reactions in which the chemical energy in glucose is converted to usable energy (as ATP) and waste products.

E An chlorophyll-containing organelle found in plants in which the reactions of photosynthesis take place.

F The cell's energy carrier.

G A by-product of many chemical reactions, this can be used to maintain body temperature.

2. Test your understanding of photosynthesis and cellular respiration by answering the questions below.

Plant cell

(a) Name this organelle: _____

Name the major process that occurs here:

_____

Write the word equation for this process:

_____

_____

_____

Write the chemical equation for this process:

_____

_____

(b) Name this organelle: _____

Name the major process that occurs here: _____

Write the word equation for this process: _____

_____

Write the chemical equation for this process: _____

_____

TEST

© 2016 **BIOZONE** International
ISBN: 978-1-927309-46-9
Photocopying Prohibited

112 Summative Assessment

▶ Brewer's yeast is a facultative anaerobe (meaning it can respire aerobically or use the anaerobic process of fermentation). It can metabolize a variety of substrates but is usually cultured with simple sugars.

▶ Brewer's yeast will preferentially use alcoholic fermentation when sugars are in excess. One would expect glucose to be the preferred substrate, as it is the starting molecule in cellular respiration.

▶ The experiment below describes an investigation into yeast fermentation in the presence of excess glucose.

The aim
To investigate alcoholic fermentation in yeast when glucose is the substrate.

The hypothesis
In the presence of excess glucose substrate, yeast fermentation will produce an increasing amount of carbon dioxide over time.

Background
The rate at which brewer's yeast (*Saccharomyces cerevisiae*) metabolizes carbohydrate substrates is influenced by factors such as temperature, solution pH, and type of carbohydrate available.

The literature describes yeast metabolism as optimal in warm (~24°C), acid (pH 4-6) environments.

High levels of sugars suppress aerobic respiration in yeast, so yeast will preferentially use the fermentation pathway in the presence of excess substrate.

The experiment
The experiment set up is shown on the right and was carried out at 24°C and pH 4.5 using glucose as a substrate. The control contained yeast solution but no substrate.

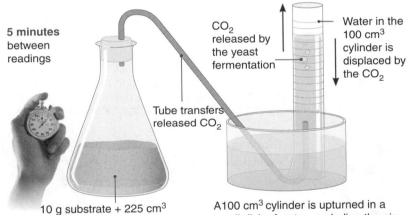

5 minutes between readings

CO_2 released by the yeast fermentation

Water in the 100 cm^3 cylinder is displaced by the CO_2

Tube transfers released CO_2

10 g substrate + 225 cm^3 water + 25 cm^3 yeast culture

A 100 cm^3 cylinder is upturned in a small dish of water, excluding the air.

The results

Substrate	Volume of carbon dioxide collected (cm^3)	
Time (min)	None	Glucose
0	0	0
5	0	0
10	0	0
15	0	0
20	0	0.5
25	0	1.2
30	0	2.8
35	0	4.2
40	0	4.6
45	0	7.4
50	0	10.8
55	0	13.6
60	0	16.1
65	0	22.0
70	0	23.8
75	0	26.7
80	0	32.5
85	0	37.0
90	0	39.9

Experimental design and results adapted from Tom Schuster, Rosalie Van Zyl, & Harold Coller , California State University Northridge 2005

1. Use the tabulated data (right) to plot an appropriate graph of the results on the grid below:

TEST

152

2. (a) Describe the results obtained on the previous page: _____

_____

_____

_____

   (b) Do you think the results supported the hypothesis? _____

   (c) Explain your reason: _____

_____

_____

3. (a) What assumptions are being made in this experimental design? _____

_____

_____

   (b) Explain why or why not you think the assumptions were reasonable? _____

_____

_____

_____

4. Why do you think yeast preferentially use an anaerobic fermentation pathway when sugars are in excess?

_____

_____

The students wanted to compare the fermentation rates of yeast on a variety of substrates to determine which was the preferred substrate. They identified four sugars to use as possible substrates (below). The students carried out some research into the sugars and found out the following information.

Glucose
• Glucose is a single sugar (monosaccharide).

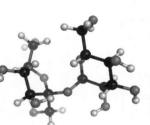

Maltose
• Maltose is a disaccharide (two sugars joined together).
• It consists of two glucose molecules.
• Hydrolyzed by maltase enzyme into two single glucose molecules.

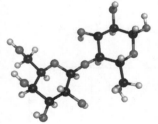

Lactose
• Lactose is a disaccharide (two sugars joined together).
• It consists of a glucose molecule and a galactose molecule.
• Hydrolyzed by lactase enzyme into glucose and galactose.

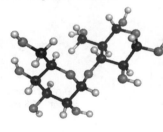

Sucrose
• Sucrose is a disaccharide (two sugars joined together).
• It consists of a glucose molecule and a fructose molecule.
• Hydrolyzed by the enzyme sucrase into glucose and fructose molecules.

5. Based on the information provided above, suggest an hypothesis for how the substrates will be used by the yeast *Saccharomyces cerevisiae*:

_____

_____

_____

_____

© 2016 **BIOZONE** International
ISBN: 978-1-927309-46-9
Photocopying Prohibited

6. The students then wanted to design an experiment to test their hypothesis. In the space below describe how an experiment could be set up to test this hypothesis:

_____

_____

_____

_____

_____

_____

_____

_____

_____

_____

_____

_____

_____

7. The students carried out their experiment and obtained the following results:

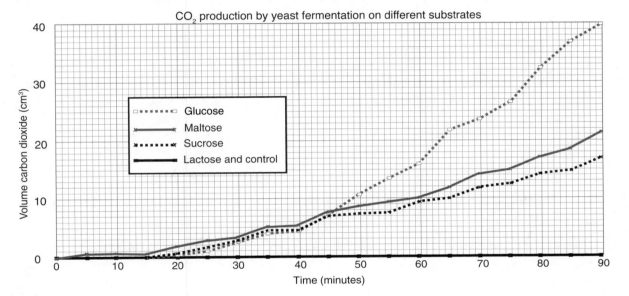

CO_2 production by yeast fermentation on different substrates

(a) Did their results support your hypothesis? Explain why or why not: _____

_____

_____

_____

(b) Why do you think CO_2 production was highest when glucose was the substrate? _____

_____

_____

(c) Suggest why fermentation rates were lower on maltose and sucrose than on glucose: _____

_____

_____

(d) Give a possible explanation for why no CO_2 was produced when lactose was the substrate? _____

_____

_____

 © 2016 **BIOZONE** International
ISBN: 978-1-927309-46-9
Photocopying Prohibited

Ecosystems: Interactions, Energy, and Dynamics

Concepts and connections
Use arrows to make your own connections between related concepts in this section of this workbook

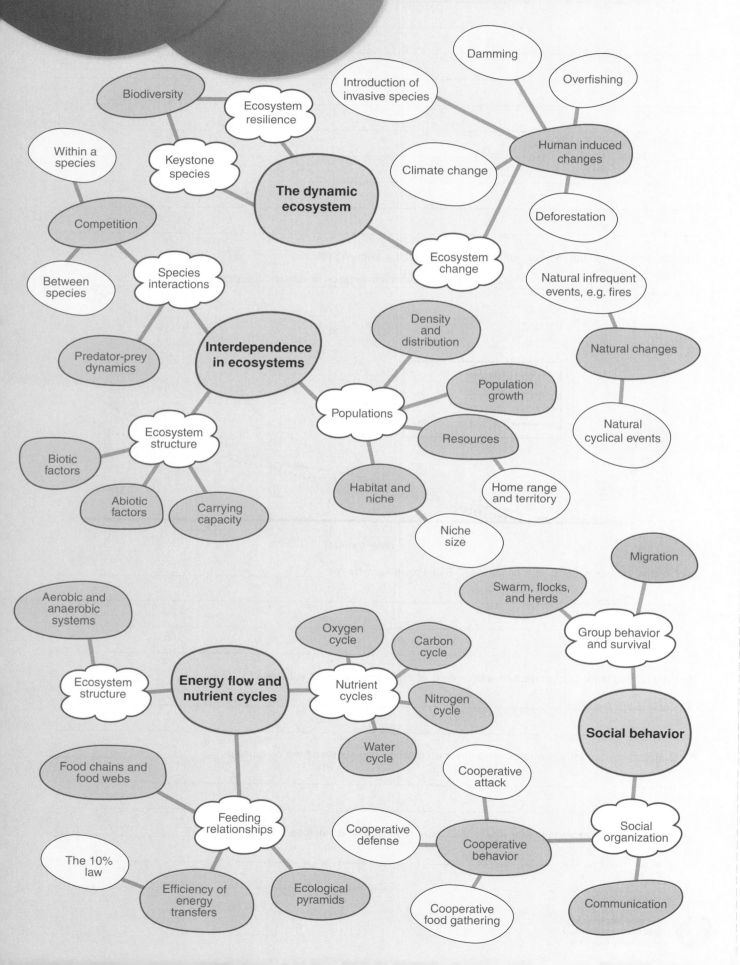

LS2.A Interdependence in Ecosystems

Key terms

abiotic factor

biotic factor

carrying capacity

competition

density

distribution

ecosystem

habitat

home range

interspecific competition

intraspecific competition

mutualism

niche

parasitism

population

population growth

predation

Disciplinary core ideas

Show understanding of these core ideas

Activity number

Species have interdependent relationships in ecosystems

☐ 1 An ecosystem includes all the living organisms and physical factors in an area. — 113

☐ 2 An organism's ecological niche (niche) describes its functional position in the ecosystem, including its habitat and its relationships with other species. — 114-116 126 127

☐ 3 Populations of organisms inhabit an ecosystem. Their distribution and density depends on factors such as competition and the availability of resources. — 117

☐ 4 Species interact in ways that may be beneficial to both species (e.g. mutualism), or harmful to at least one species (e.g. competition, predation, and parasitism). — 118 123

☐ 5 Competition occurs when species exploit the same limited resources. Intraspecific competition is usually more intense than interspecific competition. — 118-121

☐ 6 The impact of interspecific competition can be reduced when species exploit slightly different niches, e.g. foraging at different times. — 122

☐ 7 The number of organisms and populations an ecosystem can support is its carrying capacity. Abiotic and biotic factors determine the carrying capacity. — 124 125

☐ 8 The area an organism regularly occupies is its home range. Its size depends on the resources it contains. — 126 127

☐ 9 The population growth of a species is limited by the resources of its environment. — 128 131

Luc Viatour www.Lucnix.be

Crosscutting concepts

Understand how these fundamental concepts link different topics

Activity number

☐ 1 **SPQ** How a factor, such as competition or climate, affects an ecosystem's carrying capacity depends on the scale at which it occurs. — 114 120 124 125

☐ 2 **SPQ** A model of how specific factors affect biodiversity and population growth at one scale can be used to understand systems at other scales. — 128 131

☐ 3 **CE** Empirical evidence helps us make claims about cause and effect in studies of population interaction and population growth and decline. — 120-123 131

Science and engineering practices

Demonstrate competence in these science and engineering practices

Activity number

☐ 1 Use a model based on evidence to show how competition limits population size. — 121 128

☐ 2 Explain, based on evidence, how different species reduce resource competition. — 122

☐ 3 Use mathematical representations to support explanations of factors affecting carrying capacity. — 124

☐ 4 Use mathematical representations to support and revise explanations based on evidence about factors affecting population growth. — 128-131

☐ 5 Use a model to illustrate exponential population growth. — 129

☐ 6 Plan and carry out an investigation of exponential growth in bacteria. — 130

113 What is an Ecosystem?

Key Idea: An ecosystem is a natural unit encompassing all the living and non-living components in an area. These components are linked through nutrient cycles and energy flows.

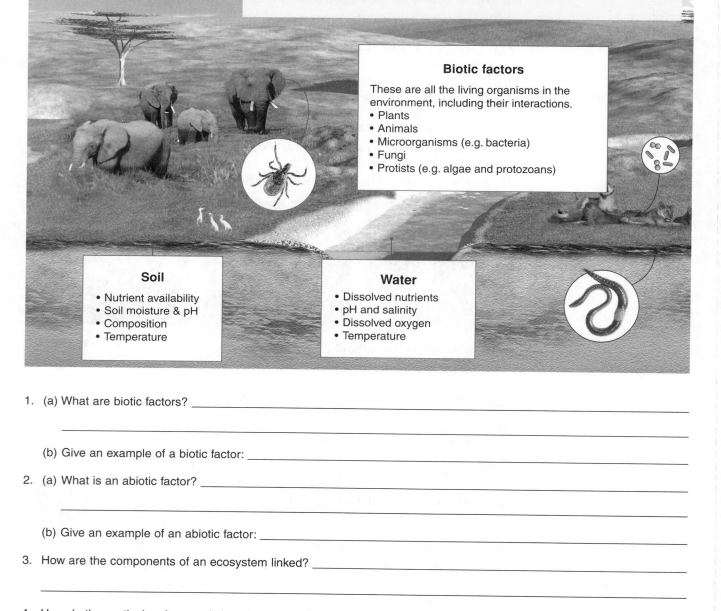

Atmosphere
- Wind speed and direction
- Humidity
- Light intensity and quality
- Precipitation
- Temperature

Ecosystems

Ecosystems are natural units made up of all the living organisms (biotic factors) and the physical (abiotic factors) in an area.

Abiotic factors (non-living chemical and physical factors) include the soil, water, atmosphere, temperature, and sunlight (SWATS). **Biotic factors** are all the living organisms (e.g. plants, animals, fungi, protists, and microorganisms).

The interactions of living organisms with each other and with the physical environment help determine the features of an ecosystem.

The components of an ecosystem are linked to each other (and to other ecosystems) through nutrient cycles and energy flows.

Biotic factors

These are all the living organisms in the environment, including their interactions.
- Plants
- Animals
- Microorganisms (e.g. bacteria)
- Fungi
- Protists (e.g. algae and protozoans)

Soil
- Nutrient availability
- Soil moisture & pH
- Composition
- Temperature

Water
- Dissolved nutrients
- pH and salinity
- Dissolved oxygen
- Temperature

1. (a) What are biotic factors? _____

    _____

   (b) Give an example of a biotic factor: _____

2. (a) What is an abiotic factor? _____

    _____

   (b) Give an example of an abiotic factor: _____

3. How are the components of an ecosystem linked? _____

    _____

4. How do the particular characteristics of an ecosystem arise?_____

    _____

© 2016 **BIOZONE** International
ISBN: 978-1-927309-46-9
Photocopying Prohibited

114 Habitat and Tolerance Range

Key Idea: The environment in which an organism lives is called its habitat. The tolerance range will determine the optimum position of organisms in their habitat.

The habitat is where an organism lives

The natural environment in which an organism lives is its **habitat**. It includes all the physical and biotic factors in that occupied area. Habitats vary widely in scale as do the physical factors that influence them.

A habitat may be vast and relatively homogeneous, as is the open ocean. Predatory barracuda (above) occur around reefs and in the open ocean.

For sessile organisms, such as this fungus, a suitable habitat may be defined by the environment in a small area, such as on this decaying log.

For microbial organisms, such as those in the ruminant gut, the habitat is defined by the chemical environment within the rumen (R) of the host, in this case, a cow.

Tolerance range determines distribution in the habitat

Each species has a **tolerance range** for factors in its environment (below). However, the members of a population are individually different and so vary in their tolerance. Organisms are usually most abundant where the conditions for their survival and reproduction are optimal. Outside this optimal ecological space, the environment is less favorable for survival and there are fewer individuals. For any species, there will also be environments that are unavailable to them.

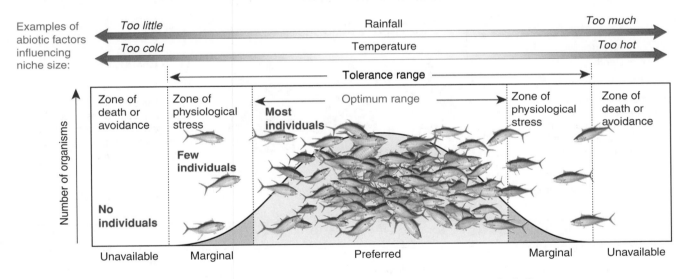

1. In which part of an organism's range will competition for resources be most intense and why? _____

_____

Species tolerant of large environmental variations tend to be more widespread than organisms with a narrow tolerance range. The Atlantic blue crab (left) is widespread along the Atlantic coast from Nova Scotia to Argentina. Adults tolerate a wide range of water salinity ranging from almost fresh to highly saline. This species is an omnivore and eats anything from shellfish to carrion and animal waste.

2. Suggest an advantage to being able to tolerate variations in a wide range of environmental factors?

_____

_____

_____

© 2016 **BIOZONE** International
ISBN: 978-1-927309-46-9
Photocopying Prohibited

CCC   WEB

SPQ 114 **KNOW**

115 The Ecological Niche

Key Idea: The niche of an organism describes its functional position in its environment. This three dimensional space can be influenced by interactions (e.g. competition) with other species.

The niche is the functional role of an organism

The **ecological niche** (or niche) of an organism describes its functional position in its environment. The full range of environmental conditions under which an organism can exist describes its **fundamental niche**.

▶ Interactions with other species, e.g. competition, usually force organisms to occupy a space that is narrower than this. This is called the **realized niche**.

▶ Central to the niche concept is the idea that two species with exactly the same niche cannot coexist, because they would compete for the same resources and one would exclude the other. More often, species compete for only some of the same resources. These competitive interactions limit population sizes and influence distributions.

The physical conditions influence the habitat. A factor may be well suited to the organism, or present it with problems to be overcome.

Adaptations enable the organism to exploit the resources of the habitat. The adaptations take the form of structural, physiological and behavioral characteristics of the organism.

Physical conditions
- Substrate
- Humidity
- Sunlight
- Temperature
- Salinity
- pH
- Exposure
- Altitude
- Depth

Resources offered by the habitat
- Food sources • Shelter • Mating sites
- Nesting sites • Predator avoidance

Adaptations for:
- Locomotion
- Activity pattern
- Tolerance to physical conditions
- Predator avoidance
- Defence
- Reproduction
- Feeding
- Competition

Resource availability is affected by the presence of other organisms and interactions with them: competition, predation, parasitism, and disease.

The habitat provides opportunities and resources for the organism. The organism may or may not have the adaptations to exploit them fully.

1. What is the niche of an organism and how is it different from the organism's habitat? _____
   _____
   _____

2. (a) In what way is the size of the realized niche flexible? _____
   _____
   _____
   _____

   (b) How does competition with another species affect the size of an organism's niche? Explain: _____
   _____
   _____
   _____

3. The diagram (right) shows the resource use curves of two bird species. They overlap in the size of the food items they exploit. On the diagram, use A, B, and C to mark:

   (a) The food item sizes most exploited by species A.
   (b) The food item sizes most exploited by species B.
   (c) The regions where competition is most intense.

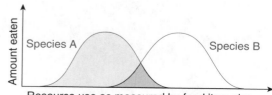

Species A   Species B

Amount eaten

Resource use as measured by food item size

116 Dingo Habitats

Key Idea: Organisms may need to exploit several different habitats to get the resources they need to survive.

Habitats provide resources

As we have seen (opposite) species may tolerate wide variations in a range of physical and biotic factors. As a result of this tolerance range, the habitat that is occupied by members of a species may be quite variable.

The important thing is that the habitat provides resources for the organisms that live there. These resources include water, food, shelter, and places to raise offspring.

Some habitats can be richer in resources than others and are usually described with reference to their main features. For example, riverine habitats (rivers and creeks containing water and thick vegetated cover) provide water, food, and cover (right).

Dingo habitats

Dingoes (right) are wild dogs found throughout Australia. The table on the far right gives information about five dingo packs at one location, including how much of their territory is made up of riverine areas. Kangaroos are the main prey for these dingoes.

Dingo pack name	Territory area (km²)	Pack size	Dingo density per 100 km²	% of territory made up of riverine areas
Pack A	113	12	10.6	10
Pack B	94	12		14
Pack C	86	3		2
Pack D	63	6		12
Pack E	45	10		14

1. Calculate the density of each of the dingo packs per 100 km² using the equation below, and record it in the table above. The first one has been done for you.

> Density = pack size ÷ territory area x 100

2. (a) Plot a scatter graph of dingo density versus how much of their territory is made up of riverine areas for each pack.

(b) Describe the relationship between dingo density and amount of riverine area:

_____

_____

_____

(c) Can you explain why this relationship might occur?

_____

_____

_____

_____

_____

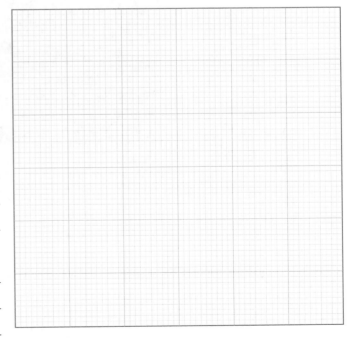

© 2016 **BIOZONE** International
ISBN: 978-1-927309-46-9
Photocopying Prohibited

KNOW

117 Population Density and Distribution

Key Idea: Population density is the number of organisms of one species per unit area or volume. Population distribution describes how those organisms are distributed relative to each other.

Population density

▶ A population refers to all the organisms of the same species in a particular area. The density of a population is the number of individuals of that species per unit area (for land organisms) or volume (for aquatic organisms).

▶ Populations can exist naturally at different densities. Social insects, such as termites exist naturally at high densities, whereas the population density of solitary or territorial species is naturally lower (below). Some species may occur at high densities at certain times of the year, e.g. during breeding.

▶ The density of populations is also affected by the availability of resources. Population density is higher where resources are plentiful and lower where they are scarce or highly variable.

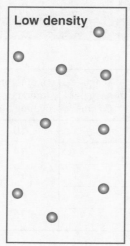

Low density

In low density populations, individuals are spaced well apart. There are only a few individuals per unit area. Highly territorial or solitary animal species, such as tigers, leopards, and bears, occur at low densities.

Plant population density is strongly correlated with the availability of water and nutrients. The density of plant populations, especially of larger species, is therefore lower in deserts, where water is limited.

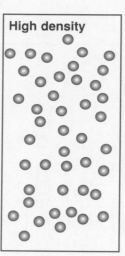

High density

In high density populations, there are many individuals per unit area. This can be a natural feature of highly social species (e.g. ants and termites) or species that reproduce asexually to form large colonies, e.g. corals.

Human populations reach their highest densities in large cities, which are often centers of commerce. Cities were originally established in areas where resources, such as water or fuel, were plentiful.

1. (a) How would you express the population density of a terrestrial species? _____

   (b) How would you express the population density of an aquatic species? _____

2. Explain how the distribution and availability of resources might influence population density? _____

   _____

   _____

© 2016 **BIOZONE** International
ISBN: 978-1-927309-46-9
Photocopying Prohibited

Population distribution

▶ Population **distribution** describes how organisms are distributed in the environment relative to each other.

▶ Three distribution patterns are usually recognized: random, clumped (or aggregated), and uniform. are described below. In the examples, the circles represent individuals of the same species.

Random distribution

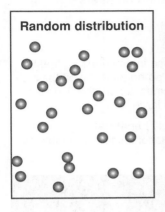

Dune grass

Oyster bed

In random distributions, the spacing between individuals is unpredictable, i.e. the position of one individual is independent of the other individuals. Random distribution is uncommon but can occur in homogeneous environments where unpredictable factors determine distribution, e.g. dandelion seeds germinating after being blown by the wind or oyster larvae settling after being carried by ocean currents.

Clumped distribution

Musk oxen herd

Elephant herd

In clumped (or aggregated) distributions, individuals are grouped in patches (often around a resource). Clumped distributions are the most common type of distribution pattern in nature and are typical of herding and other highly social species and in environments where resources are patchy.

Uniform distribution

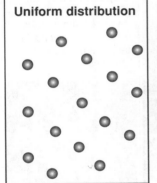

Gannet colony

Allocasuarina in Australia

In uniform (regular) distributions individuals are evenly spaced and the distance between neighboring individuals is maximized. Uniform distributions occur in territorial species, e.g. breeding colonies of seabirds, but also occurs in plants that produce chemicals to inhibit the growth of nearby plants.

3. Explain how the behavior of a species might influence the population density: _____

_____

4. What factors might influence the distribution of individuals in their environment? _____

_____

_____

5. What type of distribution pattern would you expect to see when:

   (a) Resources are not evenly spread out: _____

   (c) Animals are social: _____

   (b) Resources are evenly spread out: _____

   (d) Animals are territorial: _____

6. Why do you think random distributions are uncommon in nature? _____

_____

 © 2016 **BIOZONE** International
ISBN: 978-1-927309-46-9
Photocopying Prohibited

118 Species Interactions

Key Idea: Interactions between species may influence the size and distribution of their populations. Predation, competition, and parasitism can all act to limit the population numbers.

Species interact in ways that limit the size of populations

▶ Within ecosystems, each species interacts with others in their community. In many of these interactions, at least one of the parties in the relationship is disadvantaged. Predators eat prey, parasites and pathogens exploit their hosts, and species compete for limited resources. These interactions contribute to the abiotic factors that limit the number of organisms in a population and prevent any one population from becoming too large.

▶ Not all relationships involve exploitation. Some species form relationships that are mutually beneficial, e.g. some flowering plants have mutualistic relationships with their pollinating insects. Mutualistic relationships can enable two species to exist in greater numbers than either would alone.

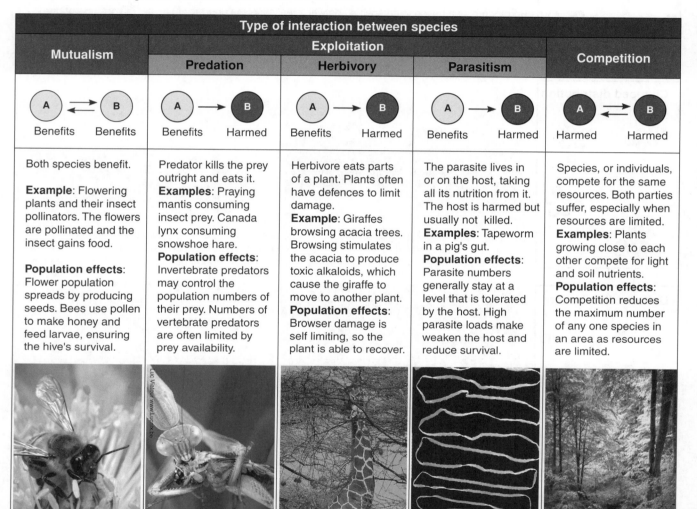

Type of interaction between species				
Mutualism	**Exploitation**			**Competition**
	Predation	**Herbivory**	**Parasitism**	
A ⇄ B Benefits Benefits	A → B Benefits Harmed	A → B Benefits Harmed	A → B Benefits Harmed	A ⇄ B Harmed Harmed
Both species benefit. **Example**: Flowering plants and their insect pollinators. The flowers are pollinated and the insect gains food. **Population effects**: Flower population spreads by producing seeds. Bees use pollen to make honey and feed larvae, ensuring the hive's survival.	Predator kills the prey outright and eats it. **Examples**: Praying mantis consuming insect prey. Canada lynx consuming snowshoe hare. **Population effects**: Invertebrate predators may control the population numbers of their prey. Numbers of vertebrate predators are often limited by prey availability.	Herbivore eats parts of a plant. Plants often have defences to limit damage. **Example**: Giraffes browsing acacia trees. Browsing stimulates the acacia to produce toxic alkaloids, which cause the giraffe to move to another plant. **Population effects**: Browser damage is self limiting, so the plant is able to recover.	The parasite lives in or on the host, taking all its nutrition from it. The host is harmed but usually not killed. **Examples**: Tapeworm in a pig's gut. **Population effects**: Parasite numbers generally stay at a level that is tolerated by the host. High parasite loads make weaken the host and reduce survival.	Species, or individuals, compete for the same resources. Both parties suffer, especially when resources are limited. **Examples**: Plants growing close to each other compete for light and soil nutrients. **Population effects**: Competition reduces the maximum number of any one species in an area as resources are limited.

1. Plants are not defenceless against herbivores. They have evolved physical and chemical defences to deter herbivores. In some cases (as in grasses) grazing stimulates growth in the plant.

(a) What is the acacia's response to giraffe browsing? _____

(b) How might this response prevent over-browsing? _____

_____

(c) How might the acacia's adaptations contribute to the sustainability of its browser population? _____

_____

_____

© 2016 **BIOZONE** International
ISBN: 978-1-927309-46-9
Photocopying Prohibited

2. Although hyenas attack and kill large animals, such as wildebeest, they will also scavenge carrion or drive other animals off their kills.

(a) Identify the interaction pictured here: _____

(b) How many species are involved in the interaction?_____

(c) Describe how each species is affected by the presence of the other (benefits/harmed/no effect):

_____

_____

3. Ticks are obligate blood feeders and must obtain blood to pass from one life stage to the next. Ticks attach to the outside of hosts (in this case a cat) where they suck blood and fluids and cause irritation.

(a) Identify this type of interaction: _____

(b) Describe how each species is affected (benefits/harmed/no effect):

_____

(c) How would the tick population be affected if the host became rare?

_____

_____

4. It is tempting to assume that large mammalian predators, such as big cats and wolves, control the numbers of their prey species. However, population studies have shown that predator-prey interactions are very complex and the outcomes depend on factors such as the food available to the prey and the availability of alternative prey for the predator. Sometimes predators can control the numbers of prey, but sometimes, the opposite is true.

How could availability of food for the prey affect the number of predators?

_____

_____

5. Many insect predators (e.g. ladybird beetles) are very effective at limiting the numbers of their invertebrate prey (e.g. aphids). As the prey become more abundant, the predators both take more prey and become more numerous. This is the basis of biological control programs in which an invertebrate predator is introduced to control the numbers of an insect pest. Once the numbers of the pest are reduced, both populations stabilize at a low level:

Why do you think a successful biological control program relies on the predator having just one prey source (the pest)?

_____

_____

_____

6. Many butterfly species breed within a relatively short time period, laying their eggs on suitable plants where the larvae will hatch and feed. Competition between the larvae will be intense as they all compete for the food available:

(a) What is the interaction occurring here? _____

_____

(b) What is the likely outcome for eggs laid too late in the season?

_____

_____

_____

© 2016 **BIOZONE** International
ISBN: 978-1-927309-46-9
Photocopying Prohibited

119 Competition for Resources

Key Idea: Species interact with other living organisms in their environment. Competition occurs when species utilize the same limited resources.

▸ No organism exists in isolation. Each organism interacts with other organisms and with the physical (abiotic) components of the environment.

▸ **Competition** occurs when two or more organisms are competing for the same limited resource (e.g. food or space).

▸ The resources available for growth, reproduction, and survival for each competitor are reduced relative to a situation of no competition. Competition therefore has a negative effect on both competitors and limits population numbers.

▸ Competition can occur between members of the same species (intraspecific competition), or between members of different species (interspecific competition).

A complex system of interactions occurs between the different species living on this coral reef in Hawaii. Population numbers will be limited by competition for limited resources, such as food and space on the reef.

Examples of limited resources

Space can be a limited resource
These sea anemones are competing for space in a tidal pool. Some species defend areas, called territories, which have resources they need.

Suitable mates can be hard to find
Within a species, individuals may compete for a mate. These male red deer are fighting to determine which one will mate with the females.

Food is usually a limited resource
In most natural systems, there is competition for food between individuals of the same species, and between different species with similar diets.

1. (a) What is competition? _____

_____

_____

(b) Why does competition occur? _____

_____

_____

(c) Why does competition have a negative effect on both competitors? _____

_____

_____

_____

© 2016 **BIOZONE** International
ISBN: 978-1-927309-46-9

120 Intraspecific Competition

Key Idea: Intraspecific competition describes competition between individuals of the same species for resources. Intraspecific competition is an important regulator of population size.

▶ Intraspecific competition occurs when individuals of the same species compete for the same limited resources. In addition to food, space, nutrients, and light, intraspecific competition also includes competition for mating partners and breeding sites.

▶ In most cases, intraspecific competition is more intense that competition between different species because individuals are all competing for the same resources (e.g. same food and mates). It is an important factor in limiting the population size of many species.

▶ Strategies such as territoriality and social hierarchies can reduce the conflict associated with resource competition and can be important in determining which individuals in the population will breed.

How does intraspecific competition limit population size?

Most resources are limited, and this is a major factor in determining how large a population can grow.

As population numbers increase, the demands on the resources are higher. The resources are used up more quickly and some individuals receive fewer resources than others. High population densities may also increase the influence of population limiting factors such as disease. Populations respond by decreasing their numbers. This occurs by:

▶ Reduced survival (more individuals die).

▶ Reduced birth rates (fewer individuals are born).

If resources increase (e.g. food increases), population numbers can increase. The relationship between resources and population numbers is shown on the right.

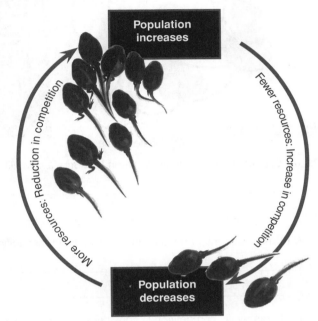

Population increases

More resources: Reduction in competition

Fewer resources: Increase in competition

Population decreases

Scramble competition

Direct competition between members of the same species for a finite resource is called scramble competition. These silkworm caterpillars are all competing for the same food. When it is insufficient, none of the individuals may survive.

Contest competition

In contest competition, there is a winner and a loser and resources are obtained completely or not at all. For example, male elephant seals fight for territory and mates. Unsuccessful males may not mate at all.

Mike Baird

Competition in social species

In some animals, strict social orders ensure that dominant individuals will have priority access to resources. Lower ranked individuals must contest what remains. If food is very limited, only dominant individuals may receive enough to survive.

1. (a) What is intraspecific competition? _____

    _____

    (b) How does scramble competition differ from contest competition? _____

    _____

    _____

Territories and limitations of population size

▶ Territoriality in birds and other animals is usually a result of intraspecific competition. A territory is a defended area containing the resources required by an individual or breeding pair to survive and reproduce. Territories space organisms out in the habitat according to the availability of resources. Those without territories usually do not breed.

▶ In the South American rufous-collared sparrow, males and females occupy small territories (below). These birds make up 50% of the population. The remaining 50% or the population, called floaters, occupy home ranges (which are undefended areas) within the territory boundaries. They are tolerated by the territory owners, but these floaters do not breed.

▶ By using tagging studies and removal of birds, researchers found that when a territory owner (male or female) dies or disappears, it is replaced by a floater of the appropriate sex. This is shown for the females in the left diagram as a darker region.

▶ Territoriality can limit population size in some circumstances. If there is no lower limit to territory size and all individuals or pairs gain a territory then the population becomes spaced out but not limited. However, if territories have a lower size limit, then only a limited number of individuals or pairs can claim a territory. Those that fail to do so must leave and this limits population numbers.

Females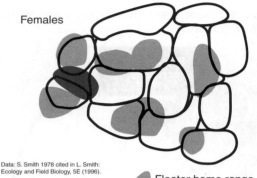

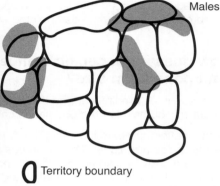 Males

Data: S. Smith 1978 cited in L. Smith:
Ecology and Field Biology, 5E (1996).

 Floater home range   Territory boundary

2. (a) Explain the relationship between resource availability and intraspecific competition: _____

_____

_____

(b) What happens to population numbers as intraspecific competition increases? _____

_____

_____

(c) Intraspecific competition is considered to be a density-dependent process. What do you think this means?

_____

_____

3. Territoriality is a way to ensure that at least some individuals have the resources to survive and reproduce. The territories are established and maintained by direct conflict and by calls and displays.

(a) Identify the benefits of possessing a territory: _____

_____

_____

(b) Can you think of a cost of having a territory? _____

_____

_____

(c) What evidence is there from the rufous-collared sparrow study to show that territoriality can limit population size?

_____

_____

_____

_____

© 2016 **BIOZONE** International
ISBN: 978-1-927309-46-9
Photocopying Prohibited

121 Interspecific Competition

Key Idea: Interspecific competition describes competition between individuals of different species for resources. It can limit the number of one or both competing species.

Interspecific competition involves individuals of different species competing for the same limited resources. They may do this by:

- Interfering directly with the ability of others to gain access to the resource.
- Exploiting the resource before other individuals can get access to it (exploitative competition).

▶ Interspecific competition is usually less intense than competition between members of the same species because competing species have different requirements for at least some of the resources (e.g. different habitat or food preferences). In other words, their niches are different even if they exploit some of the same resources.

▶ Interspecific competition can be have a role in limiting a population's size and determining the species present or their distribution. However, in naturally occurring populations, it is generally less effective at limiting population size than intraspecific competition, especially in animals. This is because each species usually has alternative resources it can exploit to avoid competition.

▶ Interspecific competition in natural plant communities is very dependent on nutrient availability and will be greater when soil nutrients are low. Fast growing plants with large, dense root systems can absorb large amounts of nitrogen, depleting soil nitrogen so that other plants cannot grow close to them. Similarly fast growing plants may quickly grow tall enough to intercept the available light and prevent the germination of plants nearby. Pest plants often have this strategy and become very difficult to control.

▶ Sometimes, humans may introduce a species with the same resource requirements as a native species. The resulting competition can lead to the decline of the native species.

In some communities, many different species may be competing for the same resource. This type of competition is called interference competition because the individuals interact directly over a scarce resource. In the example above, three species compete for what remains of a carcass.

The plant species in the forest community above compete for light, space, water, and nutrients. A tree that can grow taller than those around it will be able to absorb more sunlight, and grow more rapidly to a larger size than the plants in the shade below.

1. (a) What is interspecific competition? Describe an example: _____

   _____

   _____

   (b) Why is interspecific competition usually less intense than intraspecific competition? _____

   _____

   _____

   _____

   (c) Why is interspecific competition generally less effective at limiting population size than intraspecific competition?

   _____

   _____

   _____

© 2016 **BIOZONE** International
ISBN: 978-1-927309-46-9
Photocopying Prohibited

PRACTICES   CCC   WEB

CE   121   **KNOW**

Change in distributions of red and gray squirrels in the UK

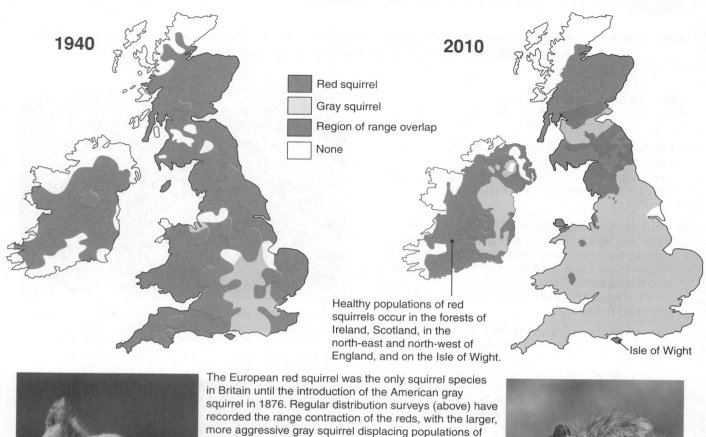

1940   **2010**

- ■ Red squirrel
- ▢ Gray squirrel
- ■ Region of range overlap
- □ None

Healthy populations of red squirrels occur in the forests of Ireland, Scotland, in the north-east and north-west of England, and on the Isle of Wight.

Isle of Wight

Paul Whippey cc 3.0

Red squirrel

BirdPhotos.com cc 3.0

Gray squirrel

The European red squirrel was the only squirrel species in Britain until the introduction of the American gray squirrel in 1876. Regular distribution surveys (above) have recorded the range contraction of the reds, with the larger, more aggressive gray squirrel displacing populations of reds over much of England. Gray squirrels can exploit tannin-rich foods, which are unpalatable to reds. In mixed woodland and in competition with grays, reds may not gain enough food to survive the winter and breed. Reds are also very susceptible to several viral diseases, including squirrelpox, which is transmitted by grays.

Whereas red squirrels once occupied a range of forest types, they are now almost solely restricted to coniferous forest. The data suggest that the gray squirrel is probably responsible for the red squirrel decline, but other factors, such as habitat loss, are also likely to be important.

2. (a) What evidence is there that competition with gray squirrels is responsible for the decline in red squirrels in the UK:

_____

_____

_____

(b) Is the evidence conclusive? If not why not?_____

_____

_____

_____

_____

3. The ability of red and gray squirrels to coexist appears to depend on the diversity of habitat type and availability of food sources (reds appear to be more successful in regions of coniferous forest). Suggest why careful habitat management is thought to offer the best hope for the long term survival of red squirrel populations in Britain:

_____

_____

_____

122 Reducing Competition Between Species

Key Idea: Competition between species for similar resources can be reduced if competing species have slightly different niches and exploit resources in different ways.

How species reduce competition

Species exploiting similar resources have adaptations (evolved features) to reduce competition. Each species exploits a smaller proportion of the entire spectrum of resources potentially available to it. For example:

▶ In a forest, each species may feed on a different part of a tree (e.g. trunk, branches, twigs, flowers, or leaves) or occupy different areas of vertical air-space (e.g. ground, understorey, sub-canopy, or canopy).

▶ Aquatic organisms may also inhabit different zones to reduce competition for resources. Some organisms will inhabit the bottom, and others may occupy surface waters.

▶ Competition may also be reduced by exploiting the same resources at a different time of the day or year (e.g. one species may feed at night and another may exploit the same resource in the morning).

Warblers of the genus *Setophaga* spend a lot of their time feeding in conifer trees. By feeding in different parts of the tree and on different food sources the warblers reduce competition. The blue areas shown on the tree indicate areas where the warblers spent 50% or more of their time feeding.

Blackburnian warblers forage in new needles and buds in upper branches of trees. They tend to move horizontally through tree feeding on insects or spiders.

Bay-breasted warblers frequent older needles and lichen near middle branches. They feed on insects, particularly the spruce budworm. These birds will also feed on berries and nectar.

Cape May warblers feed around new needles and buds near the top of tree. They hawk for flying insects and pick insects up from the tips of conifer branches. These warblers also feed on berry juice and nectar.

Black-throated warbler forage in new needles and buds as well as older needles and branches in the upper to middle parts of tree. They feed mainly on insects, sometimes hovering-(gleaning), or catching insects in flight-(hawking). Berries will occasionally be consumed.

Myrtle warblers forage on lower trunks and middle branches. These birds are insectivorous, but will readily take wax-myrtle berries in winter, a habit which gives the species its name. They make short flights in search of insects.

1. (a) How do the species of warbler avoid competition? _____

_____

   (b) What evidence is there that their adaptations for doing this are largely behavioral? _____

_____

2. Cape May warblers and Blackburnian warblers appear to occupy the same habitat in the tree. Explain how they are able to avoid competition and coexist:

_____

_____

_____

 © 2016 **BIOZONE** International
ISBN: 978-1-927309-46-9
Photocopying Prohibited

PRACTICES   CCC   WEB

CE   122   **KNOW**

Adaptations reduce competition in foraging bumblebees

Studies on bumblebee foraging have shown that when bumblebees forage in the presence of other bumblebee species they tend to spend the majority of their time on particular flower types. In many cases, the length of the corolla (the length of the flower petals) of flowers visited correlates with the length of the bumblebee's proboscis (mouthparts).

Bumblebee species in the mountains of Colorado (graph right) compete for nectar from flowers. Species with a long proboscis take nectar from flowers with long petals. Species with a short proboscis take nectar from flowers with short petals. This reduces competition for food between the bumblebee species.

Bumblebee species

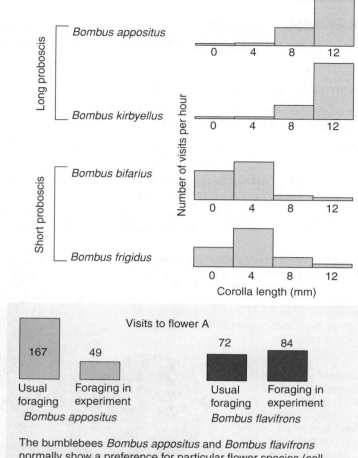

The bumblebees *Bombus appositus* and *Bombus flavifrons* normally show a preference for particular flower species (call these A and F respectively for reference). However, in the absence of competition, they will forage on either flower species. This was shown in an experiment in which visits of *Bombus appositus* to its usual forage flower A were restricted. *Bombus flavifrons*, which usually forages on flower F, responded by increasing its visits to flower A and decreasing its visits to flower F.

3. (a) How do the *Bombus* species in Colorado reduce competition for flower resources? _____

_____

_____

_____

(b) Are the differences between the Colorado species mainly structural, physiological, or behavioral? Explain:

_____

_____

_____

_____

4. What evidence is there that competition restricts bumblebee species to certain flower types: _____

_____

_____

_____

_____

123 Predator-Prey Relationships

Key Idea: Predators do not always regulate prey numbers, but predator numbers are often dependent upon prey numbers.

Do predators limit prey numbers?

▶ It was once thought that predators always limited the numbers of their prey populations. While this is often true for invertebrate predator-prey systems, very often prey species are regulated more by factors such as climate and the availability of food than by predation.

▶ In contrast, predator populations can be strongly affected by the availability of prey, especially when there is little opportunity for prey switching (hunting another prey if the preferred one becomes scarce).

▶ Predator and prey populations may settle into a stable oscillations, where the predator numbers follow those of the prey, with a time lag (right).

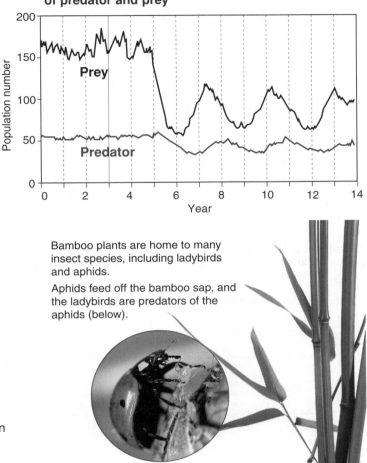

Fluctuations in hypothetical populations of predator and prey

A case study in predator-prey numbers

In some areas of Northeast India, a number of woolly aphid species colonize and feed off bamboo plants. The aphids can damage the bamboo so much that it is no longer able to be utilized by the local people for construction and textile production.

Giant ladybird beetles (*Anisolemnia dilatata*) feed exclusively off the woolly aphids of bamboo plants. There is some interest in using them as biological control agents to reduce woolly aphid numbers, and limit the damage woolly aphids do to bamboo plants.

The graph below shows the relationship between the giant ladybird beetle and the woolly aphid when grown in controlled laboratory conditions.

Bamboo plants are home to many insect species, including ladybirds and aphids.

Aphids feed off the bamboo sap, and the ladybirds are predators of the aphids (below).

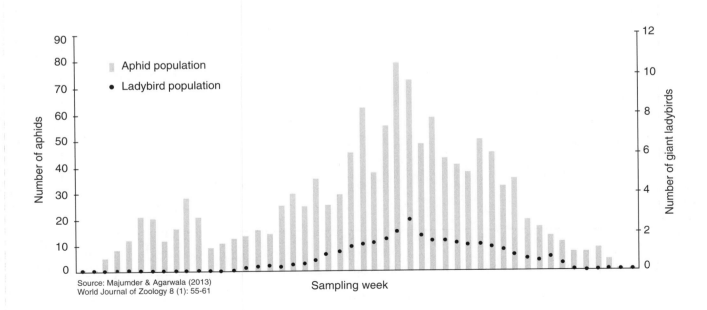

Source: Majumder & Agarwala (2013)
World Journal of Zoology 8 (1): 55-61

1. (a) On graph the above, mark the two points (using different colored pens) where the peak numbers of woolly aphids and giant ladybirds occurs.

(b) Do the peak numbers for both species occur at the same time? _____

(c) Why do you think this is? _____

_____

_____

2. (a) What is the response of the ladybird population when their prey decline? _____

_____

(b) Although this was a laboratory situation, what features of the ladybird predator suggest it would be a good choice to control woolly aphids?:

_____

_____

_____

3. A census of a deer population on an island forest reserve Indicated a population of 2000 animals in 1960. In 1961, ten wolves (natural predators of deer) were brought to the island in an attempt to control deer numbers. The numbers of deer and wolves were monitored over the next nine years. The results of these population surveys are presented right.

(a) Plot a line graph for the results. Use one scale (on the left) for numbers of deer and another scale (on the right) for the number of wolves. Use different symbols or colors to distinguish the lines and include a key.

(b) What does the plot show? _____

_____

_____

_____

Island population surveys (1961-1969)		
Year	Wolf numbers	Deer numbers
1961	10	2000
1962	12	2300
1963	16	2500
1964	22	2360
1965	28	2244
1966	24	2094
1967	21	1968
1968	18	1916
1969	19	1952

(c) Suggest a possible explanation for the pattern in the data: _____

_____

_____

© 2016 BIOZONE International
ISBN: 978-1-927309-46-9
Photocopying Prohibited

124 The Carrying Capacity of an Ecosystem

Key Idea: Carrying capacity is the maximum number of organisms of a particular species a particular environment can support indefinitely. Carrying capacity regulates population numbers.

The **carrying capacity** is the maximum number of organisms of a given species a particular environment can support indefinitely. An ecosystem's carrying capacity, and therefore the maximum population size it can sustain is limited by its resources and affected by both biotic factors (e.g. food) and abiotic factors (e.g. water, temperature). It is determined by the most limiting factor and can change over time (e.g. as a result of a change in food availability or a climate shift).

The graph, right, shows how the carrying capacity of a forest-dwelling species changes based on changes to the limiting factors:

1 A population moves into the forest and rapidly increases in numbers due to abundant resources.

2 The population overshoots the carrying capacity.

3 Large numbers damage the environment and food becomes more limited, lowering the original carrying capacity.

4 The population becomes stable at the new carrying capacity.

5 The forest experiences a drought and the carrying capacity is reduced as a result.

6 The drought breaks and the carrying capacity rises but is less than before because of habitat damage during the drought.

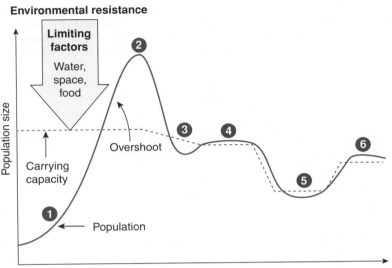

Environmental resistance

Limiting factors — Water, space, food

Carrying capacity

Overshoot

Population

Population size

Time

Factors affecting population size

Density dependent factors

The effect of these on population size is influenced by population density. They include:

▶ Competition
▶ Predation
▶ Disease

Density dependent factors tend to be biotic and are less important when population density is low.

They regulate population size by decreasing birth rates and increasing death rates.

Density independent factors

The effect of these on population size does not depend on population density. They include catastrophic events such as:

▶ Volcanic eruptions, fire
▶ Drought, flood, tsunamis
▶ Earthquakes

Density independent factors tend to be abiotic.

They regulate population size by increasing death rates.

1. What is carrying capacity? _____

_____

2. How does carrying capacity limit population numbers? _____

_____

_____

3. What limiting factors have changed at points 3, 5, and 6 in the graph above, and how have they changed?

(a) 3: _____

(b) 5: _____

(c) 6: _____

PRACTICES   CCC   WEB

SPQ   124   **KNOW**

125 A Case Study in Carrying Capacity

Key Idea: Predator-prey interactions are not always predictable because environmental factors influence the relationship.

When wolves were introduced to Coronation Island

Coronation Island is a small, 116 km² island off the coast of Alaska. In 1960, the Alaska Department of Fish and Game released two breeding pairs of wolves onto the island. Their aim was to control the black-tailed deer that had been overgrazing the land. The results (below) were not what they expected.

Introduction of the wolves initially appeared to have the desired effect. The wolves fed off the deer and successfully bred, and deer numbers fell. However, within a few years the deer numbers crashed. The wolves ran out of food (deer) and began eating each other, causing a drop in wolf numbers. Within eight years, only one wolf inhabited the island, and the deer were abundant. By 1983, there were no wolves on the island, and the deer numbers were high.

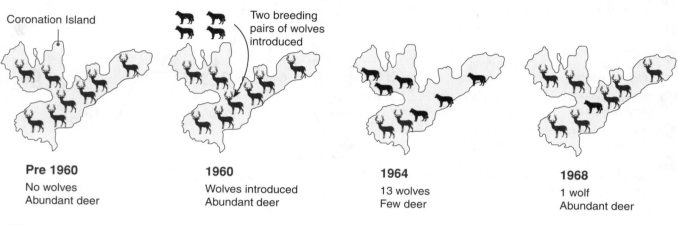

Pre 1960
No wolves
Abundant deer

1960
Wolves introduced
Abundant deer

1964
13 wolves
Few deer

1968
1 wolf
Abundant deer

What went wrong?

▶ The study showed Coronation Island was too small to sustain both the wolf and the deer populations.

▶ The deer could not easily find refuge from the wolves, so their numbers were quickly reduced.

▶ Reproductive rates in the deer may have been low because of poor quality forage following years of over-grazing. When wolves were introduced, predation and low reproductive rates, caused deer numbers to fall.

▶ The deer were the only food source for the wolves. When deer became scarce the wolves ate each other because there was no other prey available.

1. Why were wolves introduced to Coronation Island? _____

   _____

   _____

2. (a) What were some of the factors that caused the unexpected result?

   _____

   _____

   _____

   _____

   (b) What do these results tell you about the carrying capacity of Coronation Island? _____

   _____

   _____

   _____

   _____

 © 2016 **BIOZONE** International
ISBN: 978-1-927309-46-9
Photocopying Prohibited

126 Home Range Size in Dingoes

Key Idea: A home range is the area an animal normally inhabits. Home range size is influenced by the resources offered by the ecosystem.

Ecosystem and home range

The **home range** is the area where an animal normally lives and moves about in. An animal's home range can vary greatly in size. Animals that live in ecosystems rich in resources (e.g. good supply of food, water, shelter) tend to have smaller home ranges than animals that live in resource-poor ecosystems. This is because animals in a resource-poor ecosystem must cover a wider area to obtain the resources they need.

Dingo home ranges

Dingoes are found throughout Australia, in ecosystems as diverse as the tropical rainforests of the north, to the arid deserts of central Australia. The table (right) shows the home range sizes for dingo packs living in a variety of ecosystems. Some of the ecosystems in Australia in which dingoes are found are described below.

Dingo home range size in different ecosystems

	Location (study site)	Ecosystem	Range (km^2)
1	Fortescue River, North-west Australia	Semi-arid, coastal plains and hills	77
2	Simpson Desert, Central Australia	Arid, stony and sandy desert	67
3	Kapalga, Kakadu N.P., North Australia	Tropical, coastal wetlands and forests	39
4	Harts Ranges, Central Australia	Semi-arid, river catchment and hills	25
5	Kosciusko N.P., South-east Australia	Moist, cool forested mountains	21
6	Georges Creek N.R., East Australia	Moist, cool forested tablelands (plateaux)	18
7	Nadgee N.R., South-east Australia	Moist, cool coastal forests	10

Australian ecosystems

Arid: Little or no rain, and very dry. Very little, or no, vegetation grows. Often desert regions.

Semi arid: Rainfall is low, but sufficient to support some scrubby vegetation and grasses.

Cool forests: Moderate temperatures. Adequate water and abundant vegetation.

Tropical forests: Warm regions with high rainfall. Abundant lush vegetation, including large trees.

Temsabulla cc2.0

1. Using the information on dingo home range size from the table above:

   (a) Name the two regions where home ranges were largest: _____

   (b) Name the two regions where home ranges were smallest: _____

   (c) Use the information provided on Australian ecosystems to explain how ecosystem type influences the home range of the dingo packs identified in (a) and (b):

   _____

   _____

   _____

   _____

   _____

 © 2016 **BIOZONE** International
ISBN: 978-1-927309-46-9
Photocopying Prohibited

KNOW

127 Resources and Distribution

Key Idea: Some social species have exclusive core areas (territories) within a larger home range. These areas contain the best resources and are defended against competitors.

Baboon home ranges in Nairobi Park

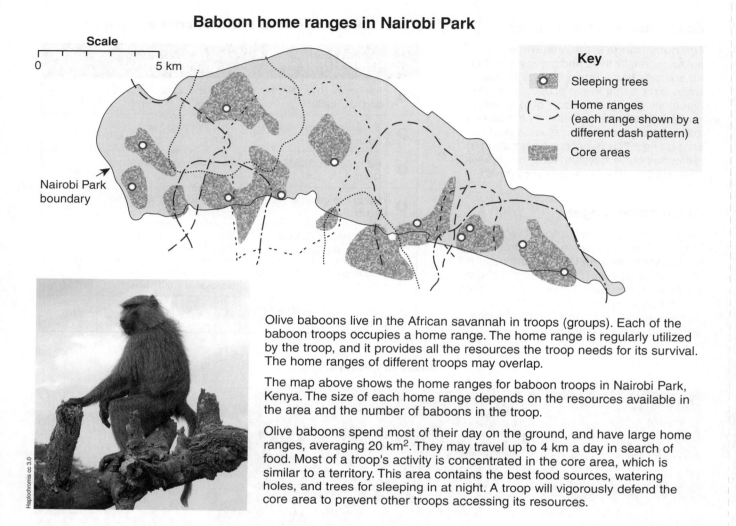

Scale

0 ——— 5 km

Nairobi Park boundary

Key
- Sleeping trees
- Home ranges (each range shown by a different dash pattern)
- Core areas

Haplochromis cc 3.0

Olive baboons live in the African savannah in troops (groups). Each of the baboon troops occupies a home range. The home range is regularly utilized by the troop, and it provides all the resources the troop needs for its survival. The home ranges of different troops may overlap.

The map above shows the home ranges for baboon troops in Nairobi Park, Kenya. The size of each home range depends on the resources available in the area and the number of baboons in the troop.

Olive baboons spend most of their day on the ground, and have large home ranges, averaging 20 km^2. They may travel up to 4 km a day in search of food. Most of a troop's activity is concentrated in the core area, which is similar to a territory. This area contains the best food sources, watering holes, and trees for sleeping in at night. A troop will vigorously defend the core area to prevent other troops accessing its resources.

1. (a) What factors influence the size of baboon home ranges? _____

    _____

   (b) Would a small baboon troop have a large or a small home range relative to a large troop (assume both areas have similar resources)?

    _____

   (c) Explain the answer you gave in 1(b): _____

    _____

    _____

   (d) what might happen to home range size during a drought? _____

2. Why do you think baboons aggressively defend their core areas? _____

    _____

3. How might the home ranges and their defended core areas operate to regulate population size in a particular region?

    _____

    _____

© 2016 **BIOZONE** International
ISBN: 978-1-927309-46-9
Photocopying Prohibited

128 Population Growth

Key Idea: Populations have the potential for continued exponential growth if resources are non-limiting. However, this rarely occurs because the carrying capacity limits population growth.

Population growth

▶ Births, deaths, immigrations (movements into the population) and emigrations (movements out of the population) are events that determine the numbers of individuals in a population. Population growth depends on the number of individuals added to the population from births and immigration, minus the number lost through deaths and emigration.

▶ Scientists usually measure the rate of these events. These rates are influenced by environmental factors, such as the availability of resources, and by the characteristics of the organisms themselves.

▶ In population studies, the per capita rate of population increase (also called the biotic potential) is often used. Ignoring migration, B – D / N gives the per capita rate of increase (denoted by the italicised letter r).

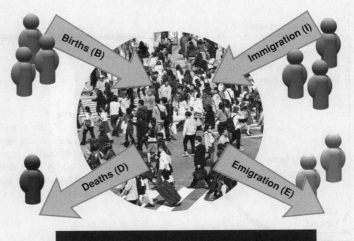

| Population growth | $= B - D + I - E$ |
| Per capita growth rate, | $r = B - D / N$ |

Population growth curves

▶ The change in population numbers over time is often presented as a population growth curve. Two basic population growth curves exist. Exponential growth (left) is unconstrained by the environment. Logistic growth (following page) is limited by carrying capacity.

Exponential growth

Exponential growth occurs when resources are unlimited (this rarely occurs). Growth is extremely rapid, but not sustainable. It produces a J-shaped growth curve.

Exponential growth is expressed mathematically as: $dN/dt = rN$

Exponential growth
$$dN/dt = rN$$

Here the number being added to the population per unit time is large. The steepness of the curve depends on the value of r.

Exponential (J) curve
Exponential growth is sustained only when there is no environmental resistance.

Early on, population growth is slow because there are so few individuals.

Lag phase

Population numbers (N)

Time (t)

In bacterial populations, each cell divides in two, so the population doubles with every generation. Bacteria growing without environmental constraint show exponential growth.

1. Using the terms, B, D, I, and E (above), construct equations to express the following (the first is completed for you):

   (a) A population in equilibrium: ___B + I = D + E___

   (b) A declining population: _____

   (c) An increasing population: _____

PRACTICES PRACTICES CCC WEB

SPQ 128 **KNOW**

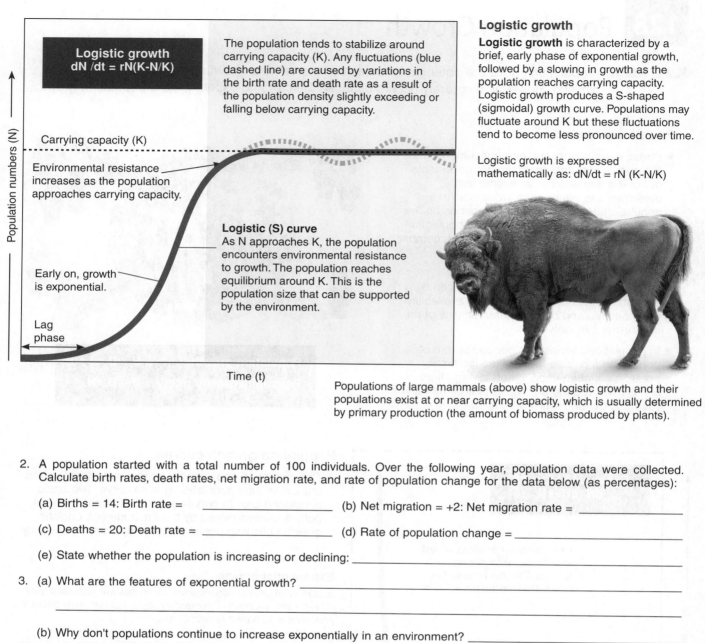

Logistic growth
$$dN/dt = rN(K-N/K)$$

The population tends to stabilize around carrying capacity (K). Any fluctuations (blue dashed line) are caused by variations in the birth rate and death rate as a result of the population density slightly exceeding or falling below carrying capacity.

Logistic growth

Logistic growth is characterized by a brief, early phase of exponential growth, followed by a slowing in growth as the population reaches carrying capacity. Logistic growth produces a S-shaped (sigmoidal) growth curve. Populations may fluctuate around K but these fluctuations tend to become less pronounced over time.

Logistic growth is expressed mathematically as: $dN/dt = rN(K-N/K)$

Carrying capacity (K)

Environmental resistance increases as the population approaches carrying capacity.

Logistic (S) curve
As N approaches K, the population encounters environmental resistance to growth. The population reaches equilibrium around K. This is the population size that can be supported by the environment.

Early on, growth is exponential.

Lag phase

Population numbers (N)

Time (t)

Populations of large mammals (above) show logistic growth and their populations exist at or near carrying capacity, which is usually determined by primary production (the amount of biomass produced by plants).

2. A population started with a total number of 100 individuals. Over the following year, population data were collected. Calculate birth rates, death rates, net migration rate, and rate of population change for the data below (as percentages):

(a) Births = 14: Birth rate = _____   (b) Net migration = +2: Net migration rate = _____

(c) Deaths = 20: Death rate = _____   (d) Rate of population change = _____

(e) State whether the population is increasing or declining: _____

3. (a) What are the features of exponential growth? _____

(b) Why don't populations continue to increase exponentially in an environment? _____

4. (a) Describe the features of logistic growth: _____

(b) What is environmental resistance and what role does it have in limiting population growth? _____

(c) Explain why a population might overshoot carrying capacity before stabilizing around carrying capacity: _____

5. What happens to population growth rate as K-N/K approaches 0? _____

6. Use a spreadsheet to demonstrate how the equations for exponential and logistic growth produce their respective curves. Watch the weblink animations provided or use the spreadsheet on the Teacher's Digital Edition if you need help.

© 2016 **BIOZONE** International
ISBN: 978-1-927309-46-9
Photocopying Prohibited

129 Plotting Bacterial Growth

Key Idea: The growth of a microbial population in non-limiting conditions can be plotted on a semi-log graph and this can be used to predict microbial cell numbers at a set time.

Bacterial growth

▶ Bacteria normally reproduce by a binary fission, a simple cell division that involves one cell dividing in two.

▶ When actively growing bacteria are inoculated into a liquid growth medium and the population is counted at intervals, a line can be plotted to show the growth of the cell population over time.

NEED HELP?
See Activity 17

One cell   Binary fission   Two cells

Time (min)	Population size
0	1
20	2
40	4
60	8
80	
100	
120	
140	
160	
180	
200	
220	
240	
260	
280	
300	
320	
340	
360	

1. Complete the table above by doubling the number of bacteria for every 20 minute interval.

2. State how many bacteria were present after: 1 hour: _____ 3 hours: _____ 6 hours: _____

3. Graph the results on the grid above. Make sure that you choose suitable scales and labels for each axis.

4. (a) Predict the number of cells present after 380 minutes: _____

   (b) Plot this value on the graph above.

5. Why is a semi-log graph used to plot microbial growth? _____

_____

PRACTICES   PRACTICES   WEB

129   **DATA**

130 Investigating Bacterial Growth

Key Idea: The growth of microbial populations over time can be measured indirectly by recording change in the absorbance of the culture with a spectrophotometer.

Background

The increase in cell numbers in a bacterial culture can be measured indirectly with a spectrophotometer as an increase in culture turbidity.

The aim

To investigate the growth rate of *E.coli* in two different liquid cultures, a minimal growth medium and a nutrient-enriched complex growth medium.

The method

Using aseptic technique, the students added 0.2 mL of a pre-prepared *E.coli* culture to two test tubes, one with 5.0 mL of a minimal growth medium and one with 5.0 mL of a complex medium. Both samples were immediately mixed, and 0.2 mL samples removed from each and added to a cuvette. The absorbance of the sample was measured using a spectrophotometer at 660 nm. This was the 'time zero' reading. The test tubes were covered with parafilm, and placed in a 37°C water bath. Every 30 minutes, the test tubes were lightly shaken and 0.2 mL samples were taken from each and the absorbance measured. The results are presented right.

A spectrophotometer (left) is an instrument used to measure transmittance of a solution and so can be used to quantify bacterial growth where an increase in cell numbers results in an increase in turbidity.

In this experiment, students measured the absorbance of the solution. Absorbance measures the amount of light absorbed by the sample.

 All bacteria should be treated as pathogenic and strict hygiene practices and aseptic techniques should be followed. This prevents infection or spread of bacteria into the environment.

1. Why is it important to follow strict hygiene precautions when working with bacteria?

_____

_____

_____

2. (a) On the grid (right) plot the results for *E.coli* growth on the two media:

   (b) What is the absorbance measuring? _____

   _____

   (c) Describe the effect of the complex medium on *E.coli* growth:

   _____

   _____

   _____

   (d) Can you see any differences in the shapes of the curves? Give a possible explanation for your observations:

   _____

   _____

Results

| Incubation time (min) | Absorbance at 660 nm | |
	Minimal medium	Complex medium
0	0.021	0.014
30	0.022	0.015
60	0.025	0.019
90	0.034	0.033
120	0.051	0.065
150	0.078	0.124
180	0.118	0.238
210	0.179	0.460
240	0.273	0.698
270	0.420	0.910
300	0.598	1.070

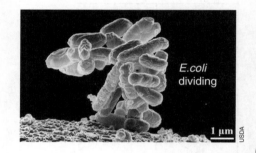

E.coli dividing

1 µm

USDA

NEED HELP? See Activities 9 & 17

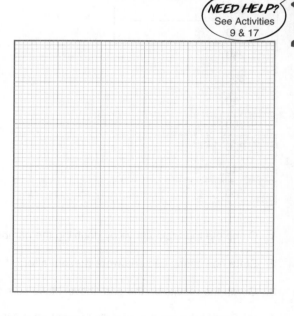

 © 2016 **BIOZONE** International
ISBN: 978-1-927309-46-9
Photocopying Prohibited

131 A Case Study in Population Growth

Key Idea: Some species show regular cycles in their population numbers. Where a predator is dependent on a single prey species, their populations may fluctuate in a similar way.

Population oscillations in a natural predator-prey system

▶ Snowshoe hares in the boreal forests of North America show cycles of population increase and decrease lasting 8-11 years. At the peak of the cycle, they can reach densities that exceed the carrying capacity of the environment.

▶ Canada lynx are very dependent on the hares for food and they have little opportunity for prey switching. Consequently, their populations also rise and fall with fluctuations in populations of snowshoe hares.

▶ Regular trapping records of lynx and hare pelts over a 90 year period showed that their population numbers oscillate with a similar periodicity, with the lynx numbers lagging behind those of the hare by 3-7 years. The lag is the response time of the lynx population to the change in hare numbers.

Oscillations in snowshoe hare and Canada lynx populations

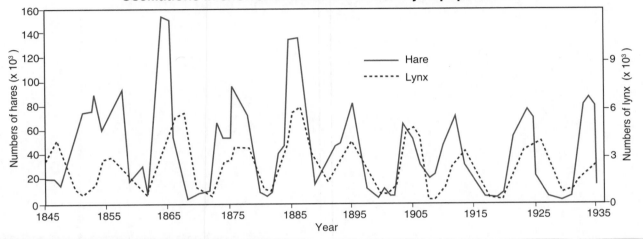

Canada lynx are largely solitary and territorial. Territory size increases when food is scarce. They prey almost exclusively on snowshoe hares, which form 60-97% of their diet and cannot maintain bodyweight successfully on alternative prey.

Snowshoe hares consume a variety of plant material depending on seasonal availability. Hare populations show regular cycles in numbers averaging, with a 2-5 year period of abundance followed by a decline following over-exploitation of food sources.

1. Why is the lynx population so dependent on the fluctuations of the hare: _____

_____

2. (a) Explain how the availability of palatable food might regulate the numbers of hares: _____

_____

_____

(b) Explain how a decline in available palatable food might affect their ability to withstand predation pressure:

_____

_____

PRACTICES   CCC   CCC   WEB

CE   SPQ   131   **KNOW**

132 Chapter Review

Summarize what you know about this topic under the headings provided. You can draw diagrams or mind maps, or write short notes to organize your thoughts. Use the introduction and the hints included to help you:

Ecosystems, habitat, and niche

HINT: Include abiotic and biotic factors. What factors affect niche size and population size?

Species interactions and their effect on population size

HINT: How do different types of interactions (e.g. predation, competition) limit population size?

Population growth and carrying capacity

HINT: Include a definition of carrying capacity and describe factors that influence it. Compare exponential and logistic growth. How does carrying capacity limit population size?

© 2016 **BIOZONE** International
ISBN: 978-1-927309-46-9
Photocopying Prohibited

133 KEY TERMS AND IDEAS: Did You Get It?

1. Test your vocabulary by matching each term to its definition, as identified by its preceding letter code.

abiotic factor	**A** The number of individuals of a species per unit area or volume.
biotic factor	**B** The functional role of an organism in its environment, including its activities and interactions with other organisms.
carrying capacity	**C** The maximum number of a species that can be supported indefinitely by the environment.
competition	**D** Exploitation in which one organism kills and eats another, usually of a different species.
density	**E** A term for any non-living part of the environment, e.g. rainfall, temperature.
ecosystem	**F** The area where an animal normally lives and moves about in.
habitat	**G** Any living component of the environment (or aspect of it) that has an effect on another organism in the environment.
home range	**H** Community of interacting organisms and the environment (both biotic and abiotic) in which they both live and interact.
interspecific competition	**I** An interaction between two species where both species benefit.
intraspecific competition	**J** Competition for a resource between members of the same species.
mutualism	**K** The physical place or environment where an organism lives. It includes all the physical and biotic factors.
niche	**L** Competition for a resource between members of different species.
population	**M** All the individuals of a particular species within a set area.
predation	**N** An interaction between organisms exploiting the same resource.

2. Study the graph of population growth for a hypothetical population below and answer the following questions:

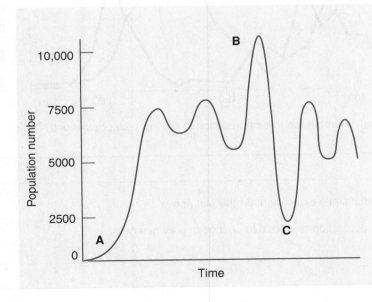

(a) Estimate the carrying capacity of the environment:

_____

_____

(b) What happened at point A?

_____

_____

_____

(c) What species interaction would be predominating at point B and why?

_____

_____

_____

(d) What happened at point C?_____

_____

(e) What factors might have contributed to the event in C? _____

_____

_____

TEST

134 Summative Assessment

Analyzing model predator-prey system

Mathematical models predict that predator and prey populations will form stable cycles of population increase and decrease. Early ecologists set out to verify these population oscillations in small model ecosystems. Two researchers, Gause and Huffaker, each worked on this question. Their results gave great insight into the nature of predatory prey interactions and the factors that control population size.

Gause's experiments

▶ Gause's experiments examined the interactions of two protists, *Paramecium* and its predator *Didinium* in simple test tube 'microcosms'. When *Didinium* was added to a culture of *Paramecium*, it quickly ate all the *Paramecium* and then died out. When sediment was placed in the microcosm, *Paramecium* could hide, *Didinium* died out and the *Paramecium* population recovered.

Huffaker's experiments

▶ Huffaker built on Gause's findings and attempted to design artificial systems that would better model a real world system. He worked on two mite species, the six spotted mite and its predator. Oranges provided both the habitat and the food for the prey.

▶ In a simple system, such as a small number of oranges grouped together, predators quickly ate all the prey and then died out.

▶ Huffaker then created a more complex system with arrays of 120 oranges (below). The amount of available food on each orange was controlled by sealing off parts of each orange with wax. Patchiness in the environment was created using balls (representing unsuitable habitat). Sticks aided dispersal of prey mites and vaseline was used to form barriers for predatory mite dispersal. In this system, the predator and prey coexisted for three full cycles (> year). In the diagram below, the arrays depict the distribution and density of the populations at the arrowed points. The circles represent oranges or balls and the dots the predatory mites.

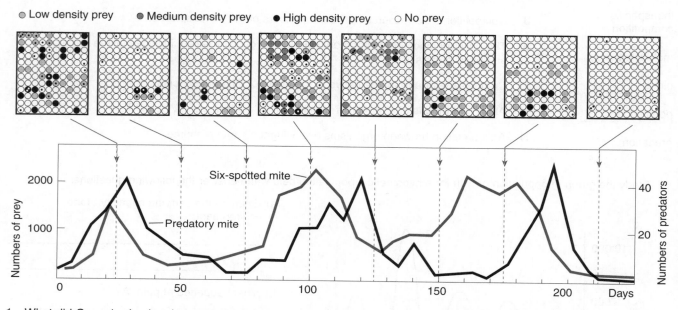

1. What did Gause's simple microcosm experiments tell us about the role of predation in limiting prey populations?

_____

_____

2. (a) Mark the three population cycles completed in Huffaker's experiment on the plot above.

   (b) In a different color, mark the lag in the predator population response to change in prey numbers.

   (c) What does the lag represent? _____

   _____

3. How well do you think Huffaker's model system approximated a real ecosystem? Use evidence from the arrays to discuss how variation in habitat makes it possible for populations to persist despite periodic declines in their numbers.

_____

_____

_____

_____

TEST

 © 2016 **BIOZONE** International
ISBN: 978-1-927309-46-9
Photocopying Prohibited

LS2.B
PS3.D
Energy Flow and Nutrient Cycles

Key terms

anaerobic

aerobic

carbon cycle

cellular respiration

consumer

ecological pyramid

food chain

food web

hydrologic cycle

nitrogen cycle

oxygen cycle

photosynthesis

producer

trophic level

Disciplinary core ideas

Show understanding of these core ideas

Energy flows through ecosystems

		Activity number
☐	1 Photosynthesis and cellular respiration provide most of the energy needed to carry out essential life processes.	135
☐	2 Aerobic respiration provides the useful energy in plants and animals. It requires oxygen to proceed.	136
☐	3 Anaerobic respiration is used by some bacteria to provide energy. It does not require oxygen to proceed.	136
☐	4 Organisms can be classified by how they obtain their energy (food). Producers, make their own food. Consumers obtain their food from eating other organisms.	137 138
☐	5 A food chain provides a model of how energy, in the form of food, passes from one organism to another. Food chains can be connected to form food webs, which show all the feeding relationships within an ecosystem.	139 140 141
☐	6 Energy and matter are transferred through trophic levels. At each level, energy is lost as heat, so ecosystems require a continuous input of light energy to sustain them. At each level in an ecosystem, matter and energy are conserved.	142 143 144
☐	7 Ecological pyramids provide a model of the number of organisms or the amount of energy or biomass.	144

Energy flows through ecosystems but matter is recycled

☐	8 Chemical elements cycle through the ecosystem. Important nutrient cycles include the hydrologic (water), carbon, oxygen, and nitrogen cycles.	145-151
☐	9 Photosynthesis and cellular respiration have important roles in carbon cycling.	147 150

Crosscutting concepts

Understand how these fundamental concepts link different topics

		Activity number
☐	1 **EM** Energy cannot be created or destroyed; it only moves between one place and another, between objects or fields, or between systems.	135 139 142 143
☐	2 **EM** Energy drives the cycling of matter within and between systems.	136 145-151
☐	3 **SSM** Models can be used to simulate flow of energy and cycling of matter.	147-150

Science and engineering practices

Demonstrate competence in these science and engineering practices

		Activity number
☐	1 Construct an explanation based on evidence for how matter cycles and energy flows in aerobic and anaerobic conditions.	136
☐	2 Use mathematical representations to describe energy transfers in ecosystems.	143 144
☐	3 Use mathematical representations to show that matter and energy are conserved as matter cycles and energy flows through ecosystems.	142 143
☐	4 Develop an evidence-based model to show how photosynthesis and cellular respiration are involved in carbon cycling.	148 150

135 Energy in Ecosystems

Key Idea: Photosynthesis and cellular respiration provide most of the energy needed for essential life processes.

Where does the energy for life processes come from?

▶ As matter and energy move through the biotic and abiotic environments, chemical elements are recombined in different ways. Each transformation results in storage and dissipation of energy into the environment as heat. Matter and energy are conserved at each transformation.

▶ The dissipation of energy as heat means that ecosystems must receive a constant input of new energy from an outside source to sustain themselves. Usually the Sun is the ultimate source of energy in an ecosystem.

▶ **Photosynthesis** and **cellular respiration** provide most of the usable energy for life's essential processes such as metabolism and growth. It is important to remember that all organisms carry out cellular respiration.

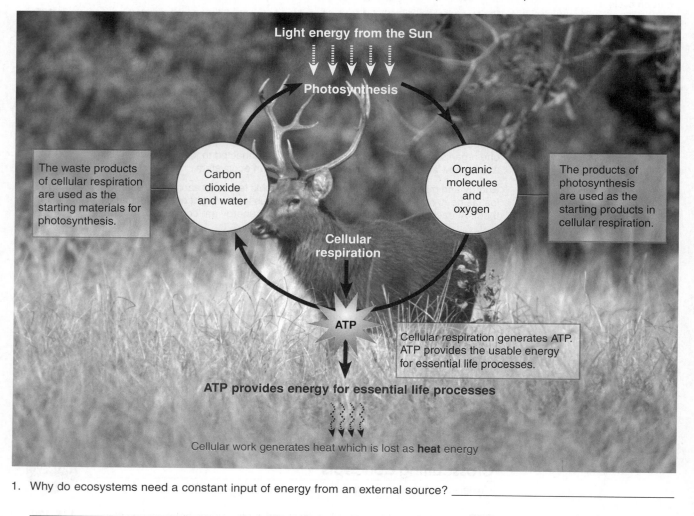

Light energy from the Sun

Photosynthesis

The waste products of cellular respiration are used as the starting materials for photosynthesis.

Carbon dioxide and water

Organic molecules and oxygen

The products of photosynthesis are used as the starting products in cellular respiration.

Cellular respiration

ATP

Cellular respiration generates ATP. ATP provides the usable energy for essential life processes.

ATP provides energy for essential life processes

Cellular work generates heat which is lost as **heat** energy

1. Why do ecosystems need a constant input of energy from an external source? _____

_____

2. How do photosynthesis and cellular respiration interact to cycle matter through an ecosystem? _____

_____

_____

_____

_____

3. What is the role of ATP in biological systems? _____

_____

_____

 © 2016 **BIOZONE** International
ISBN: 978-1-927309-46-9
Photocopying Prohibited

136 Comparing Aerobic and Anaerobic Systems

Key Idea: Aerobic and anaerobic systems produce usable energy (ATP) in different ways. Both systems are important in the cycling of matter through ecosystems.

▶ Systems that operate **aerobically** (with oxygen) and **anaerobically** (without oxygen) both produce energy as ATP for use in living systems.

▶ Most of the ATP made in respiration (aerobic or anaerobic) is generated through the transfer of electrons between electron carriers to a final electron acceptor. The energy released as a result of these transfers is captured in ATP and can be used to power essential chemical reactions in the cell.

An aerobic system: The breakdown of glucose

▶ Aerobic systems use oxygen as the final electron acceptor.

▶ Oxygen accepts electrons and joins with hydrogen to form water. Carbon dioxide is released earlier in the process.

▶ The example below shows **cellular respiration**.

An anaerobic system: Carbon dioxide to methane

▶ Anaerobic systems use a molecule other than oxygen as the terminal electron acceptor.

▶ There are many different anaerobic pathways that living organisms use to produce usable energy.

▶ **Methanogenesis** (below) is a form of anaerobic respiration found in methane-producing bacteria in the stomach of ruminants (e.g. sheep and cows).

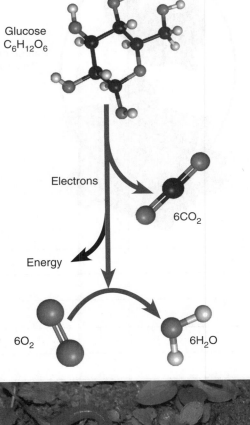

Glucose $C_6H_{12}O_6$

Electrons

$6CO_2$

Energy

$6O_2$

$6H_2O$

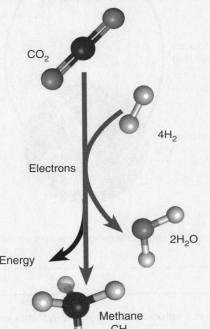

CO_2

$4H_2$

Electrons

$2H_2O$

Energy

Methane CH_4

During cellular respiration, a series of chemical reactions breaks down glucose and uses the energy released to produce the energy carrier ATP. ATP is used to carry out the work of cells. Animals cannot make their own glucose and must obtain it by consuming other organisms.

CSIRO cc3.0

Methane is a greenhouse gas and contributes to global warming. Methane-producing bacteria in ruminants use the cellulose in vegetation as a source of energy and carbon. The methane generated is breathed out. The sheep (above) are fitted with devices to measure the exhaled methane.

© 2016 **BIOZONE** International
ISBN: 978-1-927309-46-9
Photocopying Prohibited

PRACTICES   CCC   WEB

 EM 136 **KNOW**

Matter cycles through the ecosystem

▶ The total amount of matter in a closed system (and the Earth is effectively a closed system) is conserved. This means that Earth's essentially fixed supply of nutrients must be recycled to sustain life. Nutrients move between the various compartments on Earth: the atmosphere (air), hydrosphere (water), soils, and living organisms.

▶ Carbon, hydrogen, nitrogen, sulfur, phosphorus, and oxygen move through ecosystems in cycles, although the cycles may intersect and interact at various points. Energy drives the cycling of matter within and between systems.

▶ The rate of nutrient cycling can vary widely. Some nutrients (e.g. phosphorus) are cycled slowly, others (such as nitrogen) more quickly. The type of environment and diversity of an ecosystem can also have a large effect on the rate at which nutrients are cycled.

Cycling nitrogen

Nitrogen must be in a form that can be utilized by plants. Some bacteria (called nitrogen fixers) can convert nitrogen gas to nitrate. Other bacteria convert nitrates back into nitrogen gas. Both processes are anaerobic.

Cycling carbon and oxygen

Photosynthesis and cellular respiration link the cycling of carbon and oxygen. Aerobic respiration, the conversion of glucose and O_2 into CO_2 and water, is part of the carbon and oxygen cycles. The CO_2 and water are converted back into glucose and O_2 by the anaerobic process of photosynthesis.

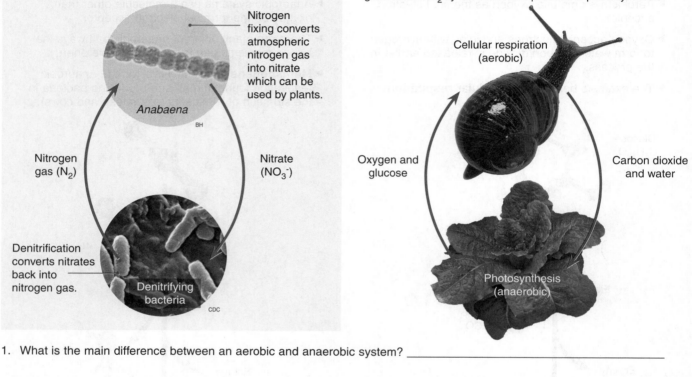

Nitrogen fixing converts atmospheric nitrogen gas into nitrate which can be used by plants.

Anabaena

BH

Nitrogen gas (N_2)

Nitrate (NO_3^-)

Denitrification converts nitrates back into nitrogen gas.

Denitrifying bacteria

CDC

Cellular respiration (aerobic)

Oxygen and glucose

Carbon dioxide and water

Photosynthesis (anaerobic)

1. What is the main difference between an aerobic and anaerobic system? _____

_____

_____

2. How are aerobic and anaerobic systems involved in nutrient cycling? _____

_____

_____

_____

_____

3. Describe how nutrient cycles can interact in the cycling of matter: _____

_____

_____

_____

_____

4. Compost heaps rely on decomposer organisms to break down plant material into a nutrient-rich soil-like humus. During the process, the ratio of carbon to nitrogen (C:N) in the heap decreases. Explain why:

_____

_____

137 Producers

Key Idea: Producers (autotrophs) make their own food. Most producers utilize the energy from the sun to do this, but some organisms use chemical energy.

What is a producer?

▶ A **producer** is an organism that can make its own food.

▶ Producers are also called **autotrophs**, which means self feeding.

▶ Plants, algae, and some bacteria are producers.

▶ Most producers are **photo**autotrophs, and use the energy in sunlight to make their food. The process by which they do this is called **photosynthesis**.

▶ Some producers are **chemo**autotrophs and use the chemical energy in inorganic molecules (e.g. hydrogen sulfide) to make their food.

Sugar (stored energy) and water

Water

Carbon dioxide gas (CO_2)

Sunlight

Oxygen gas (O_2)

Photosynthesis transforms sunlight energy into chemical energy. The chemical energy is stored as sugar (glucose), and the energy is released when the sugar undergoes further metabolic processes. The inputs and outputs of photosynthesis are shown on the leaf diagram (right).

Photosynthesis by marine algae provides oxygen and absorbs carbon dioxide. Most algae are microscopic but some, like this kelp, are large.

On land, vascular plants (plants with transport tissues) are the main producers of food.

Producers, such as grasses, make their own food, and are also the ultimate source of food and energy for consumers, such as these cows.

1. (a) What is a producer? _____

_____

(b) Name some organisms that are producers: _____

_____

2. Where do producers get their energy from? _____

_____

3. Why are producers so important in an ecosystem? _____

_____

_____

_____

© 2016 **BIOZONE** International
ISBN: 978-1-927309-46-9
Photocopying Prohibited

KNOW

138 Consumers

Key Idea: Consumers (heterotrophs) are organisms that cannot make their own food. They get their food by consuming other organisms.

What is a consumer?

Consumers (heterotrophs) are organisms that cannot make their own food and must get their food by consuming other organisms (by eating or extracellular digestion). Animals, fungi, and some bacteria are consumers. Consumers are categorized according to where they get their energy from (below).

Consumers need producers

Consumers rely on producers for survival, even if they do not consume them directly. Herbivores, such as the rabbit below, gain their energy by eating plants. Although higher level consumers, such as the eagle, may feed off herbivores, they still ultimately rely on plants to sustain them. Without the plants the rabbit would not survive, and the eagle could not eat it.

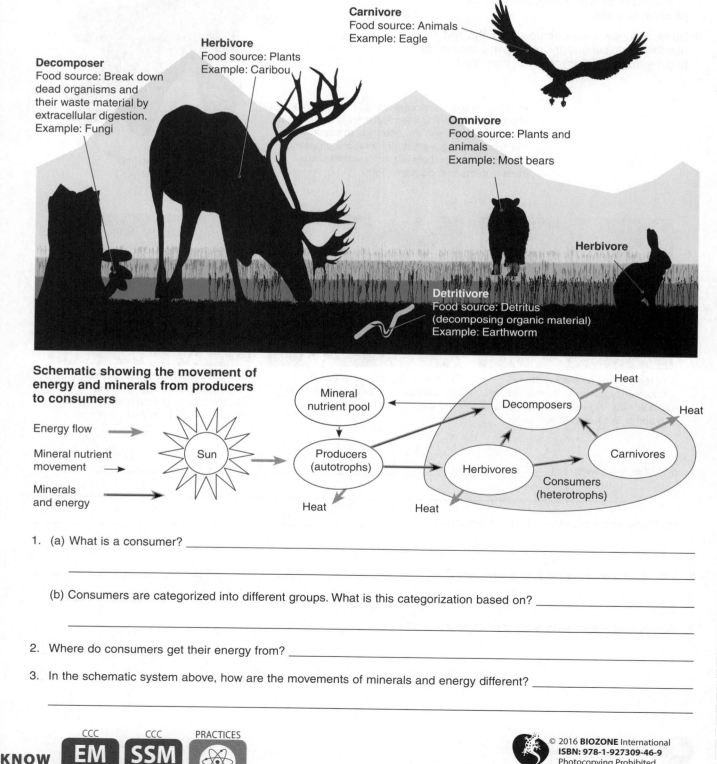

Carnivore
Food source: Animals
Example: Eagle

Herbivore
Food source: Plants
Example: Caribou

Decomposer
Food source: Break down dead organisms and their waste material by extracellular digestion.
Example: Fungi

Omnivore
Food source: Plants and animals
Example: Most bears

Herbivore

Detritivore
Food source: Detritus (decomposing organic material)
Example: Earthworm

Schematic showing the movement of energy and minerals from producers to consumers

Energy flow →
Mineral nutrient movement →
Minerals and energy →

Sun

Mineral nutrient pool

Producers (autotrophs)

Heat

Heat

Decomposers

Herbivores

Carnivores

Consumers (heterotrophs)

Heat

Heat

1. (a) What is a consumer? _____

_____

(b) Consumers are categorized into different groups. What is this categorization based on? _____

_____

2. Where do consumers get their energy from? _____

3. In the schematic system above, how are the movements of minerals and energy different? _____

_____

CCC CCC PRACTICES
EM SSM

© 2016 **BIOZONE** International
ISBN: 978-1-927309-46-9
Photocopying Prohibited

139 Food Chains

Key Idea: A food chain is a model to illustrate the feeding relationships between organisms.

Food chains

▶ Organisms in ecosystems interact in their feeding relationships. These interactions can be shown in a **food chain**, which is a simple model to illustrate how energy, in the form of food, passes from one organism to the next. Each organism in the chain is a food source for the next.

Trophic levels

▶ The levels of a food chain are called **trophic** (feeding) **levels**. An organism is assigned to a trophic level based on its position in the food chain. Organisms may occupy different trophic levels in different food chains or during different stages of their life.

▶ Arrows link the organisms in a food chain. The direction of the arrow shows the flow of energy through the trophic levels. At each link, energy is lost (as heat) from the system. This loss of energy limits how many links can be made. Most food chains begin with a producer, which is eaten by a primary consumer (herbivore). Second (and higher) level consumers eat other consumers, as shown below.

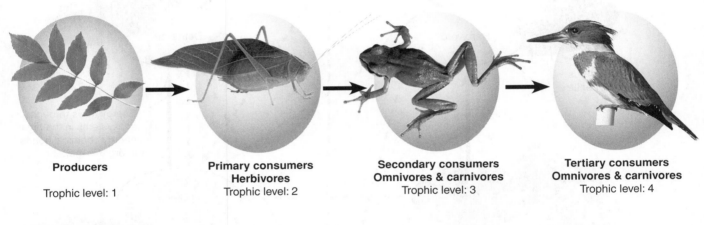

| **Producers** | **Primary consumers Herbivores** | **Secondary consumers Omnivores & carnivores** | **Tertiary consumers Omnivores & carnivores** |
| Trophic level: 1 | Trophic level: 2 | Trophic level: 3 | Trophic level: 4 |

1. What is a food chain? _____

_____

2. (a) A simple food chain for a cropland ecosystem is pictured below. Label the organisms with their trophic level and trophic status (e.g. primary consumer).

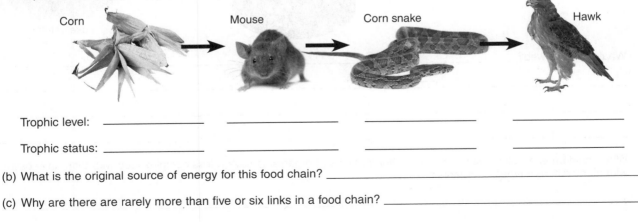

Corn   Mouse   Corn snake   Hawk

Trophic level: _____   _____   _____   _____

Trophic status: _____   _____   _____   _____

(b) What is the original source of energy for this food chain? _____

(c) Why are there are rarely more than five or six links in a food chain? _____

_____

_____

CCC SSM   CCC EM   WEB 139   KNOW

140 Food Webs

Key Idea: A food web consists of all the food chains in an ecosystem. Food webs show the complex feeding relationships between all the organisms in a community.

▶ If we show all the connections between all the food chains in an ecosystem, we can create a web of interactions called a food web. A **food web** is a model to illustrate the feeding relationships between all the organisms in a community.

▶ The complexity of a food web depends on the number of different food chains contributing to it. A simple ecosystem, with only a few organisms (and therefore only a few food chains) will have a simpler food web than an ecosystem that has many different food chains.

▶ A food web model (below) can be used to show the linkages between different organisms in a community.

A simple food web for a lake ecosystem

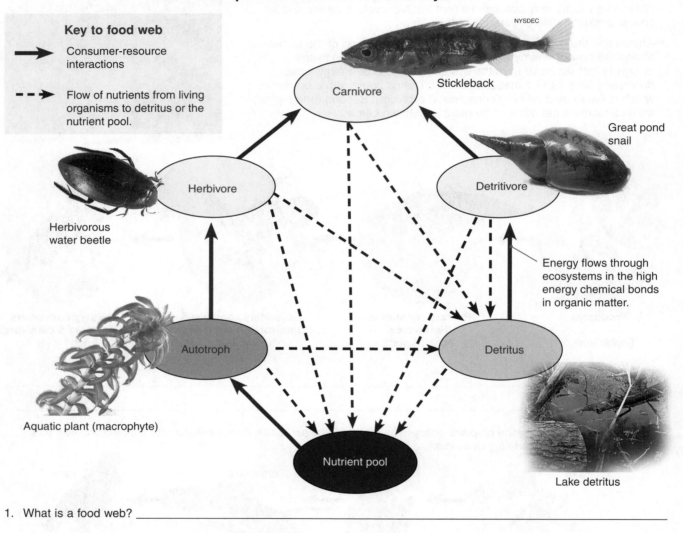

Key to food web

→ Consumer-resource interactions

⇢ Flow of nutrients from living organisms to detritus or the nutrient pool.

Carnivore

Stickleback

NYSDEC

Great pond snail

Herbivore

Detritivore

Herbivorous water beetle

Energy flows through ecosystems in the high energy chemical bonds in organic matter.

Autotroph

Detritus

Aquatic plant (macrophyte)

Nutrient pool

Lake detritus

1. What is a food web? _____

_____

_____

_____

2. Why would an ecosystem with only a few different types of organisms have a less complex food web than an ecosystem with many different types of organisms?

_____

_____

_____

 © 2016 **BIOZONE** International
ISBN: 978-1-927309-46-9
Photocopying Prohibited

141 Constructing Food Webs

Key Idea: Knowing what the inhabitants of an ecosystem feed on allows food chains to be constructed. Food chains can be used to construct a food web.

The organisms below are typical of those found in many lakes. For simplicity, only a few organisms are represented here. Real lake communities have hundreds of different species interacting together. Your task is to assemble the organisms below into a food web in a way that shows how they are interconnected by their feeding relationships.

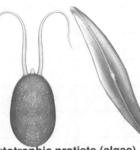

Autotrophic protists (algae)
Chlamydomonas (above left), and some diatoms (above right) photosynthesize.

Macrophytes
Aquatic green plants photosynthesize.

Detritus
Decaying organic matter.

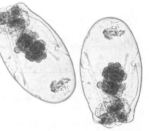

***Asplanchna* (planktonic rotifer)**
A large, carnivorous rotifer.
Diet: Protozoa and young zooplankton (e.g. *Daphnia*).

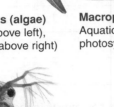

***Daphnia* (zooplankton)**
Small freshwater crustacean.
Diet: Planktonic algae.

Leech
Fluid feeding predators.
Diet: Small invertebrates, including rotifers, small pond snails, and worms.

NYSDEC

Three-spined stickleback
Common in freshwater ponds and lakes.
Diet: Small invertebrates such as *Daphnia* and insect larvae.

Diving beetle (adults and larvae)
Diet: Aquatic insect larvae and adult insects. They will also scavenge from detritus. Adults will also take fish fry. Adults and larvae voracious and may be top predators in small ponds.

Common carp
Diet: Mainly feeds on bottom living insect larvae and snails, but will also eat some plant material (not algae).

Gina Mikel

Dragonfly larva
Large aquatic insect larvae.
Diet: Small invertebrates including *Hydra*, *Daphnia*, insect larvae, and leeches.

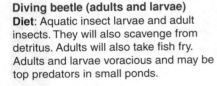

Great pond snail
Diet: Omnivorous. Main diet is macrophytes but will eat decaying plant and animal material also.

Herbivorous water beetle
Diet: Adults feed on macrophytes. Young beetle larvae are carnivorous, feeding primarily on pond snails.

Protozan (e.g. *Paramecium*)
Diet: Mainly bacteria and microscopic green algae such as *Chlamydomonas*.

Pike
Diet: Smaller fish and amphibians. They are also opportunistic predators of rodents and small birds.

Mosquito larva
Diet: Planktonic algae.

Hydra
A small, carnivorous cnidarian.
Diet: small *Daphnia* and insect larvae.

 © 2016 **BIOZONE** International
ISBN: 978-1-927309-46-9
Photocopying Prohibited

1. From the information provided for the lake food web components on the previous page, construct twelve different **food chains** to show the feeding relationships between the organisms. Some food chains may be shorter than others and most species will appear in more than one food chain. An example has been completed for you.

   Example 1: Macrophyte ⟶ Herbivorous water beetle ⟶ Carp ⟶ Pike

   (a) _____

   (b) _____

   (c) _____

   (d) _____

   (e) _____

   (f) _____

   (g) _____

   (h) _____

   (i) _____

   (j) _____

   (k) _____

   (l) _____

2. Use the food chains you created above to help you to draw up a food web for this community in the box below. Use the information supplied on the previous page to draw arrows showing the flow of energy between species (only energy **from** (not to) the detritus is required).

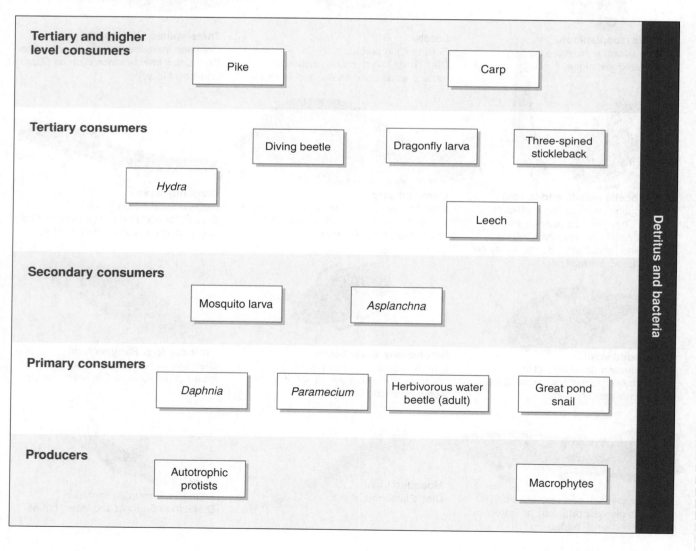

Tertiary and higher level consumers

Pike   Carp

Tertiary consumers

Diving beetle   Dragonfly larva   Three-spined stickleback

Hydra

Leech

Secondary consumers

Mosquito larva   *Asplanchna*

Primary consumers

Daphnia   *Paramecium*   Herbivorous water beetle (adult)   Great pond snail

Producers

Autotrophic protists   Macrophytes

Detritus and bacteria

 © 2016 **BIOZONE** International
ISBN: 978-1-927309-46-9
Photocopying Prohibited

142 Energy Inputs and Outputs

Key Idea: The total amount of energy captured by photosynthesis is the gross primary production. Net primary production is the amount of energy available to herbivores after respiration losses.

▶ The **gross primary production** (GPP) of any ecosystem will depend on the capacity of the producers to capture light energy and fix carbon in organic compounds.

▶ The **net primary production** (NPP) is then determined by how much of the GPP goes into plant biomass, after the respiratory needs of the producers are met. This will be the amount available to the next trophic level.

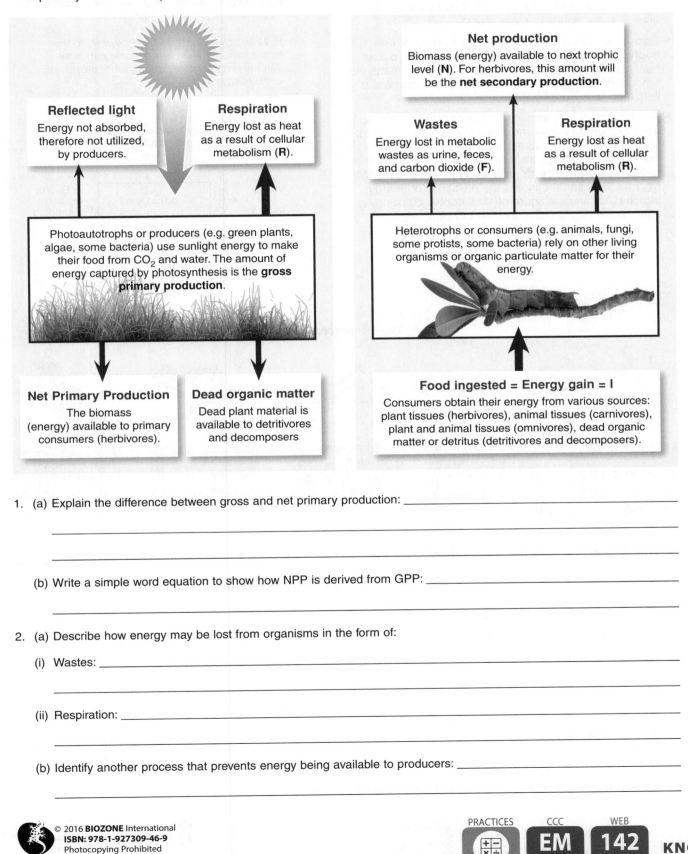

1. (a) Explain the difference between gross and net primary production: _____

_____

_____

(b) Write a simple word equation to show how NPP is derived from GPP: _____

_____

2. (a) Describe how energy may be lost from organisms in the form of:

(i) Wastes: _____

_____

(ii) Respiration: _____

_____

(b) Identify another process that prevents energy being available to producers: _____

_____

© 2016 **BIOZONE** International
ISBN: 978-1-927309-46-9
Photocopying Prohibited

PRACTICES   CCC   WEB

EM   142   **KNOW**

143 Energy Flow in Ecosystems

Key Idea: Energy flows through an ecosystem from one trophic level to the next. Only 5-20% of energy is transferred from one trophic level to the next.

Conservation of energy and trophic efficiency

▶ The Law of Conservation of Energy states that energy cannot be created or destroyed, only transformed from one form (e.g. light energy) to another (e.g. chemical energy in the bonds of molecules).

▶ Each time energy is transferred (as food) from one trophic level to the next, some energy is given out as heat, usually during cellular respiration. This means the amount of energy available to the next trophic level is less than at the previous level.

▶ Potentially, we can account for the transfer of energy from its input (as solar radiation) to its release as heat from organisms, because energy is conserved. The percentage of energy transferred from one trophic level to the next is the **trophic efficiency**. It varies between 5% and 20% and measures the efficiency of energy transfer. An average figure of 10% trophic efficiency is often used. This is called the **ten percent rule**.

Calculating available energy

The energy available to each trophic level will equal the amount entering that trophic level, minus total losses from that level (energy lost as heat + energy lost to detritus).

Heat energy is lost from the ecosystem to the atmosphere. Other losses become part of the detritus and may be utilized by other organisms in the ecosystem.

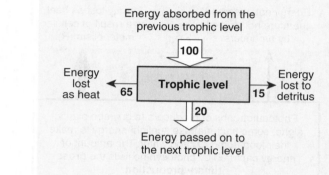

Energy absorbed from the previous trophic level

100

Energy lost as heat ← 65 | **Trophic level** | 15 → Energy lost to detritus

20

Energy passed on to the next trophic level

The ten percent rule

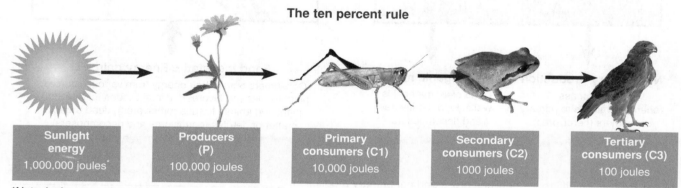

Sunlight energy	Producers (P)	Primary consumers (C1)	Secondary consumers (C2)	Tertiary consumers (C3)
1,000,000 joules*	100,000 joules	10,000 joules	1000 joules	100 joules

*Note: joules are units of energy

1. Why is the energy available to a particular trophic level less the energy in the previous trophic level? _____

_____

_____

_____

2. Why must ecosystems receive a continuous supply of energy from the Sun? _____

_____

_____

3. (a) What is trophic efficiency? _____

_____

(b) In general, how much energy is transferred between trophic levels? _____

WEB   CCC   CCC   PRACTICES

KNOW   143   EM   SSM   [+-/×÷]

© 2016 **BIOZONE** International
ISBN: 978-1-927309-46-9
Photocopying Prohibited

Calculating energy transfer and trophic efficiency in an ecosystem

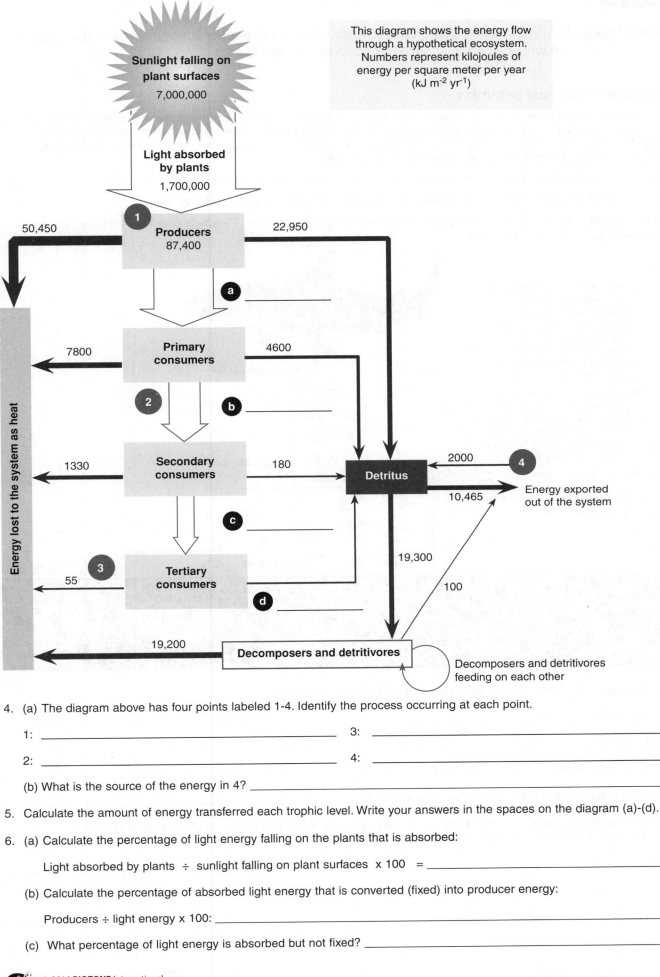

This diagram shows the energy flow through a hypothetical ecosystem. Numbers represent kilojoules of energy per square meter per year (kJ m^{-2} yr^{-1})

Sunlight falling on plant surfaces
7,000,000

Light absorbed by plants
1,700,000

1

Producers
87,400

50,450

22,950

a _____

Energy lost to the system as heat

Primary consumers

7800

4600

2

b _____

Secondary consumers

1330

180

Detritus

2000

4

Energy exported out of the system

10,465

c _____

19,300

100

3

Tertiary consumers

55

d _____

19,200

Decomposers and detritivores

Decomposers and detritivores feeding on each other

4. (a) The diagram above has four points labeled 1-4. Identify the process occurring at each point.

1: _____ 3: _____

2: _____ 4: _____

(b) What is the source of the energy in 4? _____

5. Calculate the amount of energy transferred each trophic level. Write your answers in the spaces on the diagram (a)-(d).

6. (a) Calculate the percentage of light energy falling on the plants that is absorbed:

Light absorbed by plants ÷ sunlight falling on plant surfaces x 100 = _____

(b) Calculate the percentage of absorbed light energy that is converted (fixed) into producer energy:

Producers ÷ light energy x 100: _____

(c) What percentage of light energy is absorbed but not fixed? _____

144 Ecological Pyramids

Key Idea: Ecological pyramids are used to illustrate the number of organisms, amount of energy, or amount of biomass at each trophic level in an ecosystem.

Types of ecological pyramids

The energy, biomass, or numbers of organisms at each trophic level in any ecosystem can be represented by an ecological pyramid. The first trophic level is placed at the bottom of the pyramid and subsequent trophic levels are stacked on top in their 'feeding sequence'. Ecological pyramids provide a convenient model to illustrate the relationship between different trophic levels in an ecosystem.

► Pyramid of numbers shows the numbers of individual organisms at each trophic level.

► Pyramid of biomass measures the mass of the biological material at each trophic level.

► Pyramid of energy shows the energy contained within each trophic level. Pyramids of energy and biomass are usually quite similar in appearance.

► This generalized ecological pyramid (right) shows a conventional pyramid shape, with a large number of producers at the base, and decreasing numbers of consumers at subsequent trophic levels.

► Ecological pyramids for this plankton-based ecosystem have a similar appearance regardless of whether we construct them using energy, or biomass, or numbers of organisms.

► Units refer to biomass or energy. The images provide a visual representation of the organisms present.

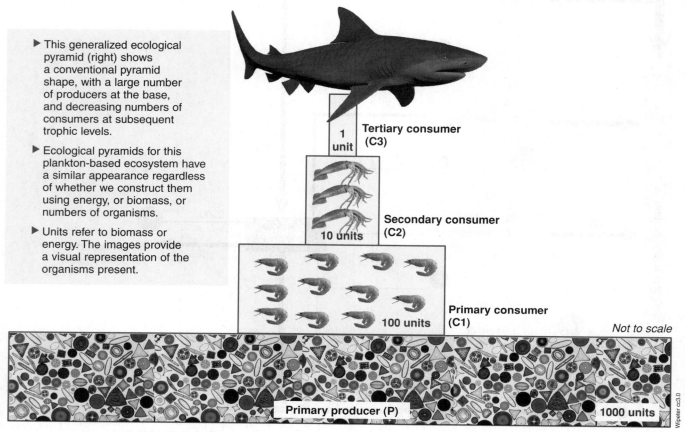

Tertiary consumer (C3) — 1 unit

Secondary consumer (C2) — 10 units

Primary consumer (C1) — 100 units

Not to scale

Primary producer (P) — 1000 units

1. What do each of the following types of ecological pyramids measure?

   (a) Number pyramid: _____

   (b) Biomass pyramid: _____

   (c) Energy pyramid: _____

2. What major group is missing from the pyramid above? _____

3. What is the advantage of using a biomass or energy pyramid rather than a pyramid of numbers to express the relationship between different trophic levels?

   _____

   _____

   _____

   _____

© 2016 **BIOZONE** International
ISBN: 978-1-927309-46-9
Photocopying Prohibited

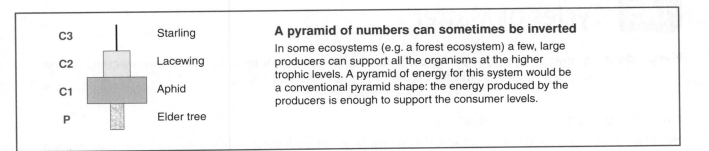

C3		Starling
C2		Lacewing
C1		Aphid
P		Elder tree

A pyramid of numbers can sometimes be inverted

In some ecosystems (e.g. a forest ecosystem) a few, large producers can support all the organisms at the higher trophic levels. A pyramid of energy for this system would be a conventional pyramid shape: the energy produced by the producers is enough to support the consumer levels.

Which pyramid provides the most useful information?

There are benefits and disadvantages to each type of pyramid.

▶ A pyramid of numbers provides information about the number of organisms at each level, but it does not account for their size, which can vary greatly. For example, elephants, rabbits, and aphids all feed on plants, but their sizes are very different and they feed on plants of very different sizes.

▶ A pyramid of biomass is often more useful because it takes into account the amount of biological material (biomass) at each level. The number of organisms at each level is multiplied by their mass to produce biomass.

▶ While number and biomass pyramids provide information about an ecosystem's structure, a pyramid of energy provides information about function (how much energy is fixed, lost, and available to the next trophic level).

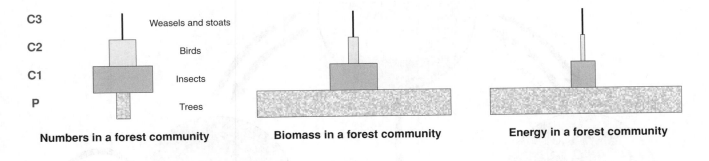

C3		Weasels and stoats
C2		Birds
C1		Insects
P		Trees

Numbers in a forest community   **Biomass in a forest community**   **Energy in a forest community**

4. (a) Why are some ecological pyramids not a typical pyramid shape? _____

_____

_____

(b) Would a pyramid of energy ever have an inverted (upturned) shape? Explain your answer: _____

_____

_____

_____

5. The pyramids below show numbers of organisms at each tropic level for two different ecosystems. Match the descriptions to the correct pyramid of numbers.

Description 1: An oak tree supports a large number of birds. The birds are hunted by foxes.

Description 2: A large cornfield is the food resource for field mice. Foxes eat the field mice.

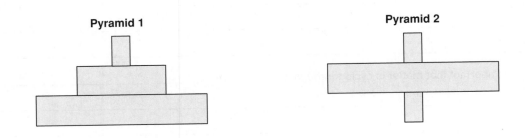

Pyramid 1   **Pyramid 2**

_____   _____

145 Cycles of Matter

Key Idea: Matter cycles through the biotic and abiotic compartments of Earth's ecosystems. These cycles are called nutrient cycles or biogeochemical cycles.

Nutrients cycle through ecosystems

▶ Nutrient cycles move and transfer chemical elements (e.g. carbon, hydrogen, nitrogen, and oxygen) through an ecosystem. Because these elements are part of many essential nutrients, their cycling is called a **nutrient cycle**, or a **biogeochemical cycle**. The term biogeochemical means that **bio**logical, **geo**logical, and **chemical** processes are involved in nutrient cycling.

▶ In a nutrient cycle, the nutrient passes through the biotic (living) and abiotic (physical) components of an ecosystem (see diagram below). Recall that energy drives the cycling of matter within and between systems. Matter is conserved throughout all these transformations, although it may pass from one ecosystem to another.

Processes in a generalized biogeochemical cycle

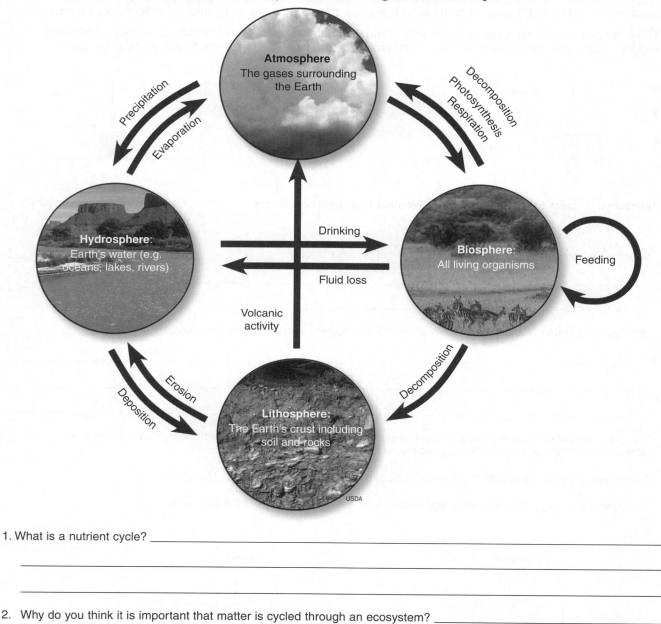

1. What is a nutrient cycle? _____

_____

_____

2. Why do you think it is important that matter is cycled through an ecosystem? _____

_____

_____

© 2016 **BIOZONE** International
ISBN: 978-1-927309-46-9
Photocopying Prohibited

146 The Hydrologic Cycle

Key Idea: The hydrologic cycle results from the cycling of water from the oceans to the land and back.

▶ About 97% of the water on Earth is stored in the oceans, which contain more than 1.3 billion cubic kilometres of water. Less than 1% of Earth's water is freely available fresh water (in lakes and streams).

▶ Water evaporates from water bodies into the atmosphere and falls as precipitation (e.g. rain). Precipitation on to land is transported back to the oceans by rivers and streams or returned to the atmosphere by evaporation or transpiration (evaporation from plant surfaces).

▶ Water can cycle very quickly if it remains near the Earth's surface, but in some circumstances it can remain locked away for hundreds or even thousands of years (e.g. in deep ice layers at the poles or in groundwater).

▶ Humans intervene in the water cycle by using water for their own needs. Irrigation from rivers and lakes changes evaporation patterns, lowers lake levels, and reduces river flows.

Water is the only substance on Earth that can be found naturally as a solid, liquid, or gas. It has the unique property of being less dense as a solid than a liquid, causing water to freeze from the top down (and float), and it has an unexpectedly high boiling point compared to other similar molecules.

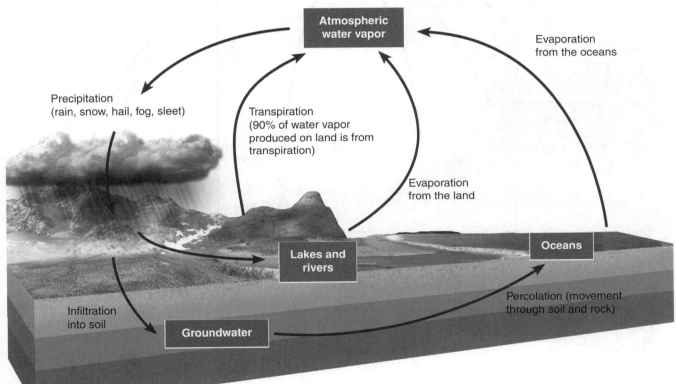

1. What is the main storage reservoir for water on Earth? _____

2. Name the two processes by which water moves from the land or oceans to the atmosphere: _____

   _____

   _____

3. Identify the feature of water that allows it to cycle as described above: _____

   _____

   _____

 © 2016 **BIOZONE** International
ISBN: 978-1-927309-46-9
Photocopying Prohibited

147 The Carbon Cycle

Key Idea: All life is carbon-based. Carbon cycles between the atmosphere, biosphere, geosphere, and hydrosphere. Photosynthesis and respiration are central to this.

▶ Carbon is the essential element of life. Its unique properties allow it to form an almost infinite number of different molecules. In living systems, the most important of these are carbohydrates, fats, nucleic acids, and proteins.

▶ Carbon in the atmosphere is found as carbon dioxide (CO_2). In rocks, it is most commonly found as either coal (mostly carbon) or limestone (calcium carbonate).

▶ The most important processes in the carbon cycle are photosynthesis and respiration.

▶ Photosynthesis removes carbon from the atmosphere and converts it to organic molecules. This organic carbon may eventually be returned to the atmosphere through respiration.

▶ Carbon cycles at different rates depending on where it is. On average, carbon remains in the atmosphere as CO_2 for about 5 years, in plants and animals for about 10 years, and in oceans for about 400 years. Carbon can remain in rocks (e.g. coal) for millions of years.

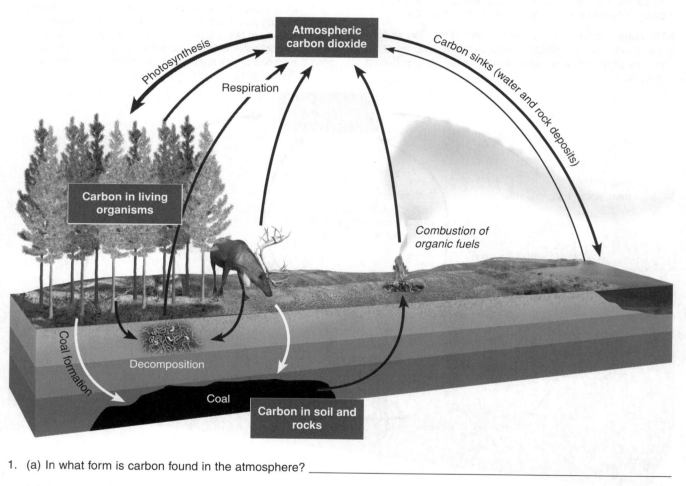

1. (a) In what form is carbon found in the atmosphere? _____

   (b) In what three important molecules is carbon found in living systems? _____

   (c) In what two forms is carbon found in rocks? _____

2. (a) Name two processes that remove carbon from the atmosphere: _____

   (b) Name two processes that add carbon to the atmosphere: _____

3. What is the effect of deforestation and burning of coal and oil on carbon cycling? _____

   _____

   _____

   _____

WEB   CCC   CCC

KNOW 147 EM SSM

© 2016 **BIOZONE** International
ISBN: 978-1-927309-46-9
Photocopying Prohibited

148 Modeling The Carbon Cycle

Key Idea: The cycling of carbon can be modeled in an artificial, closed ecosystem.

Making a closed ecosystem

▶ In this activity, a simple model was used to mimic the carbon cycle in a small closed ecosystem.

▶ A large, clear soda bottle was used as the container. It was filled almost to the top with filtered pond water.

▶ Small pebbles or rocks were added to a depth of around 2-5 cm. A dead leaf was added as detritus.

▶ The living components of the system were added. (a few stems of an aquatic plant such as *Cabomba* and four small aquatic snails). Some systems also add one small fish (e.g. feeder guppy).

▶ The system was left to stand for a day before putting the lid on. The bottle was placed in an area that gets direct sunlight for a few hours each day.

Air space
Clear bottle
Pond water
Cabomba
Snail
Dead leaf
Pebbles

1. (a) (i) Name the source of O_2 in this system:

   _____

   (ii) What process produces it? _____

   (b) (i) Name the source of CO_2 in this system:

   _____

   (ii) What process produces it? _____

   (c) What happens to the carbon in the CO_2 after it has been taken up by the plant? _____

   _____

   _____

2. In the space provided draw a simple diagram to show the how carbon cycles between *Cabomba* and the animals:

3. Predict what would happen if:

   (a) The plants were removed: _____

   _____

   _____

   (b) The animals were removed: _____

   _____

   _____

4. The pond water contains small microorganisms. What is their role in this system?

   _____

   _____

   _____

 © 2016 **BIOZONE** International
ISBN: 978-1-927309-46-9
Photocopying Prohibited

PRACTICES   CCC   CCC

SSM   EM   **KNOW**

149 The Oxygen Cycle

Key Idea: The oxygen cycle describes the movement of oxygen (O_2) through an ecosystem. The oxygen cycle is closely linked to the carbon cycle.

The importance of the oxygen cycle

▶ The **oxygen cycle** describes the movement of oxygen between the biotic and abiotic components of ecosystems. Photosynthesis is the main source of oxygen in ecosystems.

▶ Oxygen is involved to some degree in all the other biogeochemical cycles, but is closely linked to the carbon cycle in particular. This is because most producers utilize carbon dioxide in photosynthesis, and produce oxygen as a waste product. The oxygen is used in cellular respiration and carbon dioxide is produced as a waste product. This is the oxygen-carbon dioxide cycle (simplified in the diagram, right).

The link between the oxygen and carbon cycles

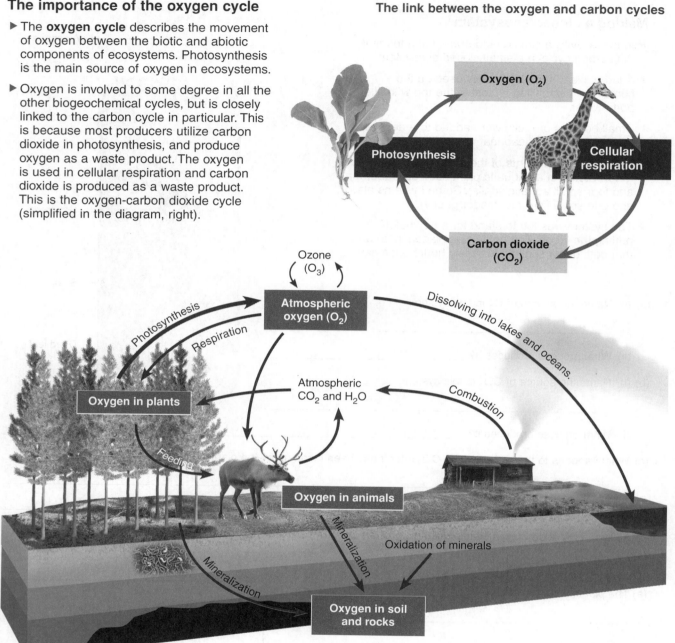

1. What is the main source of oxygen for the oxygen cycle? _____

_____

2. Why are the oxygen cycle and carbon cycle so interdependent? _____

_____

_____

_____

_____

© 2016 **BIOZONE** International
ISBN: 978-1-927309-46-9
Photocopying Prohibited

150 Role of Photosynthesis in Carbon Cycling

Key Idea: Photosynthesis removes carbon from the atmosphere and adds it to the biosphere. Respiration removes carbon from the biosphere and adds it the atmosphere.

Photosynthesis and carbon

▶ Photosynthesis removes carbon from the atmosphere by fixing the carbon in CO_2 into carbohydrate molecules. Plants use the carbohydrates (e.g. glucose) to build structures such as wood.

▶ Some carbon may be returned to the atmosphere during respiration (either from the plant or from animals). If the amount or rate of carbon fixation is greater than that released during respiration then carbon will build up in the biosphere and be reduced in the atmosphere (diagram, right).

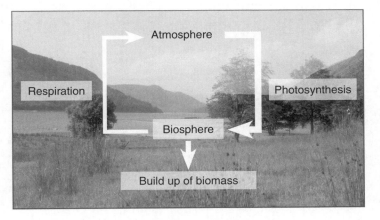

Respiration and carbon

▶ Cellular respiration releases carbon into the atmosphere as carbon dioxide as a result of the breakdown of glucose.

▶ If the rate of carbon release is greater than that fixed by photosynthesis then over time carbon may accumulate in the atmosphere (diagram bottom right). Before the Industrial Revolution, many thousands of gigatonnes (Gt) of carbon were contained in the biosphere of in the Earth's crust (e.g. coal).

▶ Deforestation and the burning of fossil fuels have increased the amount of carbon in the atmosphere.

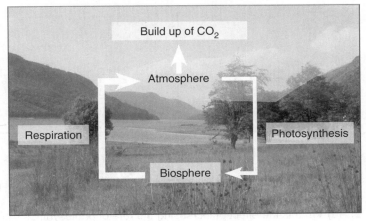

Carbon cycling simulation

Plants move about 120 Gt of carbon from the atmosphere to the biosphere a year. Respiration accounts for about 60 Gt of carbon a year. A simulation was carried out to study the effect of varying the rates of respiration and photosynthesis on carbon deposition in the biosphere or atmosphere. To keep the simulation simple, only the effects to the atmosphere and biosphere were simulated. Effects such as ocean deposition and deforestation were not studied. The results are shown in the tables right and below.

Table 1: Rate of photosynthesis equals the rate of cellular respiration.

Years	Gt carbon in biosphere	Gt carbon in atmosphere
0	610	600
20	608	600
40	608	600
60	609	598
80	612	598
100	610	596

Table 2: Rate of photosynthesis increases by 1 Gt per year.

Years	Gt carbon in biosphere	Gt carbon in atmosphere
0	610	600
20	632	580
40	651	558
60	671	538
80	691	518
100	710	498

Table 3: Rate of cellular respiration increases by 1 Gt per year.

Years	Gt carbon in biosphere	Gt carbon in atmosphere
0	610	600
20	590	619
40	570	641
60	548	664
80	528	686
100	509	703

© 2016 **BIOZONE** International
ISBN: 978-1-927309-46-9
Photocopying Prohibited

PRACTICES · PRACTICES · CCC · CCC · WEB

SSM · EM · 150 · **KNOW**

NEED HELP?
See Activity 17

1. Plot the data for tables 1,2, and 3 on the grid provided (above). Include a key and appropriate titles and axes.

2. (a) What is the effect of increasing the rate of photosynthesis on atmospheric carbon? _____

   _____

   (b) i. What is the effect of increasing the rate of photosynthesis on biospheric carbon? _____

   _____

   ii. How does this effect occur? _____

   _____

   _____

3. What is the effect of increasing the rate of cellular respiration on atmospheric and biospheric carbon? _____

   _____

   _____

4. In the real world, respiration is not necessarily increasing in comparison to photosynthesis, but many human activities cause the same effect.

   (a) Name two human activities that have the same effect on atmospheric carbon as increasing the rate of cellular respiration:

   _____

   (b) What effect does this extra atmospheric carbon have on the global climate? _____

   _____

   _____

151 The Nitrogen Cycle

Key Idea: Nitrogen is essential for building proteins. Nitrogen gas is converted to nitrates, which are taken up by plants. Animals gain nitrogen by feeding off plants or animals.

▶ Nearly eighty percent of the Earth's atmosphere is made of nitrogen gas. As a gas nitrogen is very stable and unreactive, effectively having no interaction with living systems. However, nitrogen is extremely important in the formation of amino acids, which are the building blocks of proteins.

▶ Nitrogen may enter the biosphere during lightning storms. Lightning produces extremely high temperatures in the air (around 30,000°C). At such high temperatures, nitrogen reacts with oxygen in the air to form ammonia and nitrates which dissolve in water and are washed into the soil.

▶ Some bacteria can fix nitrogen directly from the air. Some of these bacteria are associated with plants (especially legumes) and produce ammonia (NH_3). This can be converted to nitrates (NO_3^-) by other bacteria. Other bacteria produce nitrites (NO_2^-).

▶ Nitrates are absorbed and used by plants to make amino acids. Animals gain their nitrogen by feeding on plants (or on herbivores).

▶ Nitrogen is returned to the atmosphere by denitrifying bacteria which convert nitrates back into nitrogen gas.

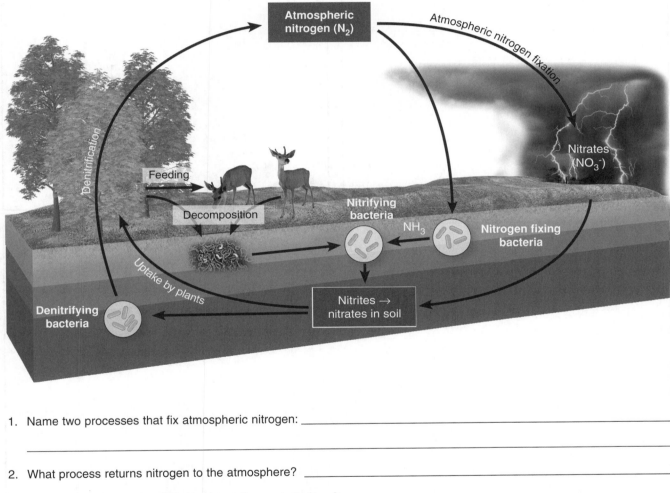

1. Name two processes that fix atmospheric nitrogen: _____

   _____

2. What process returns nitrogen to the atmosphere? _____

3. What essential organic molecule does nitrogen help form? _____

4. Why do farmers often plant legumes between cropping seasons, then plow them into the soil rather than harvest them?

   _____

   _____

   _____

5. Where do animals get their nitrogen from? _____

   _____

152 Chapter Review

Summarize what you know about this topic under the headings provided. You can draw diagrams or mind maps, or write short notes to organize your thoughts. Use the points in the introduction and the hints to help you:

Organisms are classified on how they obtain their energy

HINT: Define the terms producers and consumers and say how each obtains its energy.

Cycling of matter in ecosystems

HINT: Include reference to the carbon, oxygen, nitrogen, and hydrologic cycles. State the importance of photosynthesis and cellular respiration in the carbon cycle.

CO_2

O_2

Energy flows in ecosystems

HINT: Define trophic level and explain how energy is transferred through food chains. Describe and explain ecological efficiency.

© 2016 **BIOZONE** International
ISBN: 978-1-927309-46-9
Photocopying Prohibited

153 KEY TERMS AND IDEAS: Did You Get It?

1. Test your vocabulary by matching each term to its definition, as identified by its preceding letter code.

anaerobic _____

aerobic _____

carbon cycle _____

cellular respiration _____

consumer _____

ecological pyramid _____

food chain _____

food web _____

hydrologic cycle _____

nitrogen cycle _____

oxygen cycle _____

photosynthesis _____

producer _____

trophic level _____

A A graphical representation of the numbers, energy, or biomass at each trophic level in an ecosystem. Often pyramidal in shape.

B A metabolic process that requires oxygen in order for it to proceed.

C The processes by which nitrogen circulates between the atmosphere and the biosphere.

D The cycle describing the movement of oxygen through the ecosystem.

E A sequence of steps describing how an organism derives energy from those before it.

F Organism that is capable of making its own food, e.g. green plants.

G A complex series of interactions showing the feeding relationships between organisms in an ecosystem.

H The catabolic process in which the chemical energy in complex organic molecules is coupled to ATP production.

I The cycling of water through the environment, from lakes, seas and oceans to clouds, rain, through organisms, to ground water and rivers and back.

J Biogeochemical cycle by which carbon is exchanged among the biotic and abiotic components of the Earth..

K Any of the feeding levels that energy passes through in an ecosystem.

L Process in which carbon dioxide, water and sunlight are used to produce glucose and oxygen.

M A metabolic process that does not require oxygen to proceed.

N An organism that obtains its carbon and energy from other organisms.

2. The energy pyramid on the right is for a hypothetical plankton community.

Name the trophic levels labeled A, B, and C:

A: _____

B: _____

C: _____

Decomposers 930 kJ

C → 12 kJ

B 142 kJ

A 8690 kJ

3. For the plankton community in the above diagram, determine:

(a) The amount of energy transferred between trophic levels A and B: _____

(b) The efficiency of this energy transfer: _____

(c) The amount of energy transferred between trophic levels B and C: _____

(d) The efficiency of this energy transfer: _____

(e) Why is the amount of energy transferred from trophic level A to trophic level B considerably less than 10%:

_____

4. Why isn't all the energy transferred by sunlight available to plants? _____

_____

5. Explain why a pyramid of biomass and a pyramid of energy have similar proportions to one another: _____

_____

_____

TEST

154 Summative Assessment

The gross primary production of any ecosystem will be determined by the efficiency with which solar energy is captured by photosynthesis. The efficiency of subsequent energy transfers will determine the amount of energy available to consumers. These energy transfers can be quantified using measurements of dry mass.

Production vs productivity: What's the difference?

Strictly speaking, the primary production of an ecosystem is distinct from its productivity, which is the amount of production per unit time (a rate). However because values for production (accumulated biomass) are usually given for a certain period of time in order to be meaningful, the two terms are often used interchangeably.

In this activity, you will calculate energy and biomass transfers in real and experimental systems.

Corn field

Mature pasture

1. The energy budgets of two agricultural systems (4000 m² area) were measured over a growing season of 100 days. The results are tabulated right.

   (a) For each system, calculate the percentage efficiency of energy utilization (how much incident solar radiation is captured by photosynthesis):

   Corn: _____

   Mature pasture: _____

   (b) For each system, calculate the percentage losses to respiration:

   Corn: _____

   Mature pasture: _____

   (c) For each system, calculate the percentage efficiency of NPP:

   Corn: _____

   Mature pasture: _____

   (d) Which system has the greatest efficiency of energy transfer to biomass? _____

	Corn field	Mature pasture
	kJ x 10⁶	kJ x 10⁶
Incident solar radiation	8548	1971
Plant utilization		
Net primary production (NPP)	105.8	20.7
Respiration (R)	32.2	3.7
Gross primary production (GPP)	138.0	24.4

Estimating NPP in *Brassica rapa*

Background

Brassica rapa (right) is a fast growing brassica species, which can complete its life cycle in as little as 40 days if growth conditions are favorable. A class of students wished to estimate the gross and net primary productivity of a crop of these plants using wet and dry mass measurements made at three intervals over 21 days.

The method

▶ Seven groups of three students each grew 60 *B. rapa* plants in plant trays under controlled conditions. On day 7, each group made a random selection of 10 plants and removed them, with roots intact. The 10 plants were washed, blotted dry, and then weighed collectively (giving wet mass).

▶ The 10 plants were placed in a ceramic drying bowl and placed in a drying oven at 200°C for 24 hours, then weighed (giving dry mass).

▶ On day 14 and again on day 21, the procedure was repeated with a further 10 plants (randomly selected).

▶ The full results for group 1 are presented in Table 1 on the next page. You will complete the calculation columns.

© 2016 **BIOZONE** International
ISBN: 978-1-927309-46-9
Photocopying Prohibited

Table 1: Group 1's results for growth of 10 *B. rapa* plants over 21 days

Age in days	Wet mass of 10 plants (g)	Dry mass of 10 plants (g)	Percent biomass	Energy in 10 plants (kJ)	Energy per plant (kJ)	NPP (kJ plant^{-1} d^{-1})
7	19.6	4.2				
14	38.4	9.3				
21	55.2	15.5				

2. Calculate percent biomass using the equation: % biomass = dry mass ÷ wet mass x 100. Enter the values in Table 1.

3. Each gram of dry biomass is equivalent to 18.2 kJ of energy. Calculate the amount of energy per 10 plants and per plant for plants at 7, 14, and 21 days. Enter the values in Table 1.

4. Calculate the Net Primary Productivity per plant, i.e. the amount of energy stored as biomass per day (kJ plant^{-1} d^{-1}). Enter the values in Table 1. We are using per plant in this exercise as we do not have a unit area of harvest.

5. The other 6 groups of students completed the same procedure and, at the end of the 21 days, the groups compared their results for NPP. The results are presented in Table 2, right.

   Transfer group 1's NPP results from Table 1 to complete the table of results and calculate the mean NPP for *B. rapa*.

Table 2: Class results for NPP of *B. rapa* over 21 days

Time in days (d)	Group NPP (kJ plant^{-1} d^{-1})							Mean NPP
	1	2	3	4	5	6	7	
7		1.05	1.05	1.13	1.09	1.13	1.09	
14		1.17	1.21	1.25	1.21	1.25	1.17	
21		1.30	1.34	1.30	1.34	1.38	1.34	

6. On the grid, plot the class mean NPP vs time.

7. (a) What is happening to the NPP over time?

   _____

   _____

   (b) Explain why this is happening: _____

   _____

   _____

   _____

   _____

   _____

8. What would you need to know to determine the gross primary productivity of *B. rapa*?

   _____

   _____

   _____

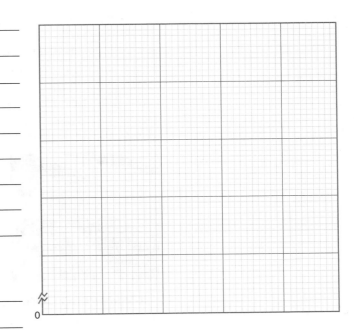

9. Net production in consumers (N), or secondary production, can be expressed as N = I - (F+R). Red meat contains approximately 700 kJ per 100 grams. If N = 20% of the energy gain (I), how much energy is lost as F and R?

   _____

10. Devise a methodology and calculations to determine the **net secondary production** and **respiratory losses** of 12 day old cabbage white caterpillars feeding on brussels sprouts for 3 days. Begin with 10 caterpillars and ~30 g brussels sprouts. How would you calculate the efficiency of energy transfer from producers to consumers? What assumptions are you making in your calculations?
    Staple your methodology to this page. You will need to know:

    Energy value of plant material: dry mass x 18.2 kJ.
    Energy value of animal material: dry mass x 23.0 kJ
    Energy value of egested waste (frass): mass x 19.87 kJ

LS2.C
ETS1.B
The Dynamic Ecosystem

Key terms

anthropogenic change
biodiversity
climate change
conservation
deforestation
ecosystem
global warming
invasive species
keystone species
overfishing
resilience
sustainability

Disciplinary core ideas

Show understanding of these core ideas

		Activity number

Ecosystems are dynamic, open systems

☐ 1 The physical and biotic environments contribute to an ecosystem's characteristics. — 155

☐ 2 In stable conditions, the numbers and types of organisms within the ecosystem are relatively stable. — 155

☐ 3 Organisms with pivotal roles in the functioning of an ecosystem are called keystone species. Their removal may cause a fundamental change in a ecosystem. — 155 158

☐ 4 Ecosystems are resilient. An ecosystem generally returns to its original state after modest fluctuations and disturbances. The resilience of an ecosystem depends on its biodiversity, health, and the frequency of disturbances. — 156 157

☐ 5 Extreme fluctuations in conditions or a population's size may challenge normal ecosystem functioning (e.g. by disturbing energy flows, species interactions, or nutrient cycling) and may cause irreversible changes. — 159 160

☐ 6 Biodiversity is the biological variety in an ecosystem. Biodiversity is an important factor in ecosystem stability and resilience. — 155 158

Anthropogenic changes can threaten species survival

☐ 7 Human (anthropogenic) activity can alter ecosystems and diminish biodiversity. — 159 161

☐ 8 Human actions that threaten biodiversity include anthropogenic causes of global warming, overfishing, and deforestation. — 160-164 166

☐ 9 Solutions to problems created by human activities must consider practical constraints, such as costs, and social, cultural, and environmental impacts. — 165 167

Jerald E. Dewey, USDA Forest Service, Bugwood.org

Crosscutting concepts

Understand how these fundamental concepts link different topics

		Activity number

☐ 1 **SPQ** The concept of orders of magnitude can be used to understand how a model of the factors affecting ecosystems can operate at different scales. — 155

☐ 2 **SC** Many aspects of science, including ecology, involve explaining how things (e.g. ecosystems) change and how they remain stable. — 155-166

Science and engineering practices

Demonstrate competence in these science and engineering practices

		Activity number

☐ 1 Analyze examples of ecosystem resilience using second-hand data. — 156 157

☐ 2 Use mathematical representations to support explanations about the factors affecting biodiversity and populations in ecosystems. — 155-158 160 164

☐ 3 Evaluate the evidence for the role of complex interactions in the stability of ecosystems and the role of changing conditions in ecosystem change. — 159 160 163 166

☐ 4 Develop and evaluate solutions for sustaining biodiversity while allowing essential human use of resources. — 161 165 167

155 Ecosystem Dynamics

Key Idea: Ecosystems are dynamic systems, responding to short-term and cyclical changes, but remaining relatively stable in the long term.

What is an ecosystem?

▶ An **ecosystem** consists of a community of organisms and their physical environment. For example a forest ecosystem consists of all the organisms within the defined area of the forest, along with the physical factors in the forest such as the temperature and the amount of wind or rain.

The dynamic ecosystem

▶ Ecosystems are dynamic in that they are constantly changing. Many ecosystem components, including the seasons, predator-prey cycles, and disease cycles, are cyclical. Some cycles may be short term, such as the change of seasons, or long term, such as the growth and retreat of deserts.

▶ Although ecosystems may change constantly over the short term, they may be relatively static over longer periods. For example, some tropical areas have wet and dry seasons, but over hundreds of years the ecosystem as a whole remains unchanged.

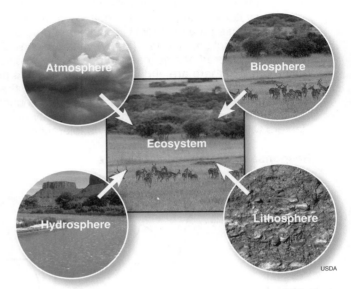

Ecosystems area a result of the interactions between biological (biotic) and physical (abiotic) factors.

An ecosystem may remain stable for many hundreds or thousands of years provided that the components interacting within it remain stable.

Small scale changes usually have little effect on an ecosystem. Fire or flood may destroy some parts, but enough is left for the ecosystem to return to is original state relatively quickly.

Large scale disturbances such as volcanic eruptions, sea level rise or large scale open cast mining remove all components of the ecosystem, changing it forever.

1. What is meant by the term dynamic ecosystem? _____

_____

_____

_____

2. (a) Describe two small scale events that an ecosystem may recover from: _____

_____

_____

(b) Describe two large scale events that an ecosystem may not recover from: _____

_____

_____

 © 2016 **BIOZONE** International
ISBN: 978-1-927309-46-9
Photocopying Prohibited

PRACTICES   CCC   CCC   WEB

SC   **SPQ**   **155**   **KNOW**

Ecosystem stability

▶ Ecosystem stability has various components, including **inertia** (the ability to resist disturbance) and **resilience** (ability to recover from external disturbances).

▶ Ecosystem stability is closely linked to biodiversity, with more diverse systems being more stable. It is hypothesized that this is because greater diversity results in a greater number of biotic interactions and few (if any) vacant niches. The system is buffered against change because it is resistant to invasions and there are enough species present to protect ecosystem functions if a species is lost. This hypothesis is supported by experimental evidence but there is uncertainty as to what level of biodiversity provides stability or what factors will stress a system beyond its tolerance.

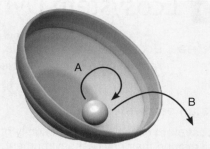

The stability of an ecosystem can be illustrated by a ball in a tilted bowl. Given a slight disturbance the ball will eventually return to its original state (**line A**). However given a large disturbance the ball will roll out of the bowl and the original state with never be restored (**line B**).

Response to environmental change

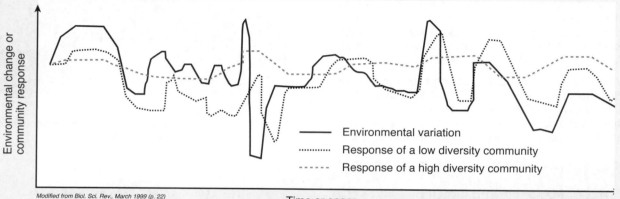

Modified from Biol. Sci. Rev., March 1999 (p. 22)

Time or space

- Environmental variation
- Response of a low diversity community
- Response of a high diversity community

▶ In models of ecosystem function, higher species diversity increases the stability of ecosystem functions such as productivity and nutrient cycling. In the graph above, note how the low diversity system varies more consistently with the environmental variation, whereas changes in high diversity system are more gradual.

▶ In any one ecosystem, some species have a disproportionate effect on ecosystem stability due to their pivotal role in some ecosystem function, e.g. nutrient recycling. These species are called **keystone** (key) **species**.

3. Why is ecosystem stability higher in ecosystems with high biodiversity than in ones with low biodiversity:

_____

_____

_____

4. The effect of changes in ecosystems can be difficult to measure in the field, so researchers often build small scale simulations of an ecosystem. The graph right shows the effect of adding nutrients to a marine ecosystem (as in nutrient runoff from land into the sea). Algal growth-promoting medium was added at 2, 10, or 20% to seawater, together with 0.1 mL of an algal mix. Two days after adding the growth medium and algae, six copepods were added to each chamber. The chambers were sealed and the population size in each chamber was measured over time.

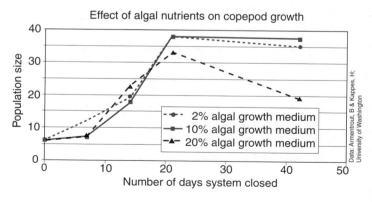

Effect of algal nutrients on copepod growth

- ● - 2% algal growth medium
- ■ - 10% algal growth medium
- ▲ - 20% algal growth medium

Data: Armentrout, B & Kappes, H; University of Washington

(a) Which chamber had the greatest environmental disturbance?

_____

(b) Which chamber(s) were able to withstand the environmental disturbance? _____

(c) What does this tell us about the stability and resilience of the system being studied? _____

_____

156 The Resilient Ecosystem

Key Idea: The resilience of a ecosystem depends on its biodiversity, health, and the frequency with which it is disturbed.

Factors affecting ecosystem resilience

Resilience is the ability of the ecosystem to recover after disturbance and is affected by three important factors: diversity, ecosystem health, and frequency of disturbance (below). Some ecosystems are naturally more resilient than others.

Ecosystem biodiversity

The greater the diversity of an ecosystem the greater the chance that all the roles (niches) in an ecosystem will be occupied, making it harder for invasive species to establish and easier for the ecosystem to recover after a disturbance.

Ecosystem health

Intact ecosystems are more likely to be resilient than ecosystems suffering from species loss or disease.

Disturbance frequency

Single disturbances to an ecosystem can be survived, but frequent disturbances make it more difficult for an ecosystem to recover. Some ecosystems depend on frequent natural disturbances for their maintenance, e.g. grasslands rely on natural fires to prevent shrubs and trees from establishing. The keystone grass species have evolved to survive frequent fires.

A study of coral and algae cover at two locations in Australia's Great Barrier Reef (right) showed how ecosystems recover after a disturbance. At Low Isles, frequent disturbances (e.g. from cyclones) made it difficult for corals to reestablish, while at Middle Reef, infrequent disturbances made it possible to coral to reestablish its dominant position in the ecosystem.

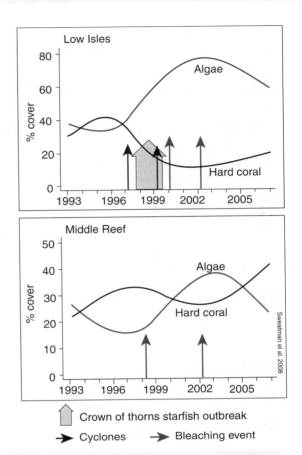

Sweatman el al 2008

 Crown of thorns starfish outbreak

→ Cyclones   → Bleaching event

Resilience and harvesting

Considering the resilience of an ecosystem is important when harvesting resources from it. For example, logging and fishing remove organisms from an ecosystem for human use. Most ecosystems will be resilient enough to withstand the removal of a certain number of individuals. However, excessive removal may go beyond the ecosystem's ability to recover. Examples include the overfishing of North Sea cod or deforestation in the Amazon basin.

Fishing   Deforestation

1. Define ecosystem resilience: _____

    _____

2. Why did the coral at Middle Reef remain abundant from 1993 to 2005 while the coral at Low Isles did not?

    _____

    _____

    _____

3. How might over-harvesting affect an ecosystem's resilience? _____

    _____

    _____

    _____

157 A Case Study in Ecosystem Resilience

Key Idea: Ecosystems fluctuate between extremes. Resilient ecosystems are able to recover from moderate fluctuations.

Spruce budworm and balsam fir

A case study of ecosystem resilience is provided by the spruce-fir forest community in northern North America. Organisms in the community include the spruce budworm, and balsam fir, spruce, and birch trees. The community fluctuates between two extremes:

▶ Between spruce budworm outbreaks the environment favors the balsam fir.

▶ During budworm outbreaks the environment favors the spruce and birch species.

USFW
Balsam fir

Jerald E. Dewey, USDA Forest Service, Bugwood.org
Spruce budworm

1 Under certain environmental conditions, the spruce budworm population grows so rapidly it overwhelms the ability of predators and parasites to control it.

2 The budworm feeds on balsam fir (despite their name), killing many trees. The spruce and birch trees are left as the major species.

3 The population of budworm eventually collapses because of a lack of food.

4 Balsam fir saplings grow back in thick stands, eventually out-competing the spruce and birch. Evidence suggests these cycles have been occurring for possibly thousands of years.

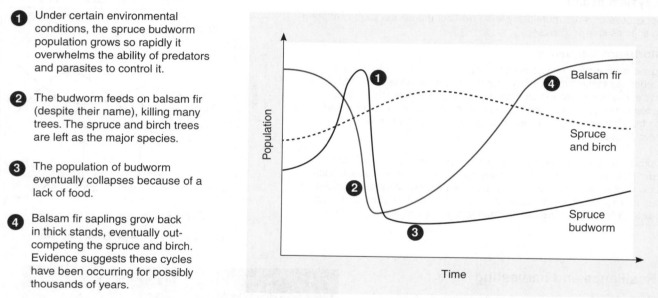

1. (a) Why could predators not control the budworm population? _____

_____

_____

_____

(b) What was the cause of the budworm population collapse after its initial rise? _____

_____

2. Under what conditions does the balsam fir out-compete the spruce and birch? _____

_____

3. In what way is the system resilient in the long term? _____

_____

_____

_____

_____

_____

© 2016 **BIOZONE** International
ISBN: 978-1-927309-46-9
Photocopying Prohibited

158 Keystone Species

Key Idea: Keystone species play a crucial role in ecosystems. Their actions are key to maintaining the dynamic equilibrium of an ecosystem.

Keystone species

▶ Some species have a disproportionate effect on the stability of an ecosystem. These species are called **keystone species** (or key species). The term keystone species comes from the analogy of the keystone in a true arch. If the keystone is removed, the arch collapses.

▶ The role of the keystone species varies from ecosystem to ecosystem, but if they are lost the ecosystem can rapidly change, or collapse completely. The pivotal role of keystone species is a result of their influence in some aspect of ecosystem functioning, e.g. as predators, prey, or processors of biological material.

Keystone

An archway is supported by a series of stones, the central one being the keystone. Although this stone is under less pressure than any other stone in the arch, the arch collapses if it is removed.

Keystone species make a difference

▶ The idea of the keystone species was first hypothesized in 1969 by Robert Paine. He studied an area of rocky seashore, noting that diversity seemed to be correlated with the number of predators present. To test this he removed the starfish from an 8 m by 2 m area of seashore. Initially the barnacle population increased rapidly before collapsing and being replaced by mussels and gooseneck barnacles. Eventually the mussels crowded out the gooseneck barnacles and the algae that covered the rocks. Limpets that fed on the algae were lost. The number of species in the study area dropped from 15 to 8.

Ochre starfish - Paine removed these in his famous study.

Elephants play a key role in maintaining the savannas by pulling down even very large trees for food. This activity maintains the grasslands.

The burrowing of prairie dogs increases soil fertility and channels water into underground stores. Their grazing promotes grass growth and diversity.

Mountain lions prey on a wide range of herbivores and to some extent dictate their home ranges and the distribution of scavenger species.

1. Define the term keystone species: _____

_____

_____

2. Prairie dogs colonies are often destroyed by ranchers who believe they compete with cattle for food. How might this lead to the collapse of prairie ecosystems?

_____

_____

_____

_____

 © 2016 **BIOZONE** International
ISBN: 978-1-927309-46-9
Photocopying Prohibited

PRACTICES   CCC   WEB

SC   158   **KNOW**

Sea otters as keystone species

▶ Sometimes the significant keystone effects of a species can be indicated when a species declines rapidly to the point of near extinction. This is illustrated by the sea otter example described below.

▶ Sea otters live along the Northern Pacific coast of North America and have been hunted for hundreds of years for their fur. Commercial hunting didn't fully begin until about the mid 1700s when large numbers were killed and their fur sold to overseas markets.

▶ The drop in sea otter numbers had a significant effect on the local marine environment. Sea otters feed on shellfish, particularly sea urchins. Sea urchins eat kelp, which provides habitat for many marine creatures. Without the sea otters to control the sea urchin population, sea urchin numbers increased and the kelp forests were severely reduced.

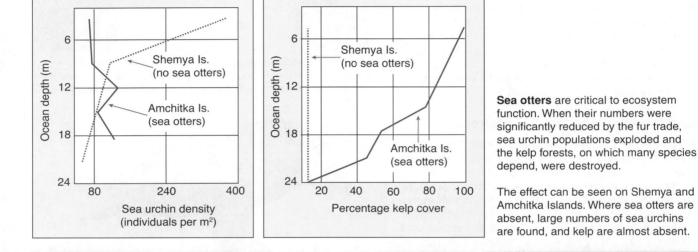

Sea otters are critical to ecosystem function. When their numbers were significantly reduced by the fur trade, sea urchin populations exploded and the kelp forests, on which many species depend, were destroyed.

The effect can be seen on Shemya and Amchitka Islands. Where sea otters are absent, large numbers of sea urchins are found, and kelp are almost absent.

Kelp are large seaweeds belonging to the brown algae. There are many forms and species of kelp. Giant kelps can grow to 45 m long.

In much the same way that forests on land provide diverse habitats for terrestrial species, kelps provide habitat, food, and shelter for a variety of marine animals.

Sea urchins kill kelp by eating the holdfast that secures the kelp to the seabed. Unchecked urchin populations can quickly turn a kelp forest into an **urchin barren**.

3. How do elephants act as a keystone species? _____

_____

4. (a) What effect do sea otters have on sea urchin numbers? _____

_____

(b) What effect do sea urchins have on kelp cover? _____

_____

(c) What evidence is there that the sea otter is a keystone species in these Northern Pacific coastal ecosystems?

_____

_____

_____

_____

© 2016 **BIOZONE** International
ISBN: 978-1-927309-46-9
Photocopying Prohibited

159 Ecosystem Changes

Key Idea: Sometimes the disturbances to an ecosystem are so extreme that the ecosystem never returns to its original state.

Ecosystems are dynamic, constantly fluctuating between particular conditions. However, large scale changes can occasionally occur to completely change the ecosystem. These include climate change, volcanic eruptions, or large scale fires.

Human influenced changes

▶ Dolly Sods is a rocky high plateau area in the Allegheny Mountains of eastern West Virginia, USA. Originally the area was covered with spruce, hemlock, and black cherry. During the 1880s, logging began in the area and virtually all of the commercially viable trees were cut down. The logging caused the underlying humus and peat to dry out. Sparks from locomotives and campfires frequently set fire to this dry peat, producing fires that destroyed almost all the remaining forest. In some areas, the fires were so intense that they burnt everything right down to the bedrock, destroying seed banks. One fire during the 1930s destroyed over 100 km² of forest. The forests have never recovered. What was once a forested landscape is now mostly open meadow.

The original dense forests of Dolly Sods included spruce, hemlock, white oak and black cherry. These species have been replaced mostly by maple, birch, beech, and low growing scrub.

Natural changes

▶ Volcanic eruptions can cause extreme and sudden changes to the local (or even global) ecosystems. The eruption of Mount St. Helens in 1980 provides a good example of how the natural event of a volcanic eruption can cause extreme and long lasting changes to an ecosystem.

Before the 1980 eruption, Mount St. Helens had an almost perfect and classic conical structure. The forests surrounding it were predominantly conifer, including Douglas-fir, western red cedar, and western white pine.

Mount St. Helens, one day before the eruption.

USGS

Schematic of eruption and recovery

The eruption covered about 600 km² (dark blue) in ash (up to 180 m deep in some areas) and blasted flat 370 km² of forest. The schematic shows the general area of bare land per decade since the eruption.

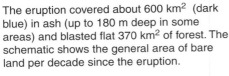

■ 1980	■ 1990
■ 2000	▫ 2010

Coldwater Lake was formed when Coldwater Creek was blocked by eruption debris.

Spirit Lake was completely emptied in the initial eruption. All life except bacteria was extinguished. The lake has since begun to recover.

N

The forests on the northern flank of the mountain have vanished, replaced by pumice plains thinly covered in low growing vegetation.

North Fork Toutle River originated from Spirit Lake before the eruption. It now originates from the mountain's crater. The river itself is now laden with sediment due to erosion.

Eruption crater
Original summit
Mt St Helens

PRACTICES   CCC   WEB
 SC 159 **KNOW**

1. (a) Identify the large scale ecosystem change that occurred at Dolly Sods: _____

    _____

    (b) What caused this change in the ecosystem? _____

    _____

    (c) Explain why the ecosystem has not been able to recover quickly after the change: _____

    _____

    _____

    _____

    _____

2. Describe the major change in the ecosystem on Mount St Helens' northern flank after the eruption:

    _____

    _____

    _____

3. Study the eruption schematic on the previous page. Why has the recovery of the ecosystem area shown in light blue (2010) been so much slower than elsewhere?

    _____

    _____

    _____

4. Describe the large scale change that occurred at Coldwater Creek after the eruption. What effect would this have had on the local ecosystem?

    _____

    _____

    _____

5. Describe the large scale change that occurred at Spirit Lake. Explain why this would cause an almost complete change in the lake ecosystem:

    _____

    _____

    _____

    _____

6. (a) What two major changes have occurred in the North Fork Toutle River?

    _____

    _____

    _____

    (b) How might this affect this riverine ecosystem?

    _____

    _____

    _____

    _____

 © 2016 **BIOZONE** International
ISBN: 978-1-927309-46-9
Photocopying Prohibited

160 Global Warming and Ecosystem Change

Key Idea: The long term warming of the Earth's atmosphere will have profound effects on the Earth's various ecosystems, such as sea level change, or rising land temperatures.

The Earth is warming

▶ There is evidence that the Earth's atmosphere is experiencing a period of accelerated warming. Fluctuations in the Earth's surface temperature as a result of climate shifts are normal and the current period of warming climate is partly explained by warming after the end of the last glacial that finished 12,000 years ago.

▶ However since the mid 20th century, the Earth's surface temperature has been increasing. This phenomenon is called **global warming** and the majority of researchers attribute it to the increase in atmospheric levels of CO_2 and other greenhouse gases emitted into the atmosphere as a result of human activity.

▶ Fifteen of the sixteen warmest years on record (since 1880) have been since the year 2000. Global surface temperatures in 2015 set a new record, 0.9°C above the long term average.

World temperatures

Rank	Year	Anomaly °C
1	2015	+0.90
2	2014	+0.74
3	2010	+0.70
4	2013	+0.66
5	2005	+0.65

Global sea level change

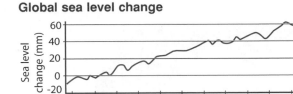

Potential effects of global warming

Hurricane Katrina damage, Mississippi

Rising sea levels: It is predicted that sea levels will rise between 300 - 1200 mm above their current mean level by 2100. The rise is due to thermal expansion (ocean water takes up more space when it is warmer) and because of melting of glaciers and ice shelves. Rising sea level will inundate coastal and low lying ecosystems and increase erosion.

The ice-albedo effect refers to the ability of ice to reflect sunlight. Cooling tends to increase ice cover, so more sunlight is reflected from the surface of the ice. Warming reduces ice cover and more solar energy is absorbed, resulting in more warming. Ice has a stabilizing effect on global climate, reflecting nearly all the sun's energy that hits it.

Weather patterns: Global warming may cause regional changes in weather patterns. High intensity hurricanes occur more frequently now than in the past, driven by higher ocean surface temperatures. Storm surge may become more frequent. A **storm surge** is a rise in sea level occurring during an intense storm. Strong storm winds push the water on shore, causing flooding.

1. What is global warming? _____

   _____

2. What is the major cause of current global warming? _____

   _____

3. How does global warming contribute to rising sea levels? _____

   _____

4. Explain how the level of ice cover can affect global climate: _____

   _____

   _____

   _____

PRACTICES   PRACTICES   CCC   WEB

SC   160   **KNOW**

Case study: Global warming and sea level rise

Low lying wet lands such as the Everglades risk being inundated by seawater. The Everglades (above) contains a variety of ecosystems including mangrove forests, sawgrass marshes, cypress swamps, and pine lands.

Studies have found that the small fish that make up the foraging base of many coastal species do better under less saline conditions. As the saltwater-freshwater interface moves inland, the production of the Everglades will decline.

Peat collapse is another effect of saltwater intrusion. The intrusion of saltwater removes the peat soil. Peat bogs support a variety of freshwater species and the plants and ecosystem they support helps filter the fresh water humans depend on.

This map shows the amount of land that would be inundated around Florida by a 1 m sea level rise, as predicted by 2100.

Florida

Land lost

The Everglades that make up a large part of southern Florida are extremely flat. In fact, moving inland they increase just 5 cm in altitude per kilometer. This means that even minor increases in sea level will have major effects on the wetland ecosystems found there.

▶ The rise in the mean sea level and inundation of the land is only part of the effect of sea level rise. As the sea level rises, ground water is intruded by saltwater, destroying freshwater habitat from underneath. In addition, storm surge becomes more important. As outlying islands, reefs, and other barriers to storm surges become closer to sea level they are less effective at preventing flooding during storms. Storm surges during severe hurricanes can be up to 15 m above the high tide level. The storm surge can therefore cause severe erosion and damage to the land far beyond the mean high tide mark.

▶ Examples of ecosystem change in the Everglades due to sea level change include the movement of mangrove forests approximately 3 km inland since the 1940s and therefore the equivalent reduction in freshwater ecosystems behind the mangroves.

5. (a) What feature of the Florida Everglades make it susceptible to flooding? _____

_____

(b) Describe the effects of sea level rise on the ecosystems of the Everglades: _____

_____

_____

_____

_____

6. In the US the American crocodile is only found in southern Florida (below). They can tolerate high temperatures and brackish water (water that is a mix of sea water and fresh water). The American crocodile is less tolerant of cold than the American alligator with which it currently coexists with in southern Florida.

(a) What could happen to the American crocodile range in Florida if sea levels rise?

_____

_____

_____

(b) How might global warming affect its range overlap with the American alligator?

_____

_____

_____

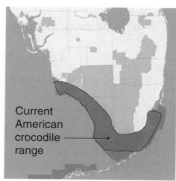

Current American crocodile range

161 Human Impact on Ecosystems

Key Idea: Human activity, either deliberate or accidental, may have an impact on ecosystems. These impacts can be so large they are irreversible.

Human activity can have major effects on ecosystems. These include polluting the air, water, or soil of ecosystems, habitat destruction including deforestation, and removal of important species and introduction of invasive species. There are few places remaining on Earth that are not directly affected by human activities. For most of human civilization these impacts on the environment have had detrimental effects. We are only now, and very slowly, beginning to minimize the environmental impact of human activities.

Detrimental impacts of human activities on ecosystems

Fragmentation of habitats
Building roads and towns in undeveloped habitat separates habitats and makes it difficult for organisms to move between them.

Introduction of new species
Some introductions may have a limited impact, but most are harmful. Introduced species can often become pests as they often have few predators in their new environment. They can quickly spread and cause huge environmental damage.

Eliminating competing species
Often native species compete for the same resources as farmed species, e.g. livestock. Natives are often actively excluded from farmed areas by fences or removal, reducing system biodiversity.

Simplifying natural ecosystems
Plowing diverse grasslands and replanting them with a low diversity pasture affects the interrelationships of thousands of species.

Overharvesting
Overgrazing of native grasslands by livestock or overharvesting of trees from forests or fish from the sea can alter the balance of species in an ecosystem and have unpredictable consequences.

Pollution
Pollution can include the deliberate dumping of waste into the environment or unintended side effects of chemical use, e.g. estrogens in the environment or emission of ozone-depleting gases.

1. Divide your class into groups. As a group discuss each of the following detrimental impacts and suggest one or more ways to reduce the harm caused. Summarize your solutions below. If required, use more paper and attach it to the page:

   (a) Creating a low diversity system: _____

   _____

   _____

   (b) Fragmentation of native habitat: _____

   _____

   (c) Pollution: _____

   _____

   _____

 © 2016 **BIOZONE** International
ISBN: 978-1-927309-46-9
Photocopying Prohibited

PRACTICES   CCC   WEB

SC   161   **KNOW**

162 The Effects of Damming

Key Idea: The increased importance of hydroelectricity and the production of water reservoirs has resulted in the damming of many rivers. This has multiple, ongoing effects on ecosystems.

Rivers are dammed for a variety of reasons including to produce electricity, for flood control, and to provide water for irrigation. Once seen as an environmentally friendly way to produce renewable electricity, the enormous damage dams can do to ecosystems both upstream and downstream of the dam is often realized too late. Most of the world's major rivers are now dammed, many more than once.

Impacts of dams

Reservoir
Displaces communities, inundates and fragments ecosystems.

Dam
Fish migration is blocked, separating spawning waters from rearing waters. Sediment accumulates in the reservoir, reducing fertility of downstream soils and leading to erosion of river deltas, which are deprived of their sediment supply.

Water quality
Water quality can be severely reduced for many years after filling. Aquatic communities in free flowing water are destroyed.

Downstream impact
Disrupted water flow and low water quality near the dam reduces biodiversity.

U.S. Bureau of Reclamation

Order of environmental impact of dams

▶ First order impacts occur immediately and are abiotic (physical). They include barrier effects (blocking water flow), and effects in water quality and sediment load.

▶ Second order effects occur soon after the establishment of the dam. These include effects related to the change in the physical and biotic environments. Rates of primary production change and the morphology of the ecosystem changes (e.g. river valley flooding and erosion).

▶ Third order impacts occur after the change in ecosystem and relate to the establishment of new ecosystem (from a flowing water system to a still water system).

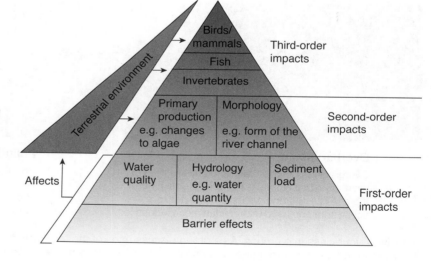

1. Identify three reasons for damming rivers: _____

_____

2. Describe how dams can affect aquatic ecosystems: _____

_____

_____

_____

_____

_____

_____

KNOW

 © 2016 **BIOZONE** International
ISBN: 978-1-927309-46-9
Photocopying Prohibited

The Yangtze River

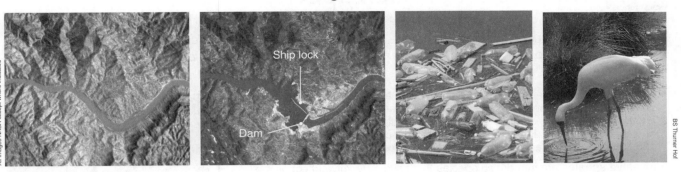

1987   2006   Pollution behind the dam   Siberian Crane

The Three Gorges Dam (above) on the Yangtze river, China, is 2.3 km wide and 101 m high, with a reservoir 660 km long. The construction of the Three Gorges Dam caused the river water level to rise by 100 m. This rise in water level flooded important wetlands where wading birds including the Siberian crane over-wintered. It also flooded 13 cities and hundreds of towns. The waste left behind in the flooded cities and towns continues to pollute the reservoir's waters. Dams reduce flood damage by regulating water flow downstream. However, this also prevents deposition of fertile silts below the dam and increases downstream erosion.

Colorado River

A number of dams are found along the Colorado River, which runs from Colorado through to Mexico. The two largest hydroelectric dams on the river are the Glen Canyon Dam and the Hoover Dam. Both dams control water flow through the Colorado River and were controversial even before their construction.

The construction of Glen Canyon Dam and the Hoover Dam effectively ended the annual flooding of the Colorado River and allowed invasive plants to establish in riparian (river side) zones. The markedly reduced flow rates of the river have also reduced the populations of many fish species downstream.

The once vast Colorado River delta has been reduced to just 5% of its original size due to the construction of various dams. The river itself no longer reaches the sea. The photo above shows the river as it is just 3 km below Morelos Dam, an irrigation diversion dam that takes almost all of the remaining river water.

3. Using the pyramid diagram, explain why the environmental impacts of dams occur in a specific order: _____

4. Describe the effects of the Three Gorges Dam on the ecosystems of the Yangtze River: _____

5. Describe the effects of the dams on the Colorado River on the river's ecosystems: _____

163 The Impact of Alien Species

Key Idea: Introduced species often have no natural predators or diseases in their new environment and so can proliferate without control and go on to cause serious environmental damage.

▶ **Introduced species** are those that have evolved at one place in the world and have been transported by humans, either intentionally or inadvertently, to another region. Some of these introductions are directly beneficial to humans and controlled in their impact, e.g. introduced agricultural plants and animals and Japanese clams and oysters (the mainstays of global shellfish industries).

▶ **Invasive species** are introduced species that have a negative effect on the ecosystems into which they have been imported. Originally these species may have been brought by humans as pets, food, ornamental specimens, or decoration, but have escaped into the wild.

▶ In their new environment, they usually lack natural predators, diseases, or other controls to limit their growth.

▶ Other species may have been accidentally transported in cargo shipments or in the ballast water of ships. Some have been deliberately introduced to control another pest species and have themselves become a problem.

▶ Some of the most destructive of all alien species are aggressive plants, e.g. mile-a-minute weed, a perennial vine from Central and South America, Miconia, a South American tree invading Hawaii and Tahiti, and Caulerpa seaweed, the aquarium strain now found in the Mediterranean. Two introductions, one unintentional and the other deliberate, are described below.

Mile-a-minute weed covering woodland in Maryland USA

Kudzu: A deliberate introduction

Kudzu (*Pueraria lobata*) is a climbing vine native to south-east Asia. It was first introduced to the United States in 1876 and promoted as a forage crop in 1908. Kudzu was widely planted during the Dust Bowl Era to try to conserve soil. In 1940, the government paid farmers almost $20 (about $345 in today's money) a hectare to plant it. However by 1953 the payments had stopped as kudzu escaped farms and invaded woodlands. By 1970, it was declared a weed and in 1997 was placed on the noxious weeds list. Half a billion dollars a year is spent trying to control it. Today, kudzu is estimated to cover 3 million ha of land in the southeastern US.

Red fire ant: An accidental introduction.

Red fire ants (*Solenopsis invicta*) were accidentally introduced into the southeastern states of the United States from South America in the 1920s and have spread north each year. Red fire ants are now resident in 17 US states where they displace populations of native insects and ground-nesting wildlife. They also damage crops and are very aggressive, inflicting a nasty sting. The USDA estimates damage and control costs for red fire ants at more than $6 billion a year.

North America

South America

1. Explain why many alien species become invasive when introduced to a new area: _____

_____

_____

_____

2. As a class, research an alien pest species in your area and discuss the impact it has had on a local ecosystem. Write a short summary of your findings and attach it to this page. You could also complete this question on your own:

164 Human Impact on Fish Stocks

Key Idea: Overfishing has caused the collapse of many fisheries. Unsustainable fishing practices continue throughout the world's oceans.

▶ Fishing is an ancient human tradition. It provides food, and is economically, socially, and culturally important. Today, it is a worldwide resource extraction industry. Decades of overfishing in all of the world's oceans has pushed commercially important species (such as cod, right) into steep decline.

▶ According to the United Nation's Food and Agriculture Organization (FAO) almost half the ocean's commercially targeted marine fish stocks are either heavily or over-exploited. Without drastic changes to the world's fishing operations, many fish stocks will soon be effectively lost.

Grand Banks fishery

Lost fishing gear can entangle all kinds of marine species. This is called **ghost fishing**.

Overfishing has resulted in many fish stocks at historic lows and fishing effort (the effort expended to catch fish) at unprecedented highs.

Huge fishing trawlers are capable of taking enormous amounts of fish at once. Captures of 400 tonnes at once are common.

Bottom trawls and dredges cause large scale physical damage to the seafloor. Non-commercial, bottom-dwelling species in the path of the net can be uprooted, damaged, or killed. An area of 8 million km^2 is bottom trawled annually.

The limited selectivity of fishing gear results in millions of marine organisms being discarded for economic, legal, or personal reasons. These organisms are defined as by-catch and include fish, invertebrates, protected marine mammals, sea turtles, and sea birds. Many of the discarded organisms die. Estimates of the worldwide by-catch is approximately 30 million tonnes per year.

Percentage of catch taken

The single largest fishery is the Northwest Pacific, taking 26% of the total global catch.

Percentage exploitation of fisheries

52% of the world's fished species are already fully exploited. Any increase in catch from these species would result in over-exploitation. 7% of the fish species are already depleted and 17% are over-exploited.

© 2016 **BIOZONE** International
ISBN: 978-1-927309-46-9
Photocopying Prohibited

PRACTICES   CCC   WEB

SC   164   **KNOW**

NOAA

Fishing techniques have become so sophisticated and efforts are on such a large scale that thousands of tonnes of fish can be caught by one vessel on one fishing cruise. Fishing vessels can reach over 100 m long. Here 360 tonnes of Chilean jack mackerel are caught in one gigantic net.

NOAA

Tuna is a popular fish type, commonly found canned in supermarkets and as part of sushi. However virtually all tuna species are either threatened or vulnerable. Demand for the fish appears insatiable with a record price of $1.7 million being paid in 2013 for a 221 kg bluefin tuna.

Derek Quinn

Illegal fishing was a major problem in the 1990s. Thousands of tonnes of catches were being unreported. International efforts have reduced this by an estimated 95%, helping the recovery of some fish stocks. Naval patrol vessels (e.g. HMNZS *Wellington* above) have helped in targeting illegal fishing vessels.

Over-fishing is only one way that humans affect fish populations. The reduction in quality habitats as a result of human activities also has an effect. The bleaching of corals, related to global warming and sea water acidification, has the potential to seriously affect fish populations.

It is estimated 8 million tonnes of plastic finds its way into the sea each year. Plastic can have severe detrimental effects on marine life, especially those that mistake plastic bags for jellyfish or other prey species. Some areas are so polluted it is dangerous to eat fish caught there.

Nicholls H

An example of the effect a fishery can have on fish stocks is the Galápagos Island sea cumber population. The commercial fishery began in 1993 and by 2004 the sea cucumber population had dropped by 98%.

1. Why is overfishing of a fish species not sustainable? _____

_____

_____

_____

2. Why is returning by-catch to the sea not always as useful as it might appear? _____

_____

_____

_____

3. What percentage of commercial fish stocks are fully exploited, overexploited, or already depleted? _____

_____

4. Identify three ways in which fish stocks are over-exploited: _____

_____

_____

_____

_____

_____

© 2016 **BIOZONE** International
ISBN: 978-1-927309-46-9
Photocopying Prohibited

165 Evaluating a Solution to Overfishing

Key Idea: The solutions to overfishing are complex and although some ideas may appear at first to provide a solution, deeper investigation often finds them to be unsustainable.

▶ There is clear evidence that many of the world's most prized fisheries are being overfished. Fisheries such as Atlantic bluefin tuna, North sea cod, and Atlantic salmon have been overfished precisely because they are prized fisheries, the fish are valuable because people like to eat them, and they were once relatively easy to catch.

▶ However as fish stocks have declined people have realized that solutions are needed to ensure the fisheries survive, both for the environment and for the economy.

▶ One of the solutions to reduce the number of fish caught in the wild while increasing the total number produced was to farm them.

▶ Aquaculture includes not just fish farming but the farming of any marine species, including mussels and prawns. It is a highly contentious issue, with arguments both for and against it (below).

Fish farming: A solution to the problem.

▶ With fish stocks plummeting and the world's population ever-increasing the demand for fish has outpaced nature's ability to keep up. Fish farming provides a way to produce fish and other marine products without the need to plunder wild populations.

▶ For example, the Atlantic salmon has been so overfished that wild caught salmon is no longer a viable fishery. Instead, Atlantic salmon are farmed in large sea pens.

▶ Around 1.5 million tonnes of salmon are produced from farms a year, compared to less than three thousand tonnes caught in wild. Returns like these from commercial fish farms help reduce the need to catch wild fish and so allow wild fish stocks to recover.

▶ Importantly, fish farming does not have to rely on specific locations. Atlantic salmon, native to the North Atlantic, is farmed in Chile, Canada, Norway, Russia, the UK, and Australia. Similarly Chinook or King salmon, native to the North Pacific, is farmed in New Zealand and Chile. New Zealand is the world's largest producer of Chinook salmon, producing more farmed salmon than the rest of the world's entire catch.

▶ Fish farming is highly efficient. In the wild, a salmon might need 10 kg of food for every 1 kg of body weight. Farmed fish require just 4 kg of food per kg of body weight. Much of this food includes fish meal.

▶ The increase in fish farming has not caused an increase in the catch of fish for fish meal. Instead, the fish meal required for fish farming has come from fish meal that was once fed to livestock such as pigs and poultry, which now use other feeds types such as grain.

Fish farming: An unsustainable disaster.

▶ Most fish farming is carried out in sea cages rather than containment facilities. This means waste from the fish farm enters the local ecosystem directly.

▶ Fish in the pens can be subject to high stocking rates and thus produce a large and concentrated amount of waste, including feces and food waste. In sites without adequate currents these wastes can build up in the area and not only pollute the water but cause disease in the farmed fish and in local populations of marine animals.

▶ Disease can be a problem in such concentrated cages, quickly transferring between individuals. Various anitfouling, antibacterial, and antiparasitic chemicals are used to reduce disease.

▶ In areas without adequate currents, heavy metals may build up in the environment and the fish, creating a hazard for human consumption. Escapes from the cages often occur, especially in high seas. If the fish being farmed are not native, they can compete with native fish. Because farmed fish have lower genetic diversity than wild fish, a large escape could decrease the overall genetic diversity of a wild population.

▶ Feeding the fish requires fish meal and other products in order to produce fish high in omega-3 fats (not made by fish but taken in with their normal diet). 50% of the world's fish oil production is fed to farmed fish.

▶ Many of the farmed fish species are carnivorous and so are fed fish meal. However this itself comes from fishing and a large percentage of the commercial fishing is dedicated to catching bait fish such as sardines specifically for fish meal. This has the potential to drive these fish to extinction.

1. Use the articles above and your own research to evaluate fish farming as a solution to overfishing. Write a short answer to the question "Can fish farming stop overfishing and produce a sustainable resource?" Include reasons and evidence to justify your answer:

_____

_____

_____

_____

_____

_____

ETS   CCC   WEB

SC   165   **KNOW**

166 Deforestation and Species Survival

Key Idea: Deforestation is the permanent removal of forest from an area. Causes of deforestation include land development for farming and plantations.

Deforestation

▶ At the end of the last glacial period, about 10,000 years ago, forests covered an estimated 6 billion hectares, about 45% of the Earth's land surface. Forests currently cover about 4 billion hectares of land (31% of Earth's surface). They include the cooler temperate forests of North and South America, Europe, China and Australasia, and the tropical forests of equatorial regions. Over the last 5000 years, the loss of forest cover is estimated at 1.8 billion hectares. 5.2 million hectares has been lost in the last 10 years alone. Temperate regions where human civilizations have historically existed the longest (e.g. Europe) have suffered the most but now the vast majority of deforestation is occurring in the tropics. Intensive clearance of forests during settlement of the most recently discovered lands has extensively altered their landscapes (e.g. in New Zealand, 75% of the original forest was lost in a few hundred years).

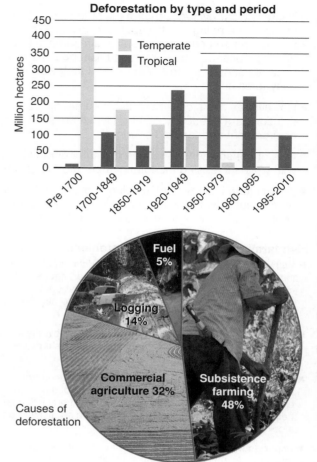

Deforestation by type and period

Y-axis: Million hectares (0 to 450)

Legend: Temperate, Tropical

X-axis periods: Pre 1700, 1700-1849, 1850-1919, 1920-1949, 1950-1979, 1980-1995, 1995-2010

Causes of deforestation

▶ **Deforestation** is the end result of many interrelated causes, which often center around socioeconomic drivers. In many tropical regions, the vast majority of deforestation is the result of subsistence farming. Poverty and a lack of secure land can be partially solved by clearing small areas of forest and producing family plots. However huge areas of forests have been cleared for agriculture, including ranching and production of palm oil plantations. These produce revenue for governments through taxes and permits, producing an incentive to clear more forest. Just 14% of deforestation is attributable to commercial logging (although combined with illegal logging it may be higher).

Causes of deforestation:
- Fuel 5%
- Logging 14%
- Commercial agriculture 32%
- Subsistence farming 48%

Tropical deforestation globally

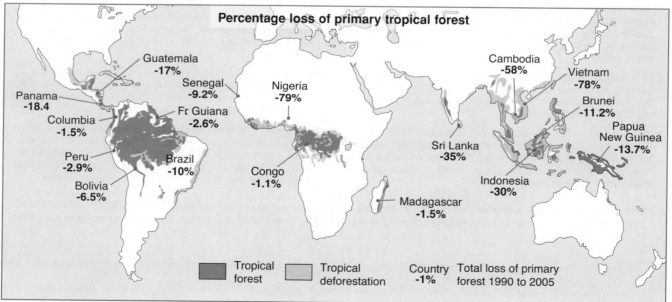

Percentage loss of primary tropical forest

- Guatemala -17%
- Senegal -9.2%
- Panama -18.4
- Columbia -1.5%
- Fr Guiana -2.6%
- Peru -2.9%
- Brazil -10%
- Bolivia -6.5%
- Nigeria -79%
- Congo -1.1%
- Madagascar -1.5%
- Cambodia -58%
- Vietnam -78%
- Brunei -11.2%
- Papua New Guinea -13.7%
- Sri Lanka -35%
- Indonesia -30%

Legend: Tropical forest | Tropical deforestation | Country -1% | Total loss of primary forest 1990 to 2005

▶ It is important to distinguish between deforestation involving primary (old growth) forest and deforestation in plantation forests. Plantations are regularly cut down and replaced and can artificially inflate a country's apparent forest cover or rate of deforestation. The loss of primary forests is far more important as these are refuges of high biodiversity, including for rare species, many of which are endemic to relatively small geographical regions (i.e. they are found nowhere else).

▶ Although temperate deforestation is still a concern, it is in equatorial regions that the pace of deforestation is accelerating (above). This is of global concern as species diversity is highest in the tropics and habitat loss puts a great number of species at risk.

© 2016 **BIOZONE** International
ISBN: 978-1-927309-46-9
Photocopying Prohibited

The Pacific Northwest and the spotted owl

▶ The forests of the Pacific Northwest of the USA include the primary forests of Oregon and Washington states. They include stands of redwood, Douglas-fir, western red cedar, and shore pine as well as alder and maple. The region is home to hundreds of species of wildlife, many dependent on the primary (old-growth) forest. One of these, the northern spotted owl, is a keystone species and an important indicator of healthy old growth forest. Listed as threatened under the Endangered Species Act, it was first protected in 1990 after nearly a century of old growth logging. In 1994, in response to the owl's status, the Northwest Forest Plan (NWFP) was implemented to govern land use on federal lands in the Pacific Northwest. In particular it called for extensive old growth reserve lands and restrictions on old growth logging.

▶ The northern spotted owl ranges over wide areas of the Pacific Northwest. It requires forests with a dense canopy of mature and old-growth trees, abundant logs, standing snags, and live trees with broken tops. Only old growth forests offer these characteristics. When forced to occupy smaller areas of less suitable habitat the birds are more susceptible to starvation, predation, and competition.

▶ Although the plight of the northern spotted owl triggered the NWFP, it is the fate of the old growth forests, the species they support, and the ecosystem services they provide that are at stake. Despite protections, the number of northern spotted owls is declining at three times the predicted rate and it is increasingly evident that competition from barred owls has a role in this.

US Fish and Wildlife Service

1. Describe the trend in temperate and tropical deforestation over the last 300 years: _____

_____

2. What are some of the causes of deforestation? _____

_____

_____

_____

3. Why is it important to distinguish between total forest loss and loss of primary forests when talking about deforestation?

_____

_____

_____

_____

4. Deforestation in temperate regions has largely stabilized and there has been substantial forest regrowth. However, these second growth forests differ in structure and composition to the forests that were lost. Why might this be of concern?

_____

_____

_____

5. **"The Pacific Northwest old growth forest should be fully protected and not logged."**
Divide your class into two groups, with one group arguing in support of this statement and one group arguing against it. Each group should present their arguments to the class as a whole. Do your own research or visit the weblinks (1-6) for this activity for more information. Present your arguments as a report and summarize the main points in the space below.

_____

_____

_____

_____

_____

© 2016 **BIOZONE** International
ISBN: 978-1-927309-46-9

167 Modeling a Solution

Key Idea: Conservation efforts are often a compromise between environmental, economic, and cultural needs.

▶ Deciding on a course of action for preserving biodiversity is not always simple. Environmental, cultural, and economic impacts must be taken into account, and compromises must often be made.

▶ The map below shows a hypothetical area of 9,300 ha (93 km²) in which two separate populations of an endangered bird species exist within a forested area of public land. A proposal to turn part of the area into a wildlife reserve has been put forward by local conservation groups. However, the area is known to have large deposits of economically viable minerals and is frequented by trampers. Hunters also spend time in the area because part of it has an established population of introduced game animals. The proposal would allow a single area of up to 1,500 ha (15 km²) to be reserved exclusively for conservation efforts.

1. Study the map below and draw on to the map where you would place the proposed reserve, taking into account economic, cultural, and environmental values. On a separate sheet, write a report justifying your decision as to where you placed the proposed reserve.

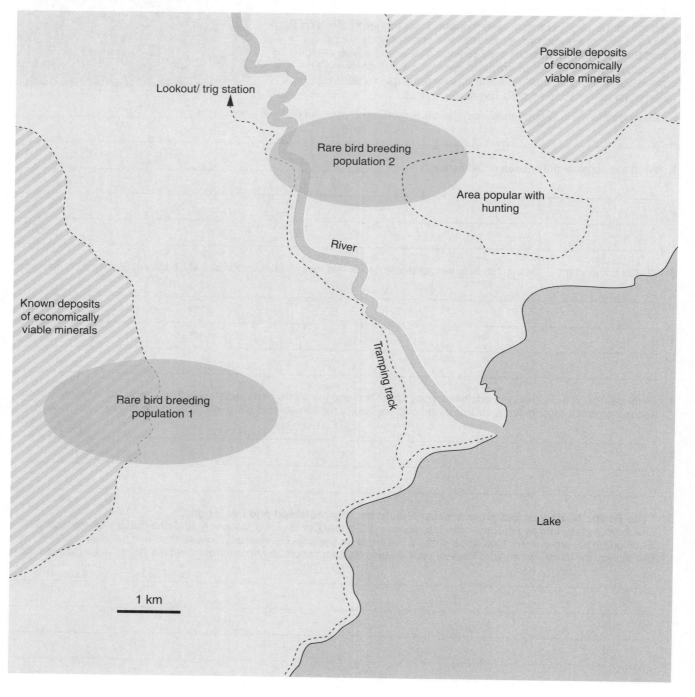

© 2016 **BIOZONE** International
ISBN: 978-1-927309-46-9
Photocopying Prohibited

168 Chapter Review

Summarize what you know about this topic under the headings provided. You can draw diagrams or mind maps, or write short notes to organize your thoughts. Use the points in the introduction and the hints to help you:

Human impacts on ecosystems

HINT: How are the resilience and sustainability of ecosystems affected by human activities?

Ecosystem change and resilience

HINT: How do ecosystems change and to what extent do the organisms in them help them recover from change?

REVISE

169 Summative Assessment

Change in winter destination, 17 US bird species

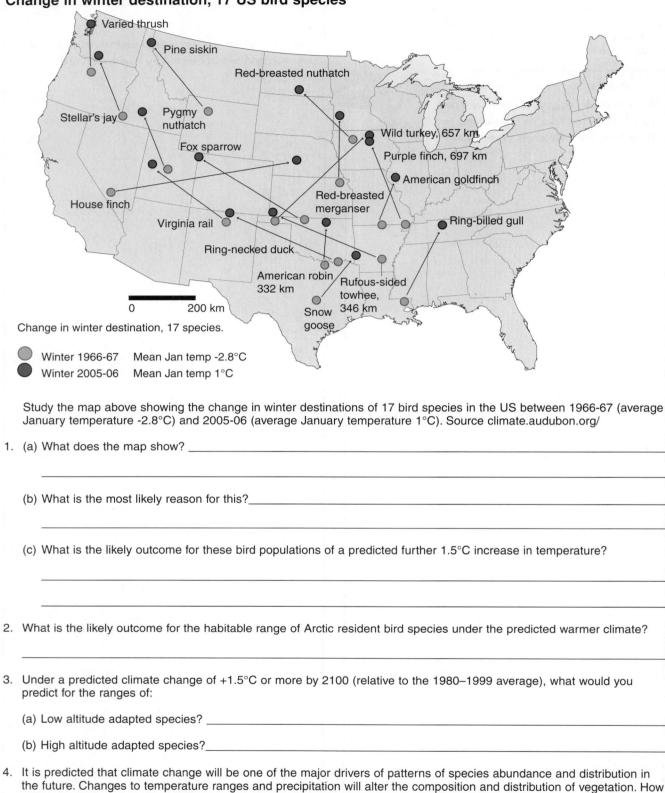

Change in winter destination, 17 species.

○ Winter 1966-67   Mean Jan temp -2.8°C
● Winter 2005-06   Mean Jan temp 1°C

Study the map above showing the change in winter destinations of 17 bird species in the US between 1966-67 (average January temperature -2.8°C) and 2005-06 (average January temperature 1°C). Source climate.audubon.org/

1. (a) What does the map show? _____

_____

(b) What is the most likely reason for this?_____

_____

(c) What is the likely outcome for these bird populations of a predicted further 1.5°C increase in temperature?

_____

_____

2. What is the likely outcome for the habitable range of Arctic resident bird species under the predicted warmer climate?

_____

3. Under a predicted climate change of +1.5°C or more by 2100 (relative to the 1980–1999 average), what would you predict for the ranges of:

(a) Low altitude adapted species? _____

(b) High altitude adapted species?_____

4. It is predicted that climate change will be one of the major drivers of patterns of species abundance and distribution in the future. Changes to temperature ranges and precipitation will alter the composition and distribution of vegetation. How might changes in vegetation patterns affect the distribution and abundance of animal species?

_____

_____

5. Species do not respond to climate change equally. On a separate sheet of paper, discuss some of the likely consequences of climate change to the composition and stability of ecosystems. What ecosystems do you think are most at risk and why? Attach your response to this page.

Key terms

altruism

cooperative behavior

flocking

herding

kin selection

migration

schooling

social group

Disciplinary core ideas

Show understanding of these core ideas

Activity number

Social interactions and group behavior increase the chances of survival

☐ 1 The organization of a group into a social structure improves the chances of survival and reproduction for members of the group. 170

☐ 2 Animals are classed as solitary, grouped (but without any social order), or grouped in close social groups. Social organization has both advantages and disadvantages. 170

☐ 3 Group behavior includes remaining as a group (e.g. schooling, herding, flocking), group attack and group defense. 171 172

☐ 4 Schooling, herding, and flocking provide benefits to the group. These include increased protection from predators and improved foraging and efficiency of locomotion (e.g. during long distance flight). 171 172

☐ 5 Social animals organize themselves in a way that divides resources and roles between members of the social group. Social behavior evolves because, on average, it improves the survival of individuals. 173 174

☐ 6 Cooperative behavior involves two or more individuals working together to achieve a common goal. Altruism and kin selection are examples of cooperative behavior. 175

☐ 7 Working together in defense and attack can help increase or maintain resources for the group, and increase the survival chances of individuals. 176 177

☐ 8 Cooperative behavior in gathering food increases the chances of foraging success and improves efficiencies. 178

Crosscutting concepts

Understand how these fundamental concepts link different topics

Activity number

☐ 1 **CE** ▶ Empirical evidence enables us to distinguish between cause and correlation and support claims about the benefits of group behavior to survival and reproduction. 171 172 174 176 177

Science and engineering practices

Demonstrate competence in these science and engineering practices

Activity number

☐ 1 Evaluate the evidence for how group behavior benefits survival during migration. 172

☐ 2 Evaluate the evidence for how the degree of cooperation (help given) can depend on relatedness of the individuals involved. 174

☐ 3 Evaluate the evidence for the benefits of group behaviors such as flocking. 171 172

☐ 4 Evaluate the evidence for the benefits to survival and reproduction of cooperative behaviors in hunting, foraging, or group defense. 175 176 177 178

170 Social Groupings

Key Idea: Animals may be solitary, form loosely associated groups, or form complex groups with clear social structures. Each behavior has its advantages and disadvantages.

▶ No animal lives completely alone. At some stage in their lives all animals must interact with others of their species (e.g. to reproduce or through competitive interactions for food or resources).

▶ Generally animals are classed as solitary, grouped together (but without any social order), or grouped together in close social groups.

Solitary animal

Non-social groups

Social groups

Solitary animals spend the majority of their lives alone, often in defended territories. They may only seek out others of their species for breeding. Offspring are often driven away shortly after they become independent.

Solitary life is often an advantage when resources are scarce or scattered over a large area. Solitary animals include many of the cat family e.g. tiger (above), bears, and various invertebrates.

Many animals form loose associations but do not interact socially. Each animal is acting directly for its own benefit with little or no direct cooperation between them. Schools of fish, flocks of birds and many herding mammals exhibit this non-social grouping.

Non-social groups provide protection from predators by reducing the possibility of being preyed upon individually. There may also be benefits during feeding and moving.

Primates form complex social structures which are usually based around a family group. Some animals that form social groups also form dominance hierarchies.

Dominance hierarchies help distribute resources and maintain social structure. In some species members of the group are divided into castes with specialized roles (e.g. ants and bees). Some produce offspring or help raise young, others may be workers or help with defense of the colony.

Advantages of large social groupings

1. Protection from physical factors and predators.
2. Assembly for mate selection.
3. Locating and obtaining food.
4. Defense of resources against other groups.
5. Division of labor amongst specialists.
6. Richer learning environment.
7. Population regulation (e.g. breeding restricted to a dominant pair).

Disadvantages of large social groups

1. Increased competition between group members for resources as group size increases.
2. Increased chance of the spread of diseases and parasites.
3. Interference with reproduction, e.g. infanticide by non-parents or cheating in parental care (so that non-parents may unknowingly raise another's offspring).

1. Give one advantage and one disadvantage of solitary living: _____

_____

2. Explain why group behavior, such as schooling, is more about individual advantage than group advantage:

_____

_____

_____

_____

3. Give one advantage and one disadvantage of a dominance hierarchy: _____

_____

_____

© 2016 **BIOZONE** International
ISBN: 978-1-927309-46-9
Photocopying Prohibited

171 Schooling, Flocking, Herding

Key Idea: Being part of a group enhances survival by providing protection from predators and by reducing energy expenditure during movement.

Dynamics in a flock, school, or herd

▶ **Schooling** by fish, **flocking** by birds, and **herding** by grazing mammals are essentially all the same behavior. Each individual is behaving in a way that helps its own survival regardless of the others within that group. Within the group, the application of a few simple rules results in apparently complex behavior.

▶ In a school, flock, or herd, three simple rules tend to apply:

- Move towards the group or others in the group.
- Avoid collision with others in the group or external objects.
- Align your movement with the movement of the others in the group.

▶ If every individual moves according to these rules, the school, flock, or herd will stay as a dynamic cohesive unit, changing according to the movement of others and the cues from the environment.

Flock of auklets (a small seabird)
D. Dibenski

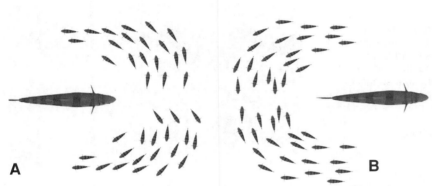

A **B**

Schooling dynamics: In schooling fish or flocking birds, every individual behaves according to a set of rules. In diagram A, each fish moves away from the predator while remaining close to each other. The school splits to avoid the predator (A), before moving close together again behind the predator (B).

In a **flash expansion**, each individual moves directly away from the predator. Collisions have never been observed, suggesting each fish is able to sense the direction of movement of the fish next to it.

Why do fish school?

▶ Schooling in predatory fish may enhance the ability of any individual to catch its prey. If the prey avoids one predator, it may get caught by the next predator. Prey fish may school for defensive reasons.

Advantages of schooling

▶ **Avoidance of predators**

- Confusion caused by the movement of the school.
- Protection by reduced probability of individual capture.
- Predator satiation (more than the predator could eat)
- Better predator detection (the many eyes effect)

▶ **Better hydrodynamics** within the school, so less energy is expended in swimming.

School of jacks

1. How do the rules for flocking and schooling help to maintain a cohesive group?

_____

_____

_____

_____

_____

_____

 © 2016 **BIOZONE** International
ISBN: 978-1-927309-46-9
Photocopying Prohibited

PRACTICES   CCC   WEB

CE   171   **KNOW**

Flocking in birds

▶ Flocking in birds follows similar rules to schooling. In flight, each individual maintains a constant distance from others and keeps flying in the average direction of the group. Flocks can be very large, with thousands of birds flying together as a loosely organized unit, such as starlings flocking in evenings or queleas, flocking over feeding or watering sites (photo A, right).

▶ Migrating flocks generally fly in a V formation (photo B, right). The V formation provides the best aerodynamics for all in the flock except the leader. Each bird gains lift from the movement of the air caused by the bird ahead of it.

Research on great white pelicans has shown that the V formation helps the birds conserve energy. As the wing moves down, air rushes from underneath the wing to above it, causing an upward moving vortex behind the wing. This provides lift to the bird flying behind and to one side, requiring it to use less effort to maintain lift. Energy savings come from increased gliding.

The wingtip vortex provides lift for the trailing bird.

A: Flocking queleas (small weaver birds)

Alastair Rae cc2.0

B: Migrating geese

Herding in mammals

▶ Herding is common in hoofed mammals, especially those on grass plains like the African savanna. A herd provides protection because while one animal has its head down feeding, another will have its head up looking for predators. In this way, each individual benefits from a continual supply of lookouts (the many eyes principle). During an attack, individuals move closer to the center of the herd, away from the predator, as those on the outside are more frequently captured. The herd moves as one group, driven by individual needs.

2. Identify some benefits of schooling to a:

   (a) Predatory fish species: _____

   (b) Prey fish species: _____

   _____

3. (a) How does flying in V formation help a bird during migration? _____

   _____

   _____

   _____

   _____

   (b) How would this help survival in the longer term? _____

   _____

   _____

   _____

   _____

   _____

4. Explain how herding in grazing mammals provides a survival advantage: _____

   _____

   _____

   _____

   _____

172 Migration

Key Idea: Migration is an energy expensive behavior. Grouping together reduces the energy used and improves survival of migrating individuals.

Migration is the long distance movement of individuals from one place to another. Migration usually occurs on a seasonal basis and for a specific purpose, e.g. feeding, breeding, or over-wintering.

▶ In order for migratory behavior to evolve, the advantages of migration must outweigh the disadvantages. Migration is an energy expensive behavior, and animals must spend a lot of time building energy stores that will fuel the effort. The destination provides enough food or shelter to enhance survival of individuals and their offspring.

▶ Although some animals migrate individually, many migrate in large groups (right).

Benefits of group migration

▶ Group migration helps navigation in what is called the "many wrongs principle" in which the combining of many inaccurate navigational compasses produces a more accurate single compass. Thus, if an animal navigates by itself with a slightly inaccurate internal compass, or inaccurately interprets environmental cues, it may arrive in the wrong location. In a group, each member can adjust its heading according to the movement of the others, thus an average direction is produced and each member is more likely to arrive in the correct place (right).

▶ Birds flying together increase aerodynamic efficiency to each other, saving energy. In schooling fish, individuals in the center of the school use less effort for movement. Flocking and schooling also provide feeding benefits and decreased risk of predation along the migration route.

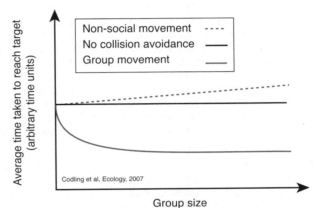

Non-social movement ----
No collision avoidance ——
Group movement ——

Average time taken to reach target (arbitrary time units)

Group size

Codling et al, Ecology, 2007

Increasing group size decreases the time taken to reach a navigational target when the group is moving as a social unit. Non-social groups take longer with increasing size because of the need to avoid others in the group.

1. What is migration? _____

_____

_____

2. Describe an advantage and a disadvantage of migration: _____

_____

_____

_____

_____

3. (a) How does grouping together increase navigational efficiency? _____

_____

_____

_____

(b) How does this enhance individual survival? _____

_____

_____

PRACTICES   CCC   WEB

 CE 172 KNOW

173 Social Organization

Key Idea: In social groups, members of the group interact regularly. Social species organize themselves in a way that divides resources and roles between group members.

Eusocial animals

▶ Eusocial animals are those in which a single female produces the offspring and non-reproductive individuals care for the young. They have the highest form of social organization. Individuals are divided into castes that carry out specific roles. In most cases there is a queen which produces the young and members of the group are normally directly related to the queen. Non-reproductive members of the group may be involved in care of the young or defense of the nest site. Examples include ants, honey bees, termites, and naked mole rats.

Honeybees   Termites

Presocial animals

▶ Presocial animals exhibit more than just sexual interactions with members of the same species, but do not have all of the characteristic of eusocial animals. They may live in large groups based around a single breeding pair. Offspring may be looked after by relations (e.g. aunts/older siblings). These groups often form hierarchies where the breeding pair are the most dominant. There may also be separate hierarchies for male and female members of the group. Examples include canine species that live in packs (e.g. wolves), many primates, and some bird species.

▶ The number of males in a social group varies between species. Wild horse herds have a single stallion, which controls a group of mares. Young males are driven away when they are old enough. Female elephants and their offspring form small groups lead by the eldest female (the matriarch). Adult male elephants only visit the group during the reproductive season.

1. (a) Identify two examples of eusocial animals: _____

_____

(b) Describe the organization of a eusocial animal group: _____

_____

_____

_____

2. What is the difference between eusocial and presocial groups? _____

_____

_____

_____

3. In eusocial animals, worker and soldier castes never breed but are normally all genetically related. How might their contribution to the group help pass their own genes to the next generation (ensure the survival or their own genes)?

_____

_____

_____

_____

 © 2016 **BIOZONE** International
ISBN: 978-1-927309-46-9
Photocopying Prohibited

174 How Social Behavior Improves Survival

Key Idea: By working together (directly or indirectly) members of a group increase each other's chances of survival. The level of help given depends on the level of relatedness.

▶ Living in a group can improve the survival of the members, e.g. improving foraging success or decreasing the chances of predation. Animals such as meerkats, ground squirrels, and prairie dogs decrease the chances of predation by using sentries, which produce alarm calls when a predator approaches.

Gunnison's prairie dogs

Gunnison's prairie dogs (right) live in large communities called towns in the grasslands of western North America. The towns are divided into territories which may include up to 20 individuals. During their foraging, above-ground individuals may produce alarm calls if a predator approaches, at which nearby prairie dogs will take cover. However, whether or not an alarm call is given depends on the relatedness of the individuals receiving the call to the individual giving it. Gunnison's prairie dogs put themselves at risk when giving an alarm call by attracting the attention of the predator. Apparently altruistic (self-sacrificing) behavior involving close relatives is called **kin selection**.

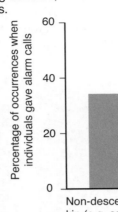

Bar graph: Percentage of occurrences when individuals gave alarm calls (y-axis, 0–60). Categories (x-axis): Non-descendant kin (e.g. cousins), Offspring, Parent / siblings, No known kin.

White fronted bee-eaters

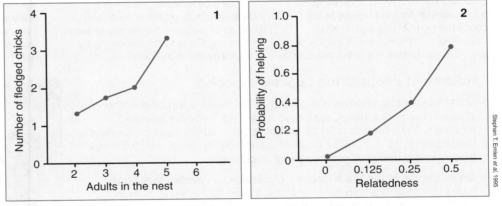

Graph 1: Number of fledged chicks (y-axis, 0–4) vs Adults in the nest (x-axis, 2–6).

Graph 2: Probability of helping (y-axis, 0–1.0) vs Relatedness (x-axis, 0, 0.125, 0.25, 0.5).

Stephen T. Emlen *et al.* 1995

White fronted bee-eaters (left) live in family groups which include a breeding pair and non-breeding pairs. All adults help provide for the chicks. Graph 1 shows the relationship between the number of adults in the nest and the number of chicks fledged. Graph 2 shows how relatedness affects the amount of help the pairs give the chicks.

1. Use an example to explain how living in a group improves survival: _____

   _____

   _____

   _____

2. (a) The level of help between group members often depends on relatedness. Using the examples above, explain how relatedness to the helper affects the level of help given:

   _____

   _____

   _____

   (b) With respect to this, what is unexpected about the prairie dog data? _____

   _____

   _____

 © 2016 **BIOZONE** International
ISBN: 978-1-927309-46-9
Photocopying Prohibited

PRACTICES   CCC   WEB

CE   174   **KNOW**

175 Cooperative Behaviors

Key Idea: Cooperative behavior is where two or more individuals work together to achieve a common goal. It increases the probability of survival for all individuals involved.

▶ **Cooperative behavior** involves behavior in which two or more individuals work together to achieve a common goal such as defense, food acquisition, or rearing young. Examples include hunting as a team (e.g. wolf packs, chimpanzee hunts), responding to the actions of others with same goal (e.g. migrating mammals), or acting to benefit others (e.g. mobbing in small birds). Cooperation occurs most often between members of the same species.

▶ Altruism is an extreme form of cooperative behavior in which one individual disadvantages itself for the benefit of another. Altruism is often seen in highly social animal groups. Most often the individual who is disadvantaged receives benefit in some non-material form (e.g. increased probability of passing genes onto the next generation).

Coordinated behavior is used by many social animals for the purpose of both attack (group hunting) and defense. Cooperation improves the likelihood of a successful outcome, e.g. a successful kill.

Animals may move en masse in a coordinated way and with a common goal, as in the mass migrations of large herbivores. Risks to the individual are reduced by the group behavior.

Kin selection is altruistic behavior towards relatives. In meerkats, individuals from earlier litters remain in the colony to care for new pups instead of breeding themselves. They help more often when more closely related.

Evidence of cooperation between species

▶ Many small birds species will cooperate to attack a larger predatory species, such as a hawk, and drive it off. This behavior is called mobbing. It is accompanied by mobbing calls, which can communicate the presence of a predator to other vulnerable species, which benefit from and will become involved in the mobbing.

▶ One example is the black-capped chickadee, a species that often forms mixed flocks with other species. When its mobbing calls in response to a screech owl were played back, at least ten other species of small bird were attracted to the area and displayed various degrees of mobbing behaviour. The interspecific communication helps to coordinate the community anti-predator mobbing behavior.

1. (a) What is altruism?_____

_____

(b) Why would altruism be more common when individuals are related? _____

_____

_____

2. How do cooperative interactions enhance the survival of both individuals and the group they are part of? _____

_____

_____

3. What evidence is there that unrelated species can act cooperatively? Why would they do this?_____

_____

_____

_____

© 2016 **BIOZONE** International
ISBN: 978-1-927309-46-9
Photocopying Prohibited

176 Cooperative Defense

Key Idea: Working together in defense decreases the risk to individuals and increases the chances of a successful defense.

▶ Group defense is a key strategy for survival in social or herding mammals. Forming groups during an attack by a predator decreases the chances of being singled out, while increasing the chances of a successful defense.

Group defense in musk oxen

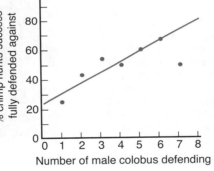

Young safely protected in center

Attack pattern by wolves

Circular defense with heads facing outwards

In the Siberian steppes, which are extensive grasslands, musk oxen must find novel ways to protect themselves from predators. There is often no natural cover, so they must make their own barrier in the form of a defensive circle. When wolves (their most common predator) attack, they shield the young inside the circle. Lone animals have little chance of surviving an attack as wolves hunt in packs.

Red colobus monkey defense

Red colobus monkeys are a common target during chimpanzee hunts. They counter these attacks by fleeing (especially females with young), hiding, or mounting a group defense. The group defense is usually the job of the males and the more defenders there are the greater the likelihood of the defense being successful.

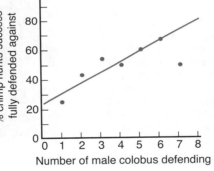

% chimp hunts success-fully defended against

Number of male colobus defending

Japanese honeybees

Japanese honeybees are often attacked by the Asian giant hornet. When a hornet scout enters the honeybee hive the honeybees mob it with more than 100 bees, forming a bee-ball. The center of the ball can reach 50°C, literally baking the scout to death.

A bee-ball formed around a hornet

1. Describe two benefits of cooperative defense: _____

   _____

   _____

2. How many colobus males are needed to effectively guarantee a successful defense against chimpanzees?

   _____

3. Sheep need to spend most of their day feeding on grass. They form mobs both naturally in the wild as well as on farms.

   (a) Explain why sheep form mobs: _____

   _____

   _____

   (b) Explain how this might enhance an individual sheep's ability to feed: _____

   _____

   _____

PRACTICES   CCC   WEB

CE   176   **KNOW**

177 Cooperative Attack

Key Idea: Working together in attack can help increase the chance of success especially if roles are distributed between the attacking members.

Adrian A. Smith

Asian giant hornet
Yasunori Koide

Lionesses hunt as a coordinated group. Several lionesses hide downwind of the prey, while others circle upwind and stampede the prey towards the lionesses in wait. Group cooperation reduces the risk of injury and increases the chance of a kill. Only 15% of hunts by a solitary lioness are successful. Those hunting in a group are successful 40% of the time.

Some ant species, known as **slavemaker ants**, raid other ant nests (called slave-raiding), killing workers and capturing grubs. The grubs are carried back to the home nest where they grow and tend the slavemaker ants' own young. Sometimes, however, the slaves rebel and can destroy the slavemaker nest.

Wasps and some species of bees (e.g. Africanized honeybees) can be extremely aggressive, although this aggressive behavior is an extension of the defense of the hive. The Asian giant hornet often attacks honeybee nests to collect grubs to feed their own young. An attack by 50 giant hornets can destroy an entire honeybee colony within a few hours.

The Gombe Chimpanzee War

Group attacks between members of the same species and even the same social groups do occur. They usually involve disputes over resources or territory, but may be due simply to rifts in social groups. One of the most well recorded and startling examples of group fighting is the Gombe Chimpanzee War. Observed by Jane Goodall, the violence began in 1974, after a split in a group of chimpanzees in the Gombe Stream National Park, in Tanzania. The group divided into two, the Kasakela in the northern part of the former territory and the Kahama in the south. Over the course of four years the Kasakela systematically destroyed the Kahama, killing all six males and one female and kidnapping three more females. The Kasakela then took over the Kahama territory. However, ironically, the territorial gains made by the Kasakela were quickly lost as their new territory bordered a larger more powerful group of chimpanzees, the Kalande. After a few violent skirmishes along this border, the Kasakela were pushed back into their former territory.

1. (a) Suggest two reasons for cooperative attacks: _____

_____

_____

(b) Suggest why cooperative attacks are more likely to be successful than individual attacks:

_____

_____

_____

_____

2. Use the data below to draw a graph of the hunting success of chimpanzees:

No. hunters	1	2	3	4	5	6
Hunt success (%)	13	29	49	72	75	42

© 2016 **BIOZONE** International
ISBN: 978-1-927309-46-9
Photocopying Prohibited

178 Cooperative Food Gathering

Key Idea: Cooperative behavior in gathering food increases the chances of foraging success and improves efficiencies.

▶ Cooperating to gather food can be much more efficient that finding it alone. It increases the chances of finding food or capturing prey.

▶ Cooperative hunting will evolve in a species if the following circumstances apply:

- If there is a sustained benefit to the hunting participants
- If the benefit for a single hunter is less than that of the benefit of hunting in a group
- Cooperation within the group is guaranteed

Worker castes in army ants

Orcas hunting

Wild beehive

Cooperative food gathering in ants often involves division of labor. Leaf-cutter ants harvest parts of leaves and use them to cultivate a fungus, which they eat. Workers that tend the fungus gardens have smaller heads than the foragers, which cut and transport the leaves. Similarly, army ants have several distinct worker castes. The smaller castes collect small prey, and larger porter ants collect larger prey. The largest workers defend the nest.

Dolphins herd fish into shallow water and trap them against the shore where they can be easily caught. When a pod of orcas (killer whales) spot a seal on an iceberg they swim towards the flow at high speed before ducking under the ice. This causes a large wave to wash over the iceberg, knocking the seal into the sea where it can be captured. If the seal fails to fall into the water, one of the whales will land itself onto the iceberg tipping the seal into the water.

Sometimes different species work together to gather food. The greater honeyguide, a bird that is found in the sub-Sahara, is notable for its behavior of guiding humans (either deliberately or not) to wild beehives. When the humans retrieve the honey, they leave behind some of the wax comb which the honeyguide eats. It is not known why this behavior evolved, because the honeyguide can enter a beehive without human help.

Army ants foraging

▶ There are two species of army ant that have quite different raiding patterns (below): *Eciton hamatum* whose columns go in many directions and *Eciton burchelli*, which is a swarm-raider, forming a broad front. Both species cache food at various points along the way.

▶ Through group cooperation, the tiny ants are able to subdue prey much larger than themselves, even managing to kill and devour animals such as lizards and small mammals. This would not be possible if they hunted as individuals.

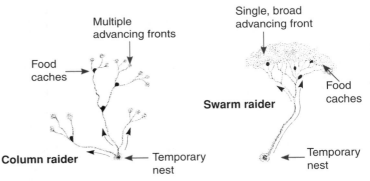

1. Using examples, describe how cooperative food gathering provides an advantage to survival or reproduction: _____

_____

_____

_____

_____

2. What conditions favor cooperative food gathering? _____

_____

_____

_____

PRACTICES WEB

178 KNOW

179 Chapter Review

Summarize what you know about this topic so far under the headings provided. You can draw diagrams or mind maps, or write short notes to organize your thoughts. Use the checklist in the introduction and the hints to help you:

Migration, flocking, schooling and herding

HINT: Include definitions for each, explain why these behaviors occur, and their benefits to survival and reproduction.

Social organization and cooperative behaviors

HINT: How does social behavior enhance survival?

© 2016 **BIOZONE** International
ISBN: 978-1-927309-46-9
Photocopying Prohibited

180 KEY TERMS AND IDEAS: Did You Get It?

1. Test your vocabulary by matching each term to its definition, as identified by its preceding letter code.

altruism ....

cooperative behavior ....

flocking ....

herding ....

kin selection ....

migration ....

schooling ....

social group ....

A A behavior in which individuals work together in order to reach a common goal, e.g. finding food.

B The long distance movement of individuals from one place to another.

C A behavior seen in hoofed mammals for protection against predators.

D Behavior that favors the reproductive success of an organism's relatives, even at a cost to their own survival and/or reproduction.

E Behavior in which an animal sacrifices its own well-being for the benefit of another animal.

F The grouping together of a large number of fish.

G The grouping together of a large number of birds.

H Examples of this may be loosely associated groups or complex groups with clear social structures.

2. Describe the difference between solitary, non-social group, and social group behavior in animals:

_____

_____

_____

3. (a) Identify the structure formed by the fish in the photo right:

_____

(b) How do the fish benefit from the formation of this structure?

_____

_____

_____

4. Draw lines to match up the first half and second half of the sentences below:

Cooperative behavior in a group …

Cooperative behavior evolves when there is a …

By displaying altruistic behavior to family members, an individual can indirectly increase …

The energetic costs of migration are outweighed …

Cooperation has been taken to the extreme by eusocial animals in which the majority of members of the group …

… sacrifice their individual reproductive chances to ensure the collective's genes (and therefore their own genes) are passed to the next generation.

… by the reproductive benefits.

… the likelihood of their own genes being passed on to the next generation.

… benefits all members of that group.

… sustained benefit to the members of the group.

5. Explain how altruistic behavior between closely related individuals benefits the survival of all participants:

_____

_____

_____

_____

© 2016 **BIOZONE** International
ISBN: 978-1-927309-46-9
Photocopying Prohibited

VOCAB

181 Summative Assessment

Cooperative hunting in chimpanzees

Chimpanzees benefit from cooperative hunting. Although they may hunt alone, they also form hunting groups of up to six members or more. Chimpanzee hunts differ from the cooperative hunting of most other animals in that each chimpanzee in the hunt has a specific role in the hunt, such as a blocker or ambusher. Studies of chimpanzee hunting show that different groups employ different hunting strategies.

The hunt information in table 1 was gathered from chimpanzees in the Tai National Park in Ivory Coast.

Number of hunters	Number of hunts	Hunting success (%)	Meat per hunt (kg)	Net benefit per hunter (kJ)
1	30	13	1.23	4015
2	34	29	0.82	1250
3	39	49	3.12	3804
4	25	72	5.47	5166
5	12	75	4.65	3471
6	12	42	3.17	1851
>6	10	90	9.27	5020

Christophe Boesch 1994

The hunt information in table 2 was gathered from chimpanzees in the Gombe Stream National Park in Tanzania.

Number of hunters	Number of hunts	Hunting success (%)	Meat per hunt (kg)	Net benefit per hunter (kJ)
1	30	50	1.23	4245
2	13	61	1.85	3201
3	9	78	1.61	1837
4	7	100	2.86	2494
5	1	100	3.00	2189
6	2	50	2.00	861

Christophe Boesch 1994

1. Use the information in the table to discuss the differences between the two groups of chimpanzee in the extent of cooperation and how it relates to hunting success. You should plot graphs to help illustrate reasons for differences:

_____

_____

_____

_____

_____

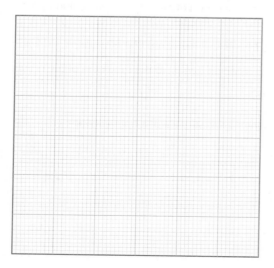

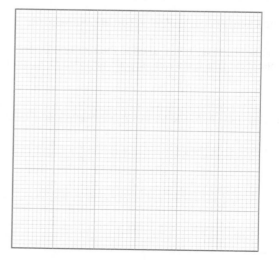

 © 2016 **BIOZONE** International
ISBN: 978-1-927309-46-9
Photocopying Prohibited

_____
_____
_____
_____
_____
_____

Sharing and bonding in chimpanzees

In Tai chimpanzees, hunting is a chance to form social bonds. Study the information below showing the number of chimpanzees taking part in a hunt and eating afterwards and the mean (average) number of bystanders during the hunt and eating afterwards.

2. Explain what the information is showing and discuss the reasons why this might occur:

Number of hunters	Mean number of hunters eating	Mean number of bystanders	Mean number of bystanders eating
1	0.7	3.5	3.0
2	1.6	3.6	2.6
3	2.5	3.6	3.0
4	2.5	2.7	2.1
5	3.5	2.7	2.3
6	4.7	2.4	2.2

Christophe Boesch 1994

_____
_____
_____
_____
_____
_____

Sentinel behavior in meerkats

Meerkats are highly social carnivores that live in mobs consisting of a dominant (alpha) breeding pair and up to 40 subordinate helpers of both sexes who do not normally breed but are usually related to the alpha pair. They are known for their sentinel behavior, watching for predators and giving alarm calls when they appear.

The graphs right show the likelihood of female or male meerkats standing sentinel when pups are either in the burrow or outside in the sentinel's group. The scale represents a statistical measure from a large number of observations. Error bars are ± SE.

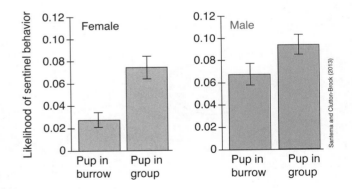

3. Discuss the evidence that meerkat sentinel behavior is altruistic in its nature: _____

_____
_____
_____
_____
_____
_____
_____

Heredity: Inheritance and Variation of Traits

Concepts and Connections
Use arrows to make your own connections between related concepts in this section of this workbook

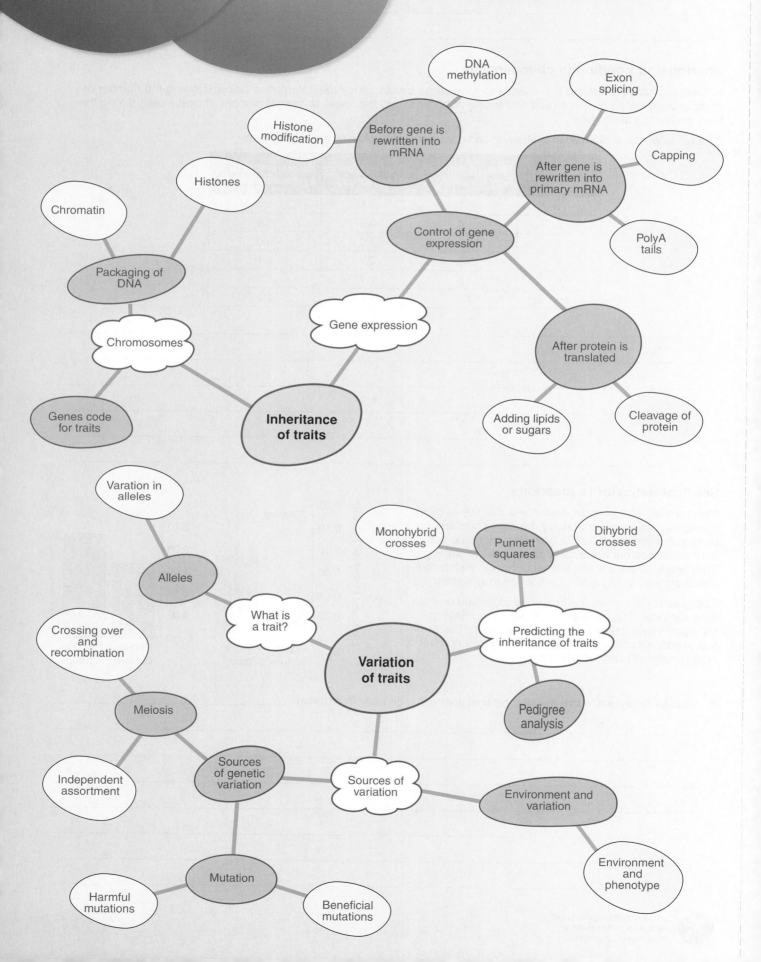

LS3.A

Inheritance of Traits

Key terms

chromosome

chromatin

DNA

gene

gene expression

histone

histone modification

transcription

translation

Disciplinary core ideas

Show understanding of these core ideas

		Activity number
Chromosomes consist of a single long DNA molecule		
☐	1 Chromosomes consist of a single, long DNA molecule. DNA is wound around histone proteins so that it can fit into a cell's nucleus. This complex of DNA and histone is called chromatin.	182
☐	2 Chromosomes contain long sections of DNA called genes. Genes code for the production of specific proteins.	182 183
Genes expressed by the cell can be regulated in different ways		
☐	3 Gene expression is the process of rewriting a gene into a protein. It involves two stages: transcription of the DNA and translation of the mRNA into protein.	186 187
☐	4 Not all DNA codes for proteins. Some segments of DNA are involved in regulatory or structural functions. Some segments have no as-yet known function.	184
☐	5 All cells in an organism have the same DNA but different types of cell regulate the expression of the DNA in different ways. Variations in the way genes are expressed (or not expressed) can cause significant differences between cells or organisms with identical DNA.	185-187

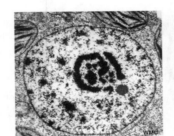 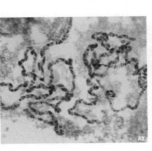 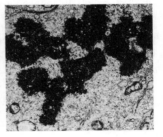

Crosscutting concepts

Understand how these fundamental concepts link different topics

		Activity number
☐	1 **CE** ▶ Empirical evidence clarified the role of chromosomes and DNA in coding for heritable traits.	183 185-187

Science and engineering practices

Demonstrate competence in these science and engineering practices

		Activity number
☐	1 Ask and evaluate questions about how the structure of DNA and chromosomes can encode the instructions for heritable traits.	182 183 185 186
☐	2 Use models to explain how many proteins can be produced from just one gene.	187

182 Chromosomes

Key Idea: A chromosome is a single long molecule of DNA coiled around histone proteins. Chromosomes contain protein-coding regions called genes.

In eukaryotes, DNA is complexed with proteins to form chromatin. The proteins in the chromatin are responsible for packaging the chromatin into discrete linear structures called chromosomes. The extent of packaging changes during the life cycle of the cell, with the chromosomes becoming visible during mitosis.

Each chromosome includes protein-coding regions called genes. Eukaryotic genes are typically preceded by segments of DNA called promoters and enhancers which are involved in starting the process of transcribing the gene into mRNA.

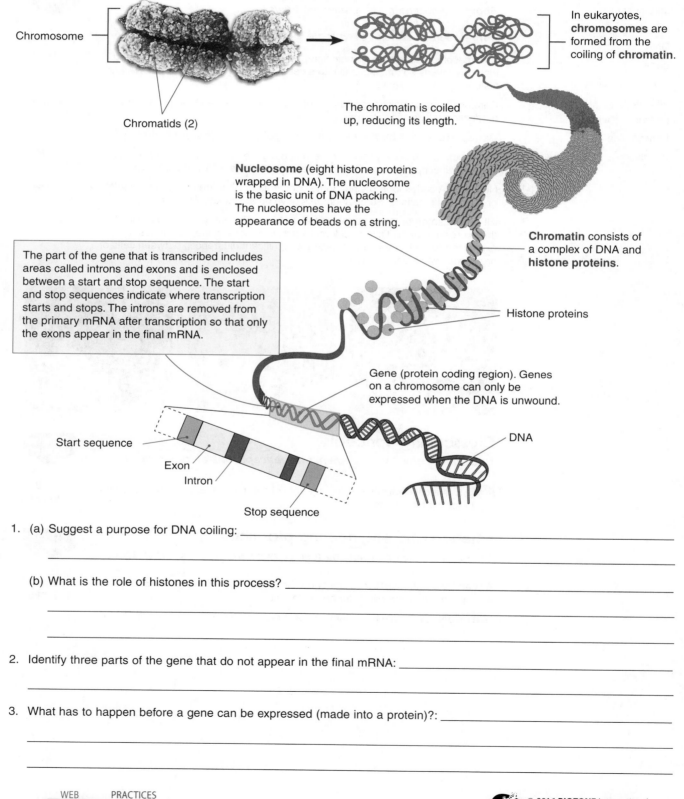

Chromosome

Chromatids (2)

In eukaryotes, **chromosomes** are formed from the coiling of **chromatin**.

The chromatin is coiled up, reducing its length.

Nucleosome (eight histone proteins wrapped in DNA). The nucleosome is the basic unit of DNA packing. The nucleosomes have the appearance of beads on a string.

Chromatin consists of a complex of DNA and **histone proteins**.

The part of the gene that is transcribed includes areas called introns and exons and is enclosed between a start and stop sequence. The start and stop sequences indicate where transcription starts and stops. The introns are removed from the primary mRNA after transcription so that only the exons appear in the final mRNA.

Histone proteins

Gene (protein coding region). Genes on a chromosome can only be expressed when the DNA is unwound.

Start sequence

Exon

Intron

Stop sequence

DNA

1. (a) Suggest a purpose for DNA coiling: _____

_____

(b) What is the role of histones in this process? _____

_____

_____

2. Identify three parts of the gene that do not appear in the final mRNA: _____

_____

3. What has to happen before a gene can be expressed (made into a protein)?: _____

_____

_____

WEB   PRACTICES

KNOW   182   ?

© 2016 **BIOZONE** International
ISBN: 978-1-927309-46-9
Photocopying Prohibited

183 DNA Carries the Code

Key Idea: The progressive studies of several scientists help to establish DNA as the heritable material responsible for the characteristics we see in organisms.

▶ Many years before Watson and Crick discovered the structure of DNA, biologists had deduced, through experimentation, that DNA carried the information that was responsible for the heritable traits we see in organisms.

▶ Prior to the 1940s, it was thought that proteins carried the code. Little was known about nucleic acids and the variety of protein structure and function suggested they could account for the many traits we see in organisms.

▶ Two early experiments, one by Griffith and another by Avery, MacLeod, and McCarty, provided important information about how traits could be passed on and what cellular material was responsible. The experiments involved strains of the bacterium *Streptococcus pneumoniae*. The S strain is pathogenic (causes disease). The R strain is harmless.

Griffith (1928)

▶ Griffith found that when he mixed killed pathogenic bacteria with living harmless cells, some of the living cells became pathogenic. Moreover, the newly acquired trait of pathogenicity was inherited by all descendants of the transformed bacteria. He concluded that the living R cells had been transformed into pathogenic cells by a heritable substance from the dead S cells.

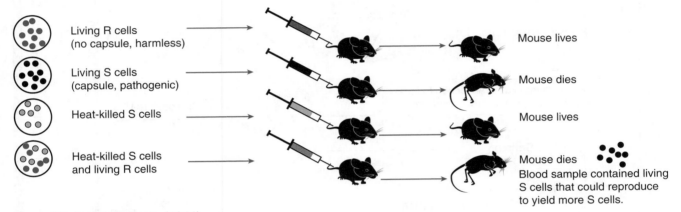

Avery-MacLeod-McCarty (1944)

▶ What was the unknown transformation factor in Griffith's experiment? Avery designed an experiment to determine if it was RNA, DNA, or protein. He broke open the heat-killed pathogenic cells and treated samples with agents that inactivated either protein, DNA, or RNA. He then tested the samples for their ability to transform harmless bacteria.

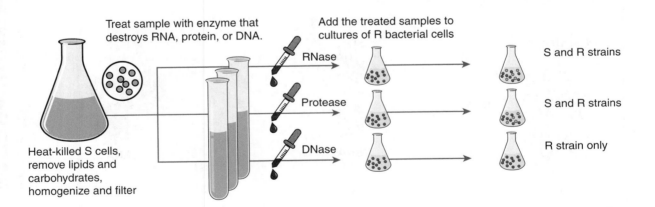

1. Griffith did not predict transformation in his experiment. What results was he expecting? Explain: _____

_____

_____

2. (a) What did Avery's experimental results show? _____

_____

(b) How did Avery's experiment build on Griffith's findings? _____

_____

_____

PRACTICES   CCC   WEB

? **CE** **183**   **KNOW**

184 Not All DNA Codes for Protein

Key Idea: Only about 2% of DNA codes for protein. Much of the non protein-coding DNA appears to have regulatory function, while other parts have as yet unknown functions.

The sequencing of the human genome, completed in 2003, revealed that humans had only 20,000-25,000 protein-coding genes, far fewer than the predictions of 100,000 or more. With improvement in sequencing technology, scientists now know that multitasking genes make more than one protein and about 98% of the genome does not code for proteins at all! These non protein-coding DNA sequences include introns and regulatory sequences and DNA that encodes RNA molecules. Among eukaryotes, greater complexity is associated with a higher proportion of non protein-coding DNA. This makes sense if the non protein-coding DNA is involved in regulating genomic function.

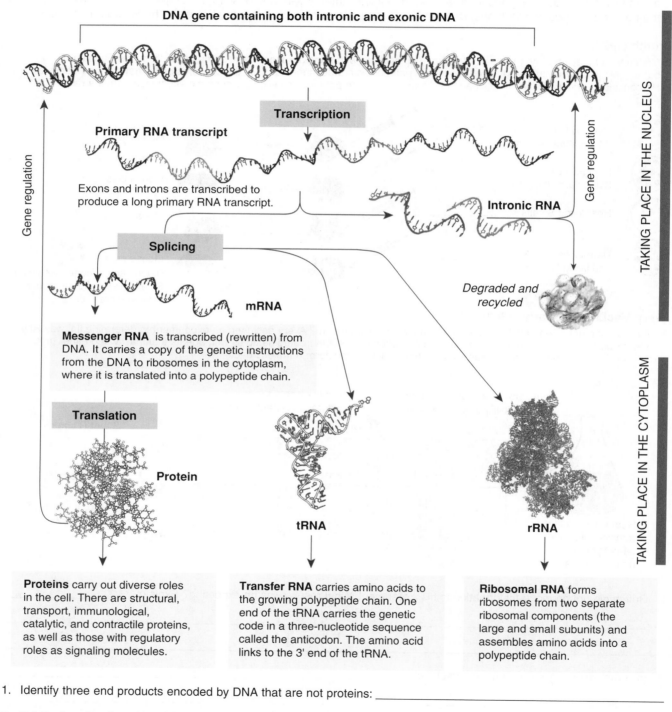

DNA gene containing both intronic and exonic DNA

Transcription

Primary RNA transcript

Exons and introns are transcribed to produce a long primary RNA transcript.

Intronic RNA

Splicing

mRNA

Degraded and recycled

Messenger RNA is transcribed (rewritten) from DNA. It carries a copy of the genetic instructions from the DNA to ribosomes in the cytoplasm, where it is translated into a polypeptide chain.

Translation

Protein

tRNA

rRNA

Gene regulation (left)
Gene regulation (right)

TAKING PLACE IN THE NUCLEUS
TAKING PLACE IN THE CYTOPLASM

Proteins carry out diverse roles in the cell. There are structural, transport, immunological, catalytic, and contractile proteins, as well as those with regulatory roles as signaling molecules.

Transfer RNA carries amino acids to the growing polypeptide chain. One end of the tRNA carries the genetic code in a three-nucleotide sequence called the anticodon. The amino acid links to the 3' end of the tRNA.

Ribosomal RNA forms ribosomes from two separate ribosomal components (the large and small subunits) and assembles amino acids into a polypeptide chain.

1. Identify three end products encoded by DNA that are not proteins: _____

2. Briefly describe the roles of these three products: _____

_____

_____

WEB

KNOW **184**

© 2016 **BIOZONE** International
ISBN: 978-1-927309-46-9
Photocopying Prohibited

185 The Outcomes of Differing Gene Expression

Key Idea: Variations in the way genes are expressed can cause significant differences between cells or organisms, even if their DNA is identical.

Same genes, different result

All the cells in an multicellular organism have identical DNA. As an organism develops from the zygote, differences in the way the DNA is expressed in developing cells cause them to differentiate into different types (right). Cells of the same type can express different proteins or amounts of protein depending on the environment they are exposed to during development or due to random gene inactivations (below).

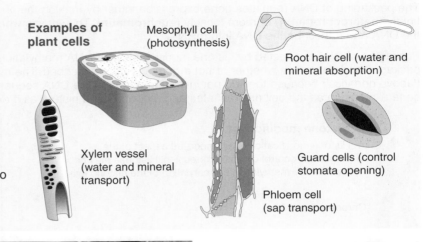

Examples of plant cells

Mesophyll cell (photosynthesis)

Root hair cell (water and mineral absorption)

Xylem vessel (water and mineral transport)

Guard cells (control stomata opening)

Phloem cell (sap transport)

Female mammals have two X chromosomes, one of which is inactivated in every cell. However, which X chromosome is inactivated is random. In cats, the gene for coat color is found on the X chromosome. Cats with two different alleles for coat color will only have one active allele per pigment cell, so each cell will produce one or the other color, giving a patchwork (calico) coat.

All worker bees and the queen bee (circled above) in a hive have the same genome, yet the queen looks and behaves very differently from the workers. Only bee larvae fed a substance called royal jelly will develop into queens. Research shows that royal jelly contains factors that silence the activation of a gene called Dnmt3, which itself silences many other genes.

Twins have the same genome but over the years of their life become different in both appearance and behavior. Studies have found that 35% of twins have significant differences in gene expression. More importantly, the older the twins, the more difference there is in the gene expression.

1. How do cells with identical DNA differentiate into different types? _____

_____

2. Why are patchwork (calico) cats always female? _____

_____

_____

_____

3. Studies on bee development focused on the Dnmt3 gene. One study switched off the Dnmt3 gene in 100 bee larvae. All the larvae developed into queens. Leaving the gene switched on in larvae causes them to develop into workers. Compare these results to feeding larvae royal jelly. Which results mimic feeding larvae royal jelly? Explain:

_____

_____

_____

_____

PRACTICES

CCC

?   **CE**   **KNOW**

186 DNA Packaging and Control of Transcription

Key Idea: Transcription and therefore the expression of genes can be controlled by regulating the extent of DNA packaging. Genes cannot be expressed when they are tightly packed.

The regulation of gene expression in eukaryotes is a complex process beginning before the DNA is even transcribed. The packaging of DNA regulates gene expression either by making the nucleosomes in the chromatin pack together tightly (**heterochromatin**) or more loosely (**euchromatin**). This affects whether or not RNA polymerase can attach to the DNA and transcribe the DNA into mRNA.

Packaging of DNA is affected by histone modification and DNA methylation. These modifications alter how the DNA is packaged and determine whether or not a gene can be transcribed. The modifications (or tags) are called **epigenetic** ('above genetics') because they do not involve changes to the DNA sequence itself. Epigenetic tags help to regulate gene expression as the cell differentiates and record a cell's history as it experiences different, changing environments.

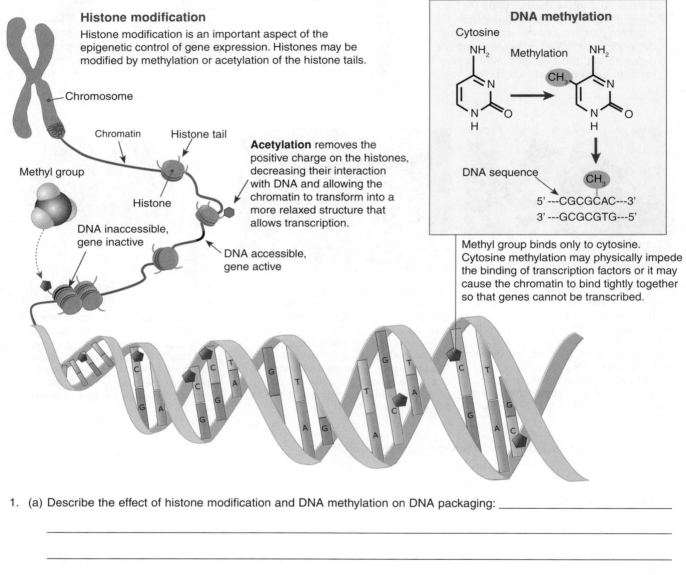

Histone modification

Histone modification is an important aspect of the epigenetic control of gene expression. Histones may be modified by methylation or acetylation of the histone tails.

Chromosome

Chromatin   Histone tail

Methyl group

Histone

Acetylation removes the positive charge on the histones, decreasing their interaction with DNA and allowing the chromatin to transform into a more relaxed structure that allows transcription.

DNA inaccessible, gene inactive

DNA accessible, gene active

DNA methylation

Cytosine

NH_2 Methylation NH_2

CH_3

DNA sequence CH_3

5' ---CGCGCAC---3'
3' ---GCGCGTG---5'

Methyl group binds only to cytosine. Cytosine methylation may physically impede the binding of transcription factors or it may cause the chromatin to bind tightly together so that genes cannot be transcribed.

1. (a) Describe the effect of histone modification and DNA methylation on DNA packaging: _____

_____

_____

(b) How do these processes affect transcription of the DNA?_____

_____

_____

2. When a zygote forms at fertilization most of the epigenetic tags are erased so that cells return to a genetic 'blank slate' ready for development to begin. However some epigenetic tags are retained and inherited. Why do you think it might be advantageous to inherit some epigenetic tags from a parent?

_____

_____

© 2016 **BIOZONE** International
ISBN: 978-1-927309-46-9
Photocopying Prohibited

187 Changes after Transcription and Translation

Key Idea: Primary mRNA molecules are modified in the nucleus before they are translated into proteins. Proteins can also be modified after translation.

▶ Human DNA contains 25,000 genes, but produces up to 1 million different proteins. Each gene must therefore produce more than one protein. This is achieved by both post transcriptional and post translational modification.

Post transcriptional modification

▶ Primary mRNA contains exons and introns. Introns are usually removed after transcription and the exons are spliced together. However, there are many alternative ways to splice the exons and these alternatives create variations in the translated proteins. In mammals, the most common method of alternative splicing involves exon skipping, in which not all exons are spliced into the final mRNA (below). Other alternative splicing options create further variants.

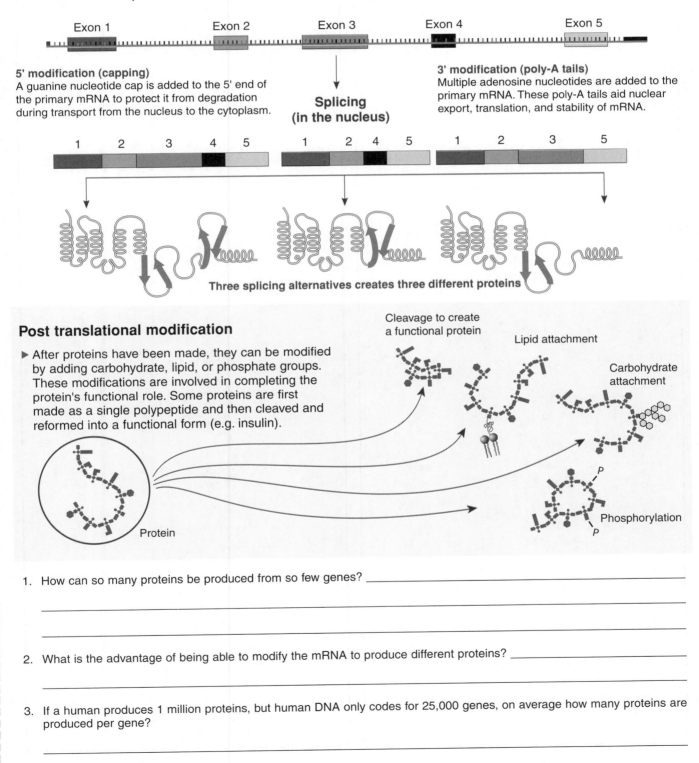

5' modification (capping)
A guanine nucleotide cap is added to the 5' end of the primary mRNA to protect it from degradation during transport from the nucleus to the cytoplasm.

Splicing (in the nucleus)

3' modification (poly-A tails)
Multiple adenosine nucleotides are added to the primary mRNA. These poly-A tails aid nuclear export, translation, and stability of mRNA.

Three splicing alternatives creates three different proteins

Post translational modification

▶ After proteins have been made, they can be modified by adding carbohydrate, lipid, or phosphate groups. These modifications are involved in completing the protein's functional role. Some proteins are first made as a single polypeptide and then cleaved and reformed into a functional form (e.g. insulin).

Protein

Cleavage to create a functional protein

Lipid attachment

Carbohydrate attachment

Phosphorylation

1. How can so many proteins be produced from so few genes? _____

_____

_____

2. What is the advantage of being able to modify the mRNA to produce different proteins? _____

_____

3. If a human produces 1 million proteins, but human DNA only codes for 25,000 genes, on average how many proteins are produced per gene?

_____

PRACTICES   CCC   WEB

CE   187   **KNOW**

188 Chapter Review

Summarize what you know about this topic so far under the headings provided. You can draw diagrams or mind maps, or write notes to organize your thoughts. Use the checklist in the introduction and the hints to help you:

The structure and organization of chromosomes

HINT: Include how the chromosome is packaged into the nucleus.

The regulation of gene expression

HINT: How does the body produce more proteins than it has protein-coding genes? Include controls over transcription, regulation after transcription, and the role of post-translational modifications.

© 2016 **BIOZONE** International
ISBN: 978-1-927309-46-9
Photocopying Prohibited

189 KEY TERMS AND IDEAS: Did You Get It?

Complete the following questions with reference to the diagram below right, complete the following questions:

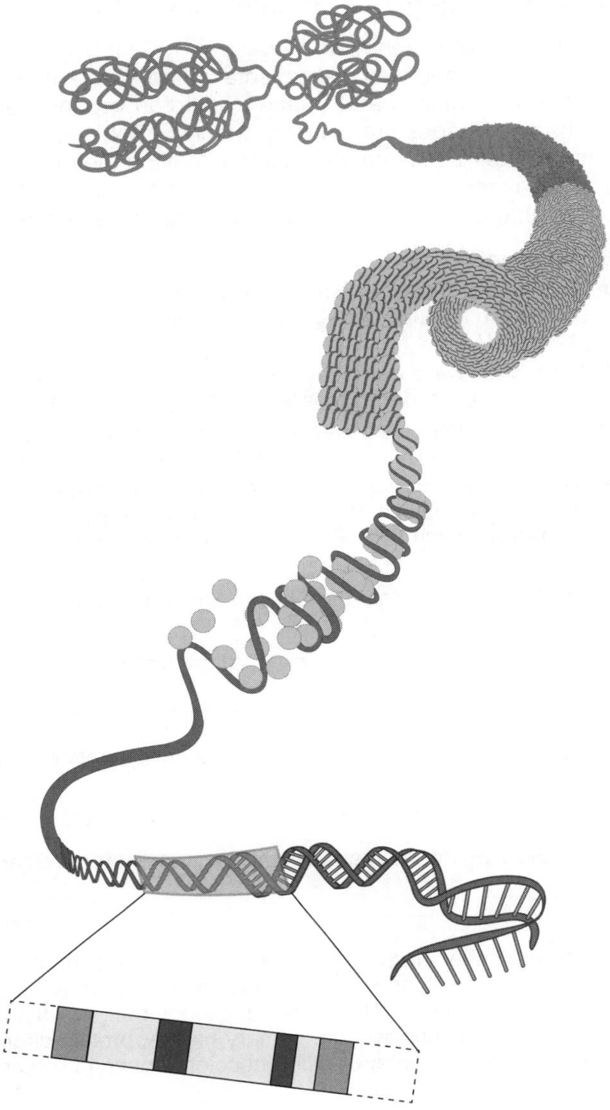

1. (a) Circle the chromosome and label the chomatids.

   (b) What is a chromatid?_____

   _____

2 (a) Label the histone proteins and the DNA.

   (b) What is the name given to the material formed by these two components?

   _____

   (c) What is the role of the histone proteins? _____

   _____

   _____

3. (a) Label the gene, the start sequence, stop sequence, intron, and exons.

   (b) Which regions of the gene are expressed?

   _____

4. (a) In what state must the DNA be in order to be transcribed?

   _____

   (b) This is achieved by (select the correct answer):

   (i) DNA methylation

   (ii) Histone acetylation

   (iii) Histone methylation

5. (a) How does the cell prevent transcription of a gene?

   _____

   (b) How is this achieved: _____

   _____

   _____

6. Test your vocabulary by matching each term to its definition, as identified by its preceding letter code.

chromosome _____

chromatin _____

DNA _____

gene _____

gene expression _____

histone _____

histone tail _____

methylation _____

A Macromolecule consisting of many millions of units, each containing a phosphate group, sugar and a base (A,T, C or G). Stores the genetic information of the cell.

B The basic unit of inheritance.

C The process of transferring the information encoded in a gene into a gene product.

D Single piece of DNA that contains many genes and associated regulatory elements and proteins.

E The addition of a CH_3 group to a DNA base, altering gene activity.

F A structure added to a histone protein that alters the way DNA binds to it.

G A protein found in the nuclei of eukaryotic cells that packages and orders the DNA into structural units called nucleosomes.

H A complex of DNA and histone proteins.

© 2016 **BIOZONE** International
ISBN: 978-1-927309-46-9
Photocopying Prohibited

TEST

190 Summative Assessment

The experiments of Hershey and Chase (1952)

▶ The work of Hershey and Chase in 1952 followed the work of Avery and his colleagues and was instrumental in the acceptance of DNA as the hereditary material. Hershey and Chase worked on viruses that infect bacteria called phages. Phages are composed only of DNA and protein. When they infect, they inject their DNA into the bacteria, leaving their protein coat outside. Hershey and Chase were able to show that DNA is the only material transferred directly from the phages into the bacteria when the bacteria are infected by the viruses.

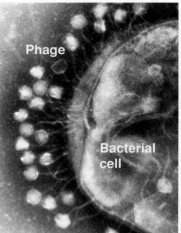

Phage

Bacterial cell

Dr Graham Beards

Experiment

▶ Batch 1 phage were grown with radioactive sulfur which was incorporated into the protein coat. Batch 2 phage were grown with radioactive phosphorus, which was incorporated into phage DNA.

▶ The radioactive phage were mixed with bacteria, which became infected. Each batch was agitated in a blender to separate the phage parts outside the bacteria from the cells.

▶ Each batch was centrifuged so that the bacteria form a pellet at the bottom of the test tube. Free phages and phage parts, which are lighter, are suspended in the liquid.

▶ When the proteins were labeled, the radioactivity was in the liquid (outside the cells). When the DNA was labeled, the radioactivity was in the pellet (the cells). The bacteria with the radioactive phage DNA also released new phages with some radioactive phosphorus.

1. (a) How did the Hershey-Chase experiment provide evidence that nucleic acids, not protein are the hereditary material?

   _____

   _____

   (b) What assumptions were made about the role of DNA in viruses and bacteria and the role of DNA in eukaryotes?

   _____

   (c) How would the results of the experiment have differed if proteins carried the genetic information? _____

   _____

2. (a) The schematic below shows the levels of control in gene expression. Fill in the boxes indicating the structures and processes, choosing from the following word list. **Word list**: 5' cap, mRNA in the nucleus, polypeptide, mRNA in the cytoplasm, DNA packing, exon, intron, functional protein, folding and assembly, poly A tail, gene, cleavage or chemical modification, primary mRNA, protein degradation, translation, transcription, exon splicing, nuclear export.
   (b) Use a highlighter or different colored pens to distinguish processes (in red) and structures (in blue).

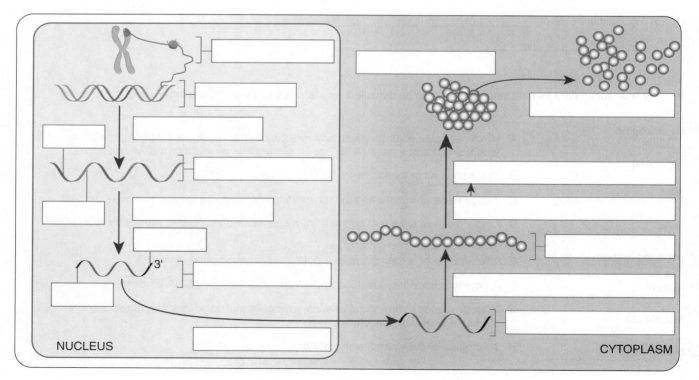

NUCLEUS

CYTOPLASM

© 2016 **BIOZONE** International
ISBN: 978-1-927309-46-9
Photocopying Prohibited

TEST

LS3.B Variation of Traits

Key terms

allele

crossing over

dihybrid

heterozygous

chromosome

gene

independent assortment

homozygous

meiosis

monohybrid

mutation

phenotype

recombination

trait

variation

Disciplinary core ideas

Show understanding of these core ideas

Activity number

Variation can result from genetic processes

☐ 1 Traits are phenotypic variants controlled by genes, which can be inherited. — 191 192

☐ 2 Variation in individuals is caused by both genetic and environmental factors. — 193-195

☐ 3 Homologous chromosomes can be heterozygous or homozygous. — 192

☐ 4 The process of meiosis gives the opportunity to produce genetic variation. Chromosomes may exchange genetic material in a process called crossing over, which results in recombination of alleles. Independent assortment allows random pairing of alleles. — 196 197

☐ 5 Mutations are the ultimate source of all new genetic variation (new alleles). Mutation can arise through copying errors during DNA replication or as a result of environmental factors. Most mutations are harmful, but some may be beneficial. — 198-202

☐ 6 Punnett squares can be used to predict the probable outcomes of monohybrid and dihybrid crosses. — 206-213

Variation can result from environmental influences

☐ 7 A phenotype is the expression of both genetic and environmental influences. — 194 203

☐ 8 The environmental influences experienced by one generation may affect subsequent generations. — 204 205

Crosscutting concepts

Understand how these fundamental concepts link different topics

Activity number

☐ 1 **CE** ▶ Empirical evidence enables us to distinguish between cause and correlation and support claims about the causes of genetic variation. — 191 198-202 205 212

☐ 2 **SPQ** ▶ Algebraic thinking is used to explain the variation and distribution of expressed traits in a population. — 195

☐ 3 **SPQ** ▶ Algebraic thinking is used to examine scientific data and predict the outcome of genetic crosses and explain the expression of traits. — 206 209 211 212

Science and engineering practices

Demonstrate competence in these science and engineering practices

Activity number

☐ 1 Use a model based on evidence to explain that heritable genetic variation arises through meiosis and mutation. — 194 197 199

☐ 2 Argue from evidence that the environment can affect the phenotype of subsequent generations. — 198 199 200 205

☐ 3 Apply concepts of statistics and probability to explain the variation and expression of traits in a population. — 206-211

☐ 4 Use evidence to build and defend the claim that variation in a population is essential for its survival and evolution. — 193 200 201 204

191 What is a Trait?

Key Idea: Traits are controlled by genes, which can be inherited and passed from one generation to the next.

Traits are inherited

▶ **Traits** are particular variants of phenotypic (observed physical) characters, e.g. blue eye color. Traits may be controlled by one gene or many genes and can show continuous variation, e.g. height in humans, or discontinuous variation, e.g. flower color in pea plants.

▶ **Gregor Mendel**, an Austrian monk (1822-1884), used pea plants to study inheritance. Using several phenotypic characteristics he was able to show that their traits were inherited in predictable ways.

Mendel's experiments

Mendel studied seven phenotypic characters of the pea plant:

- Flower color (violet or white)
- Pod color (green or yellow)
- Height (tall or short)
- Position of the flowers on the stem (axial or terminal)
- Pod shape (inflated or constricted)
- Seed shape (round of wrinkled)
- Seed color (yellow or green)

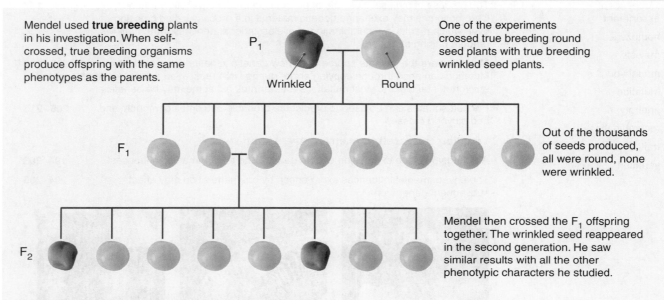

Mendel used **true breeding** plants in his investigation. When self-crossed, true breeding organisms produce offspring with the same phenotypes as the parents.

P₁ — Wrinkled / Round

One of the experiments crossed true breeding round seed plants with true breeding wrinkled seed plants.

F₁

Out of the thousands of seeds produced, all were round, none were wrinkled.

F₂

Mendel then crossed the F₁ offspring together. The wrinkled seed reappeared in the second generation. He saw similar results with all the other phenotypic characters he studied.

Three conclusions could be made from these results:

- Traits are determined by a unit that passes unchanged from parent to offspring (these units are now called genes).

- Each individual inherits one unit (gene) for each trait from each parent (each individual has two units).

- Traits may not physically appear in an individual, but the units (genes) for them can still be passed to its offspring.

1. Define a trait: _____

_____

2. Define true breeding: _____

_____

_____

3. (a) What was the ratio of smooth seeds to wrinkled seeds in the F₂ generation? _____

b) Suggest why the wrinkled seed trait did not appear in the F₁ generation: _____

_____

_____

_____

© 2016 **BIOZONE** International
ISBN: 978-1-927309-46-9
Photocopying Prohibited

192 Different Alleles For Different Traits

Key Idea: Eukaryotes generally have paired chromosomes. Each chromosome contains many genes, and each gene may have a number of versions, called alleles.

Homologous chromosomes

In sexually reproducing organisms, chromosomes are generally found in pairs. Each parent contributes one chromosome to the pair. The pairs are called **homologues** or **homologous pairs**. Each homologue carries an identical assortment of genes, but the version of the gene (the **allele**) from each parent may differ. This diagram shows the position of three different genes on the same chromosome that control three different traits (A, B and C).

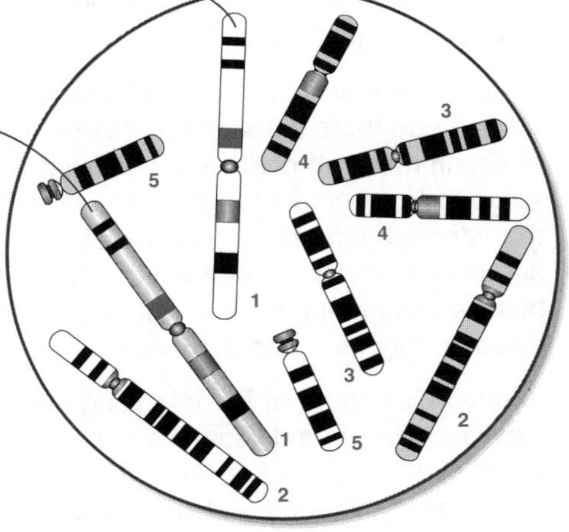

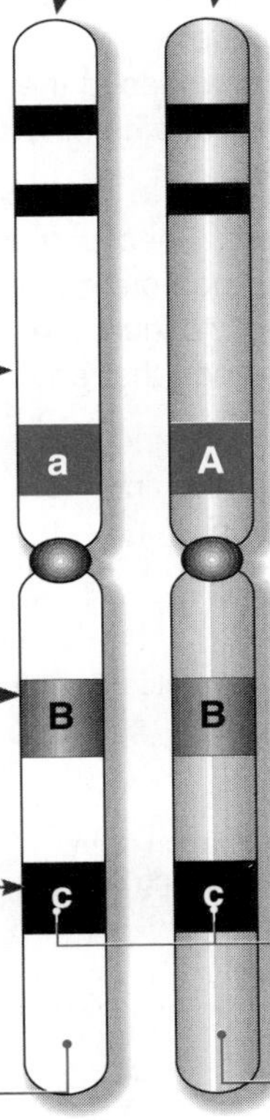

Having two different versions of gene A is a **heterozygous** condition. Only the dominant allele (A) will be expressed.

The diagram above shows the complete chromosome complement for a hypothetical organism. It has a total of ten chromosomes, as five, nearly identical pairs (each pair is numbered).

When both chromosomes have identical copies of the dominant allele for gene B the organism is **homozygous dominant** for that gene.

When both chromosomes have identical copies of the recessive allele for gene C the organism is said to be **homozygous recessive** for that gene.

Genes occupying the same position or **locus** on a chromosome code for the same characteristic (e.g. ear lobe shape).

Maternal chromosome originating from the egg of this person's mother.

Paternal chromosome originating from the sperm of this person's father.

1. Define the following terms describing the allele combinations of a gene in a sexually reproducing organism:

   (a) Heterozygous: _____

   _____

   (b) Homozygous dominant: _____

   _____

   (c) Homozygous recessive: _____

   _____

2. For a gene given the symbol 'A', write down the alleles present in an organism that is:

   (a) Heterozygous:_____   (b) Homozygous dominant: _____   (c) Homozygous recessive: _____

3. What is a homologous pair of chromosomes? _____

   _____

   _____

   _____

© 2016 **BIOZONE** International
ISBN: 978-1-927309-46-9
Photocopying Prohibited

KNOW

193 Why is Variation Important?

Key Idea: Variation in a population or species is important in a changing environment. Both sexually and asexually reproducing species have strategies to increase variation.

▸ **Variation** refers to the diversity of phenotypes or genotypes within a population or species. Variation helps organisms survive in a changing environment.

▸ Sexual reproduction produces variability, which provides ability to adapt to a changing physical environment. However, environments can change very slowly, it may take millions of years for a mountain range to rise from the seabed. This is more than enough time for even asexually reproducing species to acquire the variability needed to adapt. However changes in the biotic environment, such as the appearance of new strains of disease, require a fast response.

▸ Variation is important for defending against disease. Species that evolve to survive a disease flourish. Those that do not, die out. It is thought that sexual reproduction is an adaptation to increase variability in offspring and so provide a greater chance that any one of the offspring will survive a given disease.

▸ Even species that reproduce asexually for much of the time can show a large amount of variation within a population.

Aphids can reproduce sexually and asexually. Females hatch in spring and give birth to clones. Many generations are produced asexually. Just before fall, the aphids reproduce sexually. The males and females mate and the females produce eggs which hatch the following spring. This increases variability in the next generation.

Diagrams to show how three beneficial mutations could be combined through sexual or asexual reproduction.

Variation by sexual reproduction

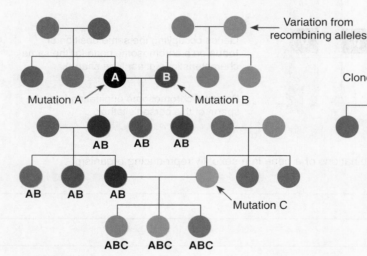

During meiosis, alleles are recombined in new combinations. Some combinations of alleles may be better suited to a particular environment than others. This variability is produced without the need for mutation. Beneficial mutations in separate lineages can be quickly combined through sexual reproduction.

Variation by asexual reproduction

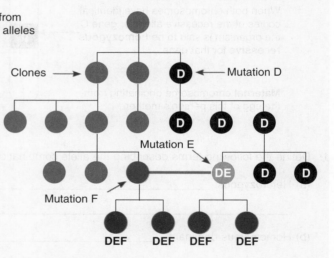

Some asexually reproducing organisms are able to exchange genes occasionally. Bacteria exchange genes with other bacteria during a process called conjugation (thicker blue line). This allows mutations that arise in one lineage to be passed to another.

1. Why is variation important in populations or species? _____

_____

_____

_____

© 2016 **BIOZONE** International
ISBN: 978-1-927309-46-9
Photocopying Prohibited

KNOW

194 Sources of Variation

Key Idea: Variation may come from changes to the genetic material (mutation), through sexual reproduction, and as a result of the effects of the environment.

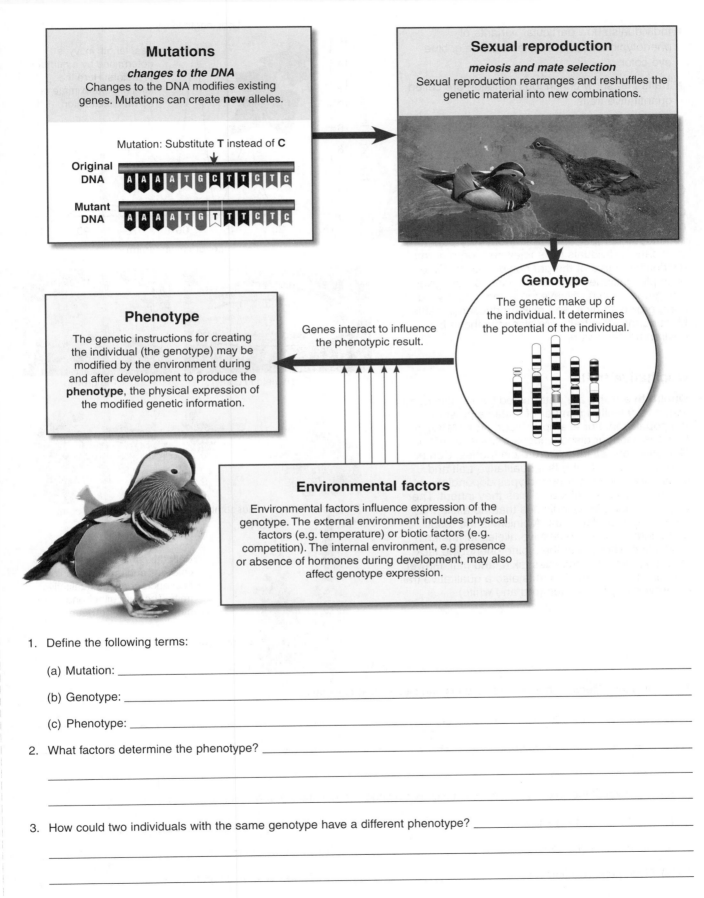

Mutations

changes to the DNA
Changes to the DNA modifies existing genes. Mutations can create **new** alleles.

Mutation: Substitute **T** instead of **C**

Original DNA: A A A A T G C T T C T C

Mutant DNA: A A A A T G T T T C T C

Sexual reproduction

meiosis and mate selection
Sexual reproduction rearranges and reshuffles the genetic material into new combinations.

Genotype

The genetic make up of the individual. It determines the potential of the individual.

Genes interact to influence the phenotypic result.

Phenotype

The genetic instructions for creating the individual (the genotype) may be modified by the environment during and after development to produce the **phenotype**, the physical expression of the modified genetic information.

Environmental factors

Environmental factors influence expression of the genotype. The external environment includes physical factors (e.g. temperature) or biotic factors (e.g. competition). The internal environment, e.g presence or absence of hormones during development, may also affect genotype expression.

1. Define the following terms:

   (a) Mutation: _____

   (b) Genotype: _____

   (c) Phenotype: _____

2. What factors determine the phenotype? _____

   _____

3. How could two individuals with the same genotype have a different phenotype? _____

   _____

PRACTICES   WEB

194   **KNOW**

195 Examples of Genetic Variation

Key Idea: Genetic variation may be continuous as a result of quantitative traits (e.g. leaf length) or it may be discontinuous as a result of qualitative traits (e.g. biological sex).

▶ Individuals show particular variants of phenotypic characters called traits, e.g. blue eye color.

▶ Traits that show continuous variation are called quantitative traits.

▶ Traits that show discontinuous variation are called qualitative traits.

Quantitative traits

Quantitative traits are determined by a large number of genes. For example, skin color has a continuous number of variants from very pale to very dark. Individuals fall somewhere on a normal distribution curve of the phenotypic range. Other examples include height in humans for any given age group, length of leaves in plants, grain yield in corn, growth in pigs, and milk production in cattle. Most quantitative traits are also influenced by environmental factors.

Qualitative traits

Qualitative traits are determined by a single gene with a very limited number of variants present in the population. For example, blood type (ABO) in humans has four discontinuous traits A, B, AB or O. Individuals fall into discrete categories. Comb shape in poultry (right) is a qualitative trait and birds have one of four phenotypes depending on which combination of four alleles they inherit. The dash (missing allele) indicates that the allele may be recessive or dominant. Albinism is the result of the inheritance of recessive alleles for melanin production. Those with the albino phenotype lack melanin pigment in the eyes, skin, and hair. Flower color in snapdragons (right) is also a qualitative trait determined by two alleles (red and white).

Leaf length in ivy

Leaf length in ivy is determined by a number of factors. Here the lengths approximate a normal distribution.

Number of leaves / Length of leaf (mm)

Grain yield in corn

Pig growth

Single comb **rrpp**   Walnut comb **R_P_**   Pea comb **rrP_**   Rose comb **R_pp**

Both the comb shape in chickens and the flower color in snapdragons are qualitative traits. They are either one or the other, but not in-between.

Snapdragons
C^W C^W   C^R C^R

1. What is the difference between continuous and discontinuous variation? _____

_____

_____

_____

2. Identify each of the following phenotypic traits as continuous (quantitative) or discontinuous (qualitative):

(a) Wool production in sheep: _____   (d) Albinism in mammals: _____

(b) Hand span in humans: _____   (e) Body weight in mice: _____

(c) Blood groups in humans: _____   (f) Flower color in snapdragons: _____

© 2016 **BIOZONE** International
ISBN: 978-1-927309-46-9
Photocopying Prohibited

196 Meiosis

Key Idea: Meiosis is a reduction division that produces haploid cells for the purposes of sexual reproduction.

Meiosis

▶ **Meiosis** is a special type of cell division necessary for the production of sex cells (gametes) for the purpose of sexual reproduction.

▶ DNA replication precedes meiosis. If genetic mistakes (mutations) occur here, they will be passed on (inherited by the offspring).

▶ Meiosis involves a single chromosomal duplication followed by two successive nuclear divisions, and it halves the diploid chromosome number.

▶ An overview of meiosis is shown on the right. Meiosis occurs in the sex organs of plants and animals.

Meiosis produces variation

▶ During meiosis, a process called **crossing over** may occur when homologous chromosomes may exchange genes. This further adds to the variation in the gametes.

▶ Meiosis is an important way of introducing genetic variation. The assortment of chromosomes into the gametes (the proportion from the father or mother) is random and can produce a huge number of possible chromosome combinations.

The process of meiosis

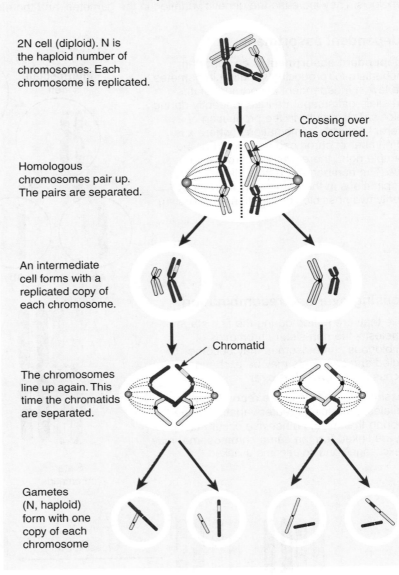

2N cell (diploid). N is the haploid number of chromosomes. Each chromosome is replicated.

Crossing over has occurred.

Homologous chromosomes pair up. The pairs are separated.

An intermediate cell forms with a replicated copy of each chromosome.

Chromatid

The chromosomes line up again. This time the chromatids are separated.

Gametes (N, haploid) form with one copy of each chromosome

1. (a) What is the purpose of meiosis? _____

   (b) Where does meiosis take place? _____

   _____

   _____

   _____

2. Describe how variation can arise during meiosis: _____

   _____

   _____

   _____

 © 2016 **BIOZONE** International
ISBN: 978-1-927309-46-9
Photocopying Prohibited

197 Meiosis and Variation

Key Idea: Meiosis produces variation. Independent assortment and crossing over are two important ways of introducing variation into the gametes formed during meiosis.

▶ Independent assortment and crossing over (leading to recombination of alleles) are mechanisms that occur during meiosis. They increase the genetic variation in the gametes, and therefore the offspring.

Independent assortment

Independent assortment is an important mechanism for producing variation in gametes. The law of independent assortment states that allele pairs separate independently during meiosis. This results in the production of 2^x different possible combinations (where x is the number of chromosome pairs). For the example right, there are two chromosome pairs. The number of possible allele combinations in the gametes is therefore $2^2 = 4$ (only two possible combinations are shown).

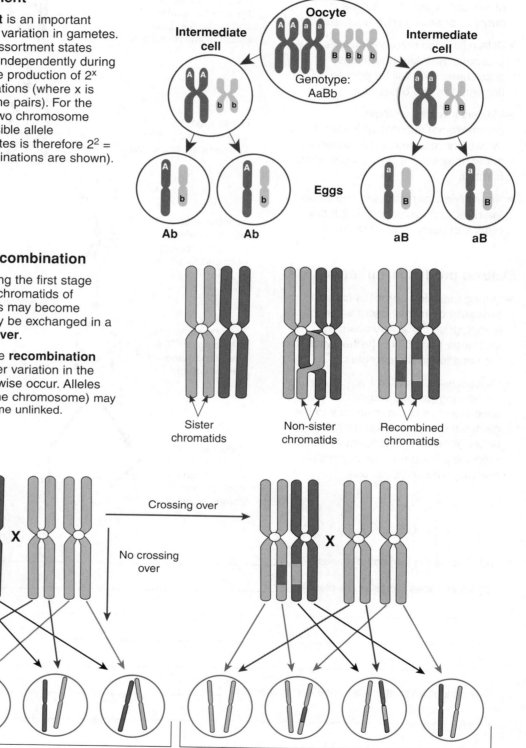

Crossing over and recombination

While they are paired during the first stage of meiosis, the non-sister chromatids of homologous chromosomes may become tangled and segments may be exchanged in a process called **crossing over**.

Crossing over results in the **recombination** of alleles, producing greater variation in the offspring than would otherwise occur. Alleles that are linked (on the same chromosome) may be exchanged and so become unlinked.

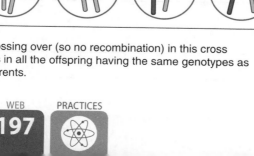

No crossing over (so no recombination) in this cross results in all the offspring having the same genotypes as the parents.

Crossing over results in recombination. Although half of the offspring are the same as the parents, half have a new genetic combination.

© 2016 **BIOZONE** International
ISBN: 978-1-927309-46-9
Photocopying Prohibited

1. (a) Using the diagram on independent assortment (previous page) draw the other two gamete combinations not shown in the diagram:

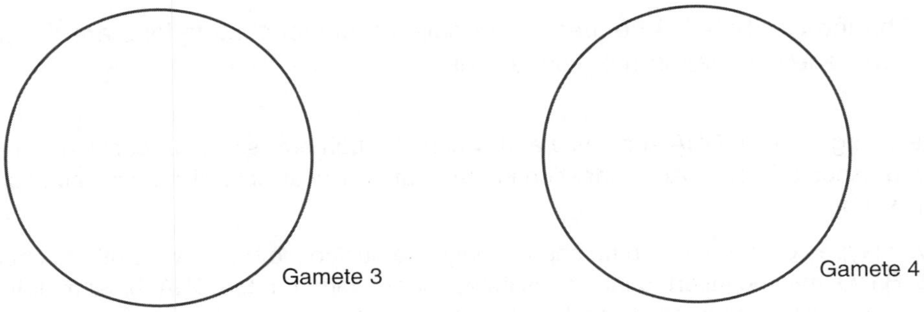

Gamete 3

Gamete 4

(b) For each of the following chromosome numbers, calculate the number of possible gamete combinations:

i. 8 chromosomes: _____

ii. 24 chromosomes: _____

iii 64 chromosomes: _____

2. What are sister and non-sister chromatids? _____

_____

_____

_____

3. (a) What is crossing over? _____

_____

_____

_____

(b) How does crossing over increase the variation in the gametes (and hence the offspring)? _____

_____

_____

_____

4. Crossing over occurs at a single point between the chromosomes below.

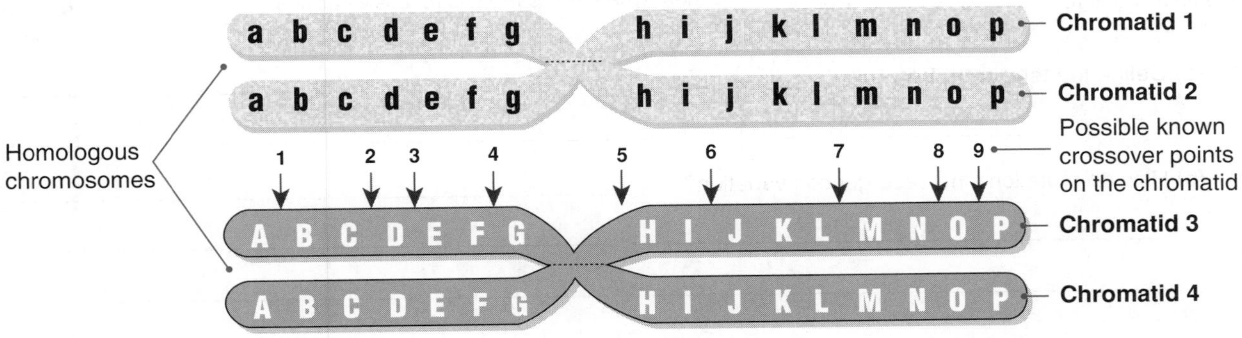

| a b c d e f g | h i j k l m n o p | — **Chromatid 1** |
| a b c d e f g | h i j k l m n o p | — **Chromatid 2** |

Homologous chromosomes

1   2 3   4   5   6   7   8 9 • — Possible known crossover points on the chromatid

| A B C D E F G | H I J K L M N O P | — **Chromatid 3** |
| A B C D E F G | H I J K L M N O P | — **Chromatid 4** |

(a) Draw the gene sequences for the four chromatids (above), after crossing over has occurred at **crossover point 2**:

1

2

3

4

(b) Which genes have been exchanged between the homologous chromosomes?

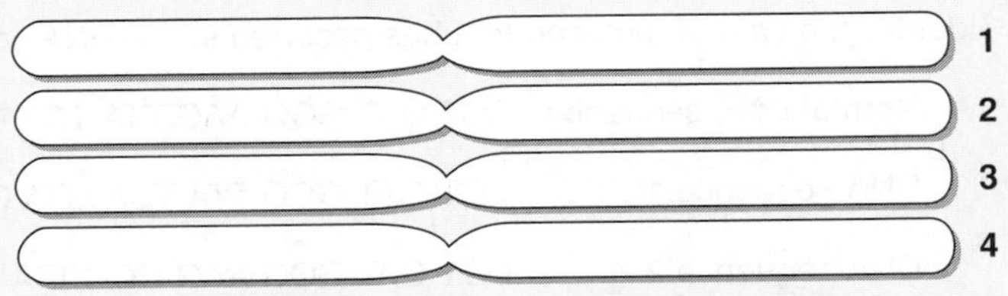

198 Mutations

Key Idea: Changes to the DNA sequence are called mutations. Mutations are the ultimate source of new genetic variation (i.e. new alleles).

▶ **Mutations** are changes to the DNA sequence and occurs through errors in DNA copying. Changes to the DNA modifies existing genes and can create variation in the form of new alleles. Ultimately, mutations are the source of all new genetic variation.

▶ There are several types of mutation. Some change only one nucleotide base, while others change large parts of chromosomes. Bases may be inserted into, substituted, or deleted from the DNA. Most mutations are harmful (e.g. those that cause cancer) but very occasionally a mutation can be beneficial.

▶ An example of a mutation producing a new allele is described below. This mutation causes the most common form of genetic hearing loss (called NSRD) in children. The mutation occurs in the gene coding for a protein called connexin 26.

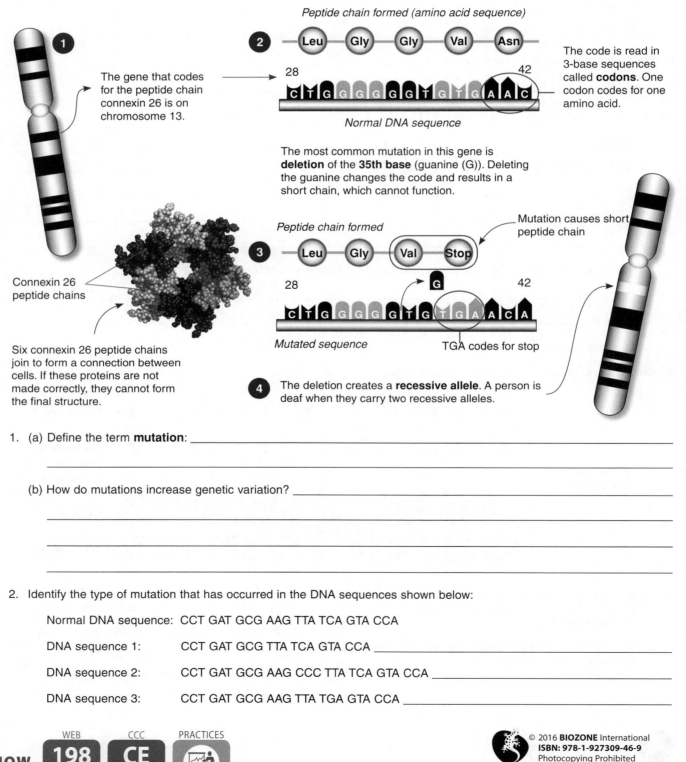

Peptide chain formed (amino acid sequence)

1

2 Leu—Gly—Gly—Val—Asn

The gene that codes for the peptide chain connexin 26 is on chromosome 13.

28   42
C T G G G G G G T G T G A A C

The code is read in 3-base sequences called **codons**. One codon codes for one amino acid.

Normal DNA sequence

The most common mutation in this gene is **deletion** of the **35th base** (guanine (G)). Deleting the guanine changes the code and results in a short chain, which cannot function.

Peptide chain formed

Connexin 26 peptide chains

3 Leu—Gly—Val—Stop

Mutation causes short peptide chain

G

28   42
C T G G G G G T G T G A A C A

Mutated sequence

TGA codes for stop

Six connexin 26 peptide chains join to form a connection between cells. If these proteins are not made correctly, they cannot form the final structure.

4 The deletion creates a **recessive allele**. A person is deaf when they carry two recessive alleles.

1. (a) Define the term **mutation**: _____

   _____

   (b) How do mutations increase genetic variation? _____

   _____

   _____

   _____

2. Identify the type of mutation that has occurred in the DNA sequences shown below:

   Normal DNA sequence: CCT GAT GCG AAG TTA TCA GTA CCA

   DNA sequence 1:   CCT GAT GCG TTA TCA GTA CCA _____

   DNA sequence 2:   CCT GAT GCG AAG CCC TTA TCA GTA CCA _____

   DNA sequence 3:   CCT GAT GCG AAG TTA TGA GTA CCA _____

© 2016 **BIOZONE** International
ISBN: 978-1-927309-46-9
Photocopying Prohibited

199 The Effects of Mutations

Key Idea: Most mutations produce harmful effects, but some can produce beneficial effects, and others have no effect at all.

▶ Most mutations have a harmful effect on the organism. This is because changes to the DNA sequence of a gene can potentially change the amino acid chain encoded by the gene. Proteins need to fold into a precise shape to function properly. A mutation may change the way the protein folds and prevent it from carrying out its usual biological function.

▶ However, sometimes a mutation can be beneficial. The mutation may result in a more efficient protein, or produce an entirely different protein that can improve the survival of the organism.

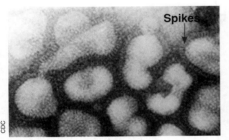

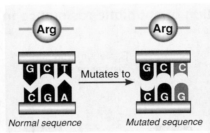

Beneficial mutations

Some mutations aid survival. In viruses (e.g. Influenzavirus above) genes coding for the glycoprotein spikes (arrowed) are constantly mutating, producing new strains that avoid detection by the host's immune system.

Silent mutations

Silent mutations do not change the amino acid sequence nor the final protein. In the genetic code, several codons may code for the same amino acid. Silent mutations may be neutral if they do not alter an organism's fitness.

Harmful mutations

Most mutations cause harmful effects, usually because they stop or alter the production of a protein (often an enzyme). Albinism (above) is one of the more common mutations in nature, and leaves an animal with no pigmentation.

Beneficial mutations in *E. coli*

An experiment known as the *E.coli* long term evolution experiment has incubated 12 lines of *E. coli* bacteria for more than 20 years.

After 31,000 generations a mutation in one of the *E. coli* populations enabled it to feed off citrate, a component of the medium they were grown in (*E. coli* are usually unable to do this). This ability gave these *E. coli* an advantage because they could use another food source.

The mutation was noticed when the optical density (cloudiness) of the flask containing the *E. coli* increased (right), indicating an increase in bacteria numbers.

Optical density of flask containing *E. coli*

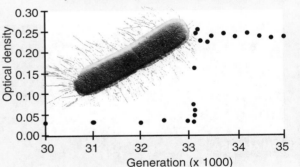

1. How might a mutation cause a beneficial effect on protein function? _____

_____

2. How might a mutation have a harmful effect on protein function? _____

_____

_____

3. Why does a silent mutation have no apparent affect on an organism? _____

_____

_____

4. How would the citrate mutation in *E. coli* give it an advantage over other *E. coli* populations? _____

_____

_____

200 Evolution of Antibiotic Resistance

Key Idea: Resistance to antibiotics can arise in bacteria by mutation. Antibiotic resistant bacteria can pass this resistance on to the next generation and to other populations.

▶ Antibiotic resistance arises when a genetic change allows bacteria to tolerate levels of an antibiotic that might normally kill it or stop its growth. This resistance may arise spontaneously through mutation or by transfer of genetic material between microbes.

▶ Genomic analyses from 30,000 year old permafrost sediments show that antibiotic resistant genes are not new. They have long been present in the bacterial genome, pre-dating the modern selective pressure of antibiotic use. In the current selective environment, these genes have proliferated and antibiotic resistance has spread.

The evolution of antibiotic resistance in bacteria

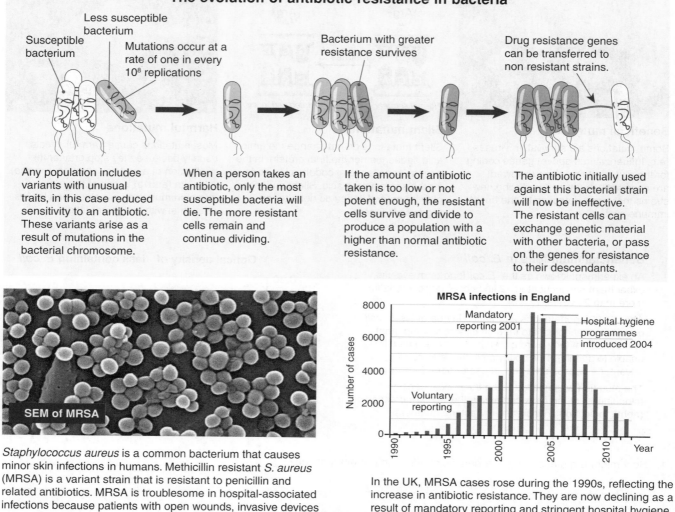

Susceptible bacterium

Less susceptible bacterium

Mutations occur at a rate of one in every 10^8 replications

Bacterium with greater resistance survives

Drug resistance genes can be transferred to non resistant strains.

Any population includes variants with unusual traits, in this case reduced sensitivity to an antibiotic. These variants arise as a result of mutations in the bacterial chromosome.

When a person takes an antibiotic, only the most susceptible bacteria will die. The more resistant cells remain and continue dividing.

If the amount of antibiotic taken is too low or not potent enough, the resistant cells survive and divide to produce a population with a higher than normal antibiotic resistance.

The antibiotic initially used against this bacterial strain will now be ineffective. The resistant cells can exchange genetic material with other bacteria, or pass on the genes for resistance to their descendants.

SEM of MRSA

Staphylococcus aureus is a common bacterium that causes minor skin infections in humans. Methicillin resistant *S. aureus* (MRSA) is a variant strain that is resistant to penicillin and related antibiotics. MRSA is troublesome in hospital-associated infections because patients with open wounds, invasive devices (e.g. catheters), or poor immunity are at greater risk of infection.

MRSA infections in England

Number of cases

Mandatory reporting 2001

Hospital hygiene programmes introduced 2004

Voluntary reporting

Year

In the UK, MRSA cases rose during the 1990s, reflecting the increase in antibiotic resistance. They are now declining as a result of mandatory reporting and stringent hospital hygiene programs. A similar pattern has been observed in the USA.

1. (a) Why can bacterial strains such as MRSA be so harmful? _____

   _____

   (b) How do bacterial strains such as MRSA arise? _____

   _____

   _____

2. How can the resistance become widespread? _____

   _____

© 2016 **BIOZONE** International
ISBN: 978-1-927309-46-9
Photocopying Prohibited

201 Beneficial Mutations in Humans

Key Idea: Beneficial mutations increase the fitness of the organisms that possess them. Beneficial mutations are relatively rare.

▶ **Beneficial mutations** are mutations that increase the fitness of the organisms that possess them. Although beneficial mutations are rare compared to those that are harmful, there are a number of well documented beneficial mutations in humans.

▶ Some of these mutations are not very common in the human population. This is because the mutations have been in existence for a relatively short time, so the mutations have not had time to become widespread in the human population.

▶ Scientists often study mutations that cause disease. By understanding the genetic origin of various diseases it may be possible to develop targeted medical drugs and therapies against them.

The Village of Limone, Italy

Apolipoprotein A1-Milano is a well documented mutation to apolipoprotein A1 that helps transport cholesterol through the blood. The mutation causes a change to one amino acid and increases the protein's effectiveness by ten times, dramatically reducing incidence of heart disease. The mutation can be traced back to its origin in Limone, Italy, in 1644. Another mutation to a gene called PCSK9 has a similar effect, lowering the risk of heart disease by 88%.

Lactose is a sugar found in milk. All infant mammals produce an enzyme called lactase that breaks the lactose into the smaller sugars glucose and galactose. As mammals become older, their production of lactase declines and they lose the ability to digest lactose. As adults, they become lactose intolerant and feel bloated after drinking milk. About 10,000 years ago a mutation appeared in humans that maintained lactase production into adulthood. This mutation is now carried in people of mainly European, African and Indian descent.

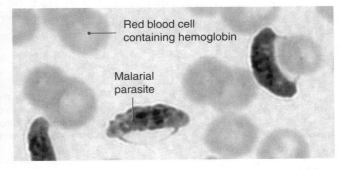
Red blood cell containing hemoglobin

Malarial parasite

Malaria resistance results from a mutation to the hemoglobin gene (Hb^S) that also causes sickle cell disease. This mutation in beneficial in regions where malaria is common. A less well known mutation (Hb^C) to the same gene, discovered in populations in Burkina Faso, Africa, results in a 29% reduction in the likelihood of contracting malaria if the person has one copy of the mutated gene and a 93% reduction if the person has two copies. In addition, the anemia that person suffers as a result of the mutation is much less pronounced than in the Hb^S mutation.

1. Why is it that many of the recent beneficial mutations recorded in humans have not spread throughout the entire human population?

_____

_____

2. What selection pressure could act on Apolipoprotein A1-Milano to help it spread through a population? _____

_____

_____

3. Why would it be beneficial to be able to digest milk in adulthood? _____

_____

_____

PRACTICES   CCC   WEB

CE   201   **KNOW**

202 Harmful Effects of Mutations in Humans

Key Idea: Many mutations are harmful. Changes to the DNA coding for the protein may prevent the protein from functioning correctly.

Cystic fibrosis

▶ Cystic fibrosis (CF) is an inherited disorder caused by a mutation of the CFTR gene. It is one of the most common lethal autosomal recessive conditions affecting white skinned people of European descent.

▶ The CFTR gene's protein product is a membrane-based protein that regulates chloride transport in cells. The ΔF508 mutation produces an abnormal CFTR protein, which cannot take its position in the plasma membrane (below, right) or perform its transport function. The ΔF508 mutation is the most common mutation causing CF.

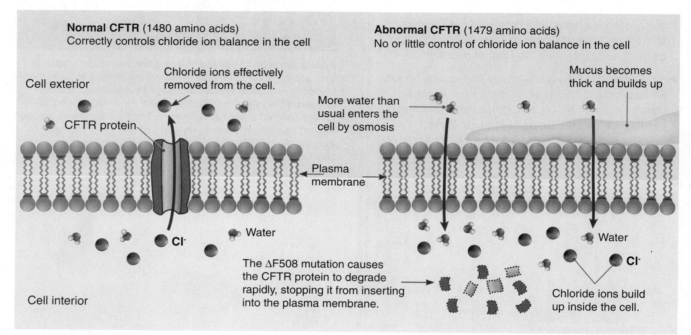

Normal CFTR (1480 amino acids)
Correctly controls chloride ion balance in the cell

Abnormal CFTR (1479 amino acids)
No or little control of chloride ion balance in the cell

Chloride ions effectively removed from the cell.

Cell exterior

CFTR protein

Plasma membrane

Water

Cl⁻

Cell interior

More water than usual enters the cell by osmosis

Mucus becomes thick and builds up

The ΔF508 mutation causes the CFTR protein to degrade rapidly, stopping it from inserting into the plasma membrane.

Chloride ions build up inside the cell.

Water

Cl⁻

Huntington's disease

▶ Huntington's disease is a progressive genetic disorder in which nerve cells in certain parts of the brain waste away, or degenerate. Symptoms include shaky hands and an awkward gait.

▶ Huntington's disease is caused by a mutation of the HTT gene on chromosome 4. The HTT gene has a repeating base sequence CAG. Normally this section repeats between 10 and 28 times, but in people with Huntington's disease this sequence repeated between 36 to 120 times. The greater the number of repeats, the greater the effects appear to be and the earlier the onset of the disease.

▶ Woody Guthrie (right) was an influential folk singer-songwriter who died in 1967 due to complications related to Huntington's disease.

Al Aumuller, NY World Telegram and the Sun, Public Domain

1. How does the ΔF508 mutation affect the amino acid sequence for the CFTR protein? _____

_____

2. (a) What causes Huntington's disease? _____

_____

(b) How does the extent of the mutation affect the symptoms and onset of the disease: _____

_____

_____

_____

 © 2016 **BIOZONE** International
ISBN: 978-1-927309-46-9
Photocopying Prohibited

203 Influences on Phenotype

Key Idea: An organism's phenotype is influenced by the effects of the environment during and after development, even though the genotype remains unaffected.

▶ The phenotype encoded by genes is a product not only of the genes themselves, but of their internal and external environment and the variations in the way those genes are controlled (epigenetics).

▶ Even identical twins have minor differences in their appearance due to epigenetic and environmental factors such as diet and intrauterine environment. Genes, together with epigenetic and environmental factors determine the unique phenotype that is produced.

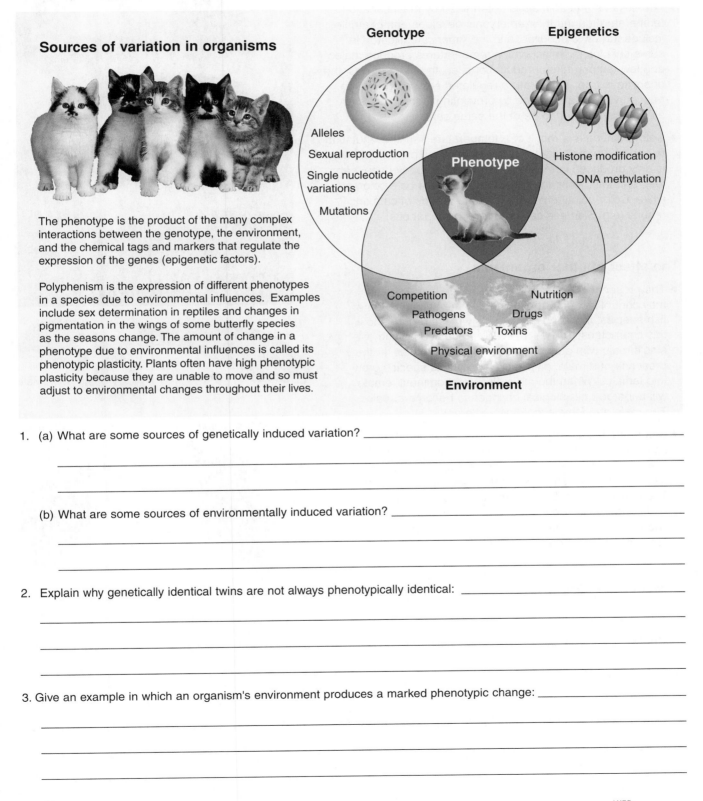

Sources of variation in organisms

The phenotype is the product of the many complex interactions between the genotype, the environment, and the chemical tags and markers that regulate the expression of the genes (epigenetic factors).

Polyphenism is the expression of different phenotypes in a species due to environmental influences. Examples include sex determination in reptiles and changes in pigmentation in the wings of some butterfly species as the seasons change. The amount of change in a phenotype due to environmental influences is called its phenotypic plasticity. Plants often have high phenotypic plasticity because they are unable to move and so must adjust to environmental changes throughout their lives.

1. (a) What are some sources of genetically induced variation? _____

_____

(b) What are some sources of environmentally induced variation? _____

_____

2. Explain why genetically identical twins are not always phenotypically identical: _____

_____

3. Give an example in which an organism's environment produces a marked phenotypic change: ___

_____

204 Environment and Variation

Key Idea: The environment can have a large effect on an organism's phenotype without affecting its genotype.

▶ Environmental factors can modify the phenotype encoded by genes without changing the genotype. This can occur both during development and later in life. Environmental factors that affect the phenotype of plants and animals include nutrients or diet, temperature, altitude or latitude, and the presence of other organisms.

The effect of temperature

▶ The sex of some animals is determined by the incubation temperature during their embryonic development. Examples include turtles, crocodiles, and the American alligator. In some species, high incubation temperatures produce males and low temperatures produce females. In other species, the opposite is true. Temperature regulated sex determination may provide an advantage by preventing inbreeding (since all siblings will tend to be of the same sex).

▶ Color-pointing is a result of a temperature sensitive mutation to one of the melanin-producing enzymes. The dark pigment is only produced in the cooler areas of the body (face, ears, feet, and tail), while the rest of the body is a pale color, or white. Color-pointing is seen in some breeds of cats and rabbits (e.g. Siamese cats and Himalayan rabbits).

The effect of other organisms

▶ The presence of other individuals of the same species may control sex determination for some animals. Some fish species, including Sandager's wrasse (right), show this characteristic. The fish live in groups consisting of a single male with attendant females and juveniles. In the presence of a male, all juvenile fish of this species grow into females. When the male dies, the dominant female will undergo physiological changes to become a male. The male and female look very different.

Female   Male

Non-helmeted *Daphnia*   Helmeted *Daphnia*

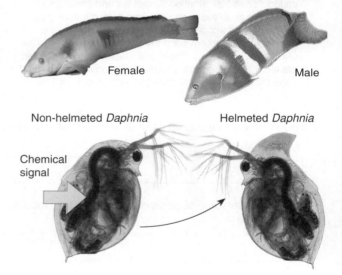

Chemical signal

▶ Some organisms respond to the presence of other, potentially harmful, organisms by changing their body shape. Invertebrates, such as some *Daphnia* species, grow a helmet when invertebrate predators are present. The helmet makes *Daphnia* more difficult to attack and handle. Such changes are usually in response to chemicals produced by the predator (or competitor) and are common in plants as well as animals.

1. (a) Give two examples of how temperature affects a phenotypic characteristic in an organism: _____

   _____

   (b) Why are the darker patches of fur in color-pointed cats and rabbits found only on the face, paws, and tail? _____

   _____

   _____

2. How is helmet development in *Daphnia* an adaptive response to environment? _____

   _____

   _____

PRACTICES

KNOW

© 2016 **BIOZONE** International
ISBN: 978-1-927309-46-9
Photocopying Prohibited

The effect of altitude

Severe stunting

Growth to genetic potential

Cline

Increasing altitude can stunt the phenotype of plants with the same genotype. In some conifers, e.g. Engelmann spruce, plants at low altitude grow to their full genetic potential, but growth becomes progressively more stunted as elevation increases. Growth is gnarled and bushy at the highest, most severe sites. Gradual change in phenotype over an environmental gradient is called a **cline**.

The effect of chemical environment

The chemical environment can influence the phenotype in plants and animals. The color of hydrangea flowers varies with soil pH. Blue flowers (due to the presence of aluminium compounds in the flowers) occur in more acidic soils (pH 5.0-5.5) in which aluminium is more readily available. In less acidic soils (pH 6.0-6.5) the flowers are pink.

3. (a) What is a **cline**? _____

_____

(b) What physical factors associated with altitude could affect plant phenotype? _____

_____

4. Describe an example of how the chemical environment of a plant can influence phenotype: _____

_____

_____

_____

_____

_____

5. Vegetable growers can produce enormous vegetables for competition. How could you improve the chance that a vegetable would reach its maximum genetic potential?

_____

_____

_____

_____

6. Two different species of plant (A and B) were found growing together on a windswept portion of a coast, Both have a low growing (prostrate) phenotype. One of each plant type was transferred to a greenhouse where "ideal" conditions were provided to allow maximum growth. In this controlled environment, species B continued to grow in its original prostrate form, but species A changed its growing pattern and became erect in form. Identify the cause of the prostrate phenotype in each of the coastal grown plant species and explain your answer:

Plant species A: _____

_____

_____

_____

Plant species B: _____

_____

_____

205 Genes and Environment Interact

Key Idea: The environment or experiences of an individual can affect the development of following generations.

▶ Studies of heredity have found that the environment or lifestyle of an ancestor can have an effect on future generations. Certain environments or diets can affect the methylation and packaging of the DNA (rather than the DNA itself) determining which genes are switched on or off and so affecting the development of the individual. These effects can be passed on to offspring, and even on to future generations. It is thought that these inherited effects may provide a rapid way to adapt to particular environmental situations.

▶ The destruction of New York's Twin Towers on September 11, 2001, traumatized thousands of people. In those thousands were 1700 pregnant women. Some of them suffered (often severe) post-traumatic stress disorder, others did not. Studies on the mothers who developed PTSD found very low levels of the stress-related hormone cortisol in their saliva. The children of these mothers also had much lower levels of cortisol than those whose mothers had not suffered PTSD. The environment of the mother had affected the offspring.

Rats and environmental effects

The effect of the environment and diet of mothers on later generations exposed to a breast cancer trigger (a carcinogenic chemical) was investigated in rats fed a high fat diet or a diet high in estrogen. The length of time taken for breast cancer to develop in later generations after the trigger for breast cancer was given was recorded and compared. The data is presented below. F_1= daughters, F_2= granddaughters, F_3 = great granddaughters.

	Cumulative percentage rats with breast cancer (high fat diet (HFD))					
	F_1%		F_2%		F_3%	
Weeks since trigger	HFD	Control	HFD	Control	HFD	Control
6	0	0	5	0	3	0
8	15	0	20	5	3	20
10	22	8	30	5	10	25
12	22	18	50	20	20	30
14	22	18	50	30	25	40
16	29	18	60	30	25	40
18	29	18	60	40	40	42
20	40	18	65	40	50	60
22	80	60	79	50	50	60

Data source: S. De Assis: Nature Communications 3 (Article 1053) (2012)

	Cumulative percentage rats with breast cancer (high estrogen diet (HED))					
	F_1%		F_2%		F_3%	
Weeks since trigger	HED	Control	HED	Control	HED	Control
6	5	0	10	0	0	0
8	10	0	10	0	15	10
10	30	15	15	20	30	20
12	38	19	30	30	40	20
14	50	22	30	40	50	20
16	50	22	30	40	50	30
18	60	35	40	40	75	40
20	60	42	50	50	80	45
22	80	55	50	50	80	60

© 2016 **BIOZONE** International
ISBN: 978-1-927309-46-9
Photocopying Prohibited

1. Use the data on the previous page to complete the graphs below. The first graph is done for you:

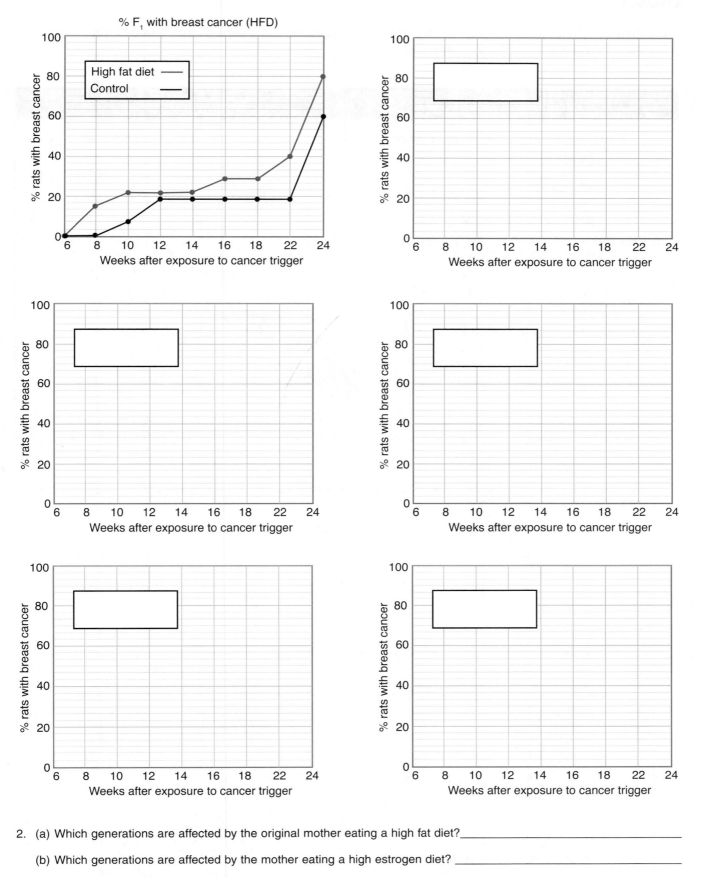

% F_1 with breast cancer (HFD)

High fat diet ——
Control ——

2. (a) Which generations are affected by the original mother eating a high fat diet?_____

   (b) Which generations are affected by the mother eating a high estrogen diet? _____

   (c) Which diet had the longest lasting effect? _____

3. What do these experiments show with respect to diet and generational effects? _____

   _____

206 Predicting Traits: The Monohybrid Cross

Key Idea: The outcome of a cross depends on the parental genotypes. A true breeding parent is homozygous for the gene involved.

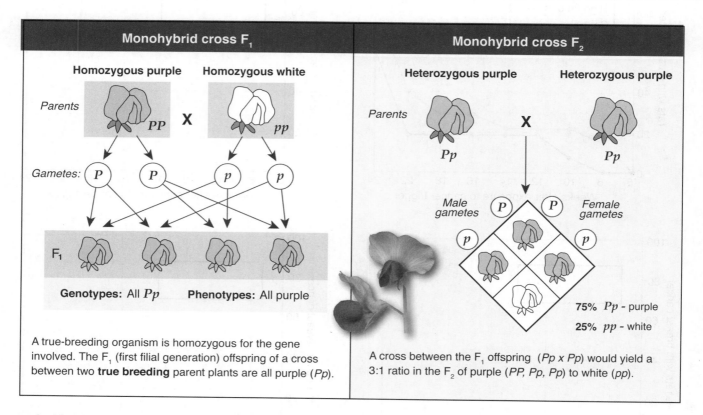

Monohybrid cross F₁	Monohybrid cross F₂

Monohybrid cross F₁

Homozygous purple   **Homozygous white**

Parents   *PP*   X   *pp*

Gametes: P   P   p   p

F₁

Genotypes: All *Pp*   **Phenotypes:** All purple

A true-breeding organism is homozygous for the gene involved. The F₁ (first filial generation) offspring of a cross between two **true breeding** parent plants are all purple (*Pp*).

Monohybrid cross F₂

Heterozygous purple   **Heterozygous purple**

Parents   *Pp*   X   *Pp*

Male gametes   P   P   *Female gametes*

p   p

75% *Pp* - purple

25% *pp* - white

A cross between the F₁ offspring (*Pp x Pp*) would yield a 3:1 ratio in the F₂ of purple (*PP, Pp, Pp*) to white (*pp*).

1. Study the diagrams above and explain why white flower color does not appear in the F₁ generation but reappears in the F₂ generation:

_____

_____

_____

_____

2. Complete the crosses below:

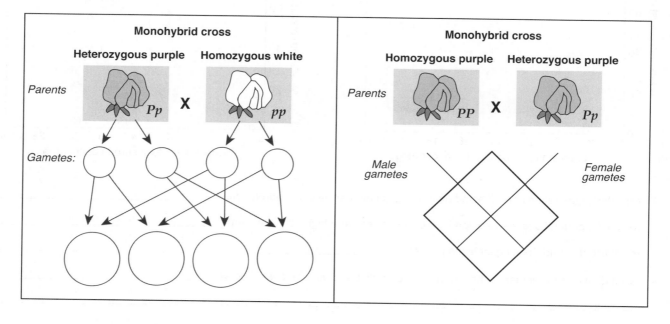

Monohybrid cross

Heterozygous purple   **Homozygous white**

Parents   *Pp*   X   *pp*

Gametes:

Monohybrid cross

Homozygous purple   **Heterozygous purple**

Parents   *PP*   X   *Pp*

Male gametes   *Female gametes*

© 2016 **BIOZONE** International
ISBN: 978-1-927309-46-9
Photocopying Prohibited

207 Predicting Traits: The Test Cross

Key Idea: If an individual's genotype is unknown, it can be determined using a test cross.

▶ It is not always possible to determine an organism's genotype by its appearance because gene expression is complicated by patterns of dominance and by gene interactions. The **test cross** was developed by Mendel as a way to establish the genotype of an organism with the dominant phenotype for a particular trait.

▶ The principle is simple. The individual with the unknown genotype is bred with a homozygous recessive individual for the trait(s) of interest. The homozygous recessive can produce only one type of allele (recessive), so the phenotypes of the offspring will reveal the genotype of the unknown parent (below). The test cross can be used to determine the genotype of single genes or multiple genes.

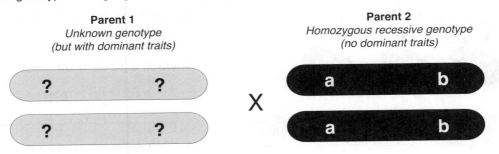

Parent 1
Unknown genotype
(but with dominant traits)

Parent 2
Homozygous recessive genotype
(no dominant traits)

The common fruit fly (*Drosophila melanogaster*) is often used to illustrate basic principles of inheritance because it has several genetic markers whose phenotypes are easily identified. Once such phenotype is body color. Wild type (normal) *Drosophila* have yellow-brown bodies. The allele for yellow-brown body color (E) is dominant. The allele for an ebony colored body (e) is recessive. The test crosses below show the possible outcomes for an individual with homozygous and heterozygous alleles for ebony body color.

A. A homozygous recessive female (ee) with an ebony body is crossed with a homozygous dominant male (EE).

B. A homozygous recessive female (ee) with an ebony body is crossed with a heterozygous male (Ee).

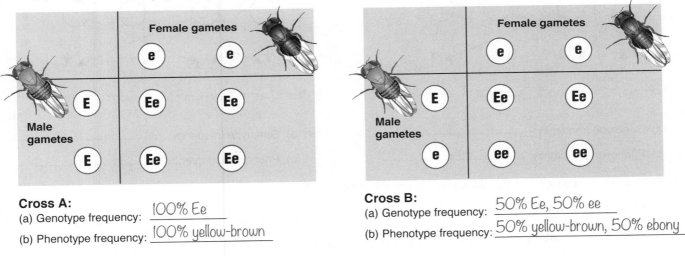

Cross A:
(a) Genotype frequency: 100% Ee
(b) Phenotype frequency: 100% yellow-brown

Cross B:
(a) Genotype frequency: 50% Ee, 50% ee
(b) Phenotype frequency: 50% yellow-brown, 50% ebony

1. In *Drosophila*, the allele for brown eyes (b) is recessive, while the red eye allele (B) is dominant. Set up and carry out a test cross to determine the genotype of a male who has red eyes:

2. 50% of the resulting progeny have red eyes, and 50% have brown eyes. What is the genotype of the male *Drosophila*?

© 2016 **BIOZONE** International
ISBN: 978-1-927309-46-9
Photocopying Prohibited

PRACTICES
WEB

207

KNOW

282

208 Practicing Monohybrid Crosses

Key Idea: A monohybrid cross studies the inheritance pattern of one gene. The offspring of these crosses occur in predictable ratios.

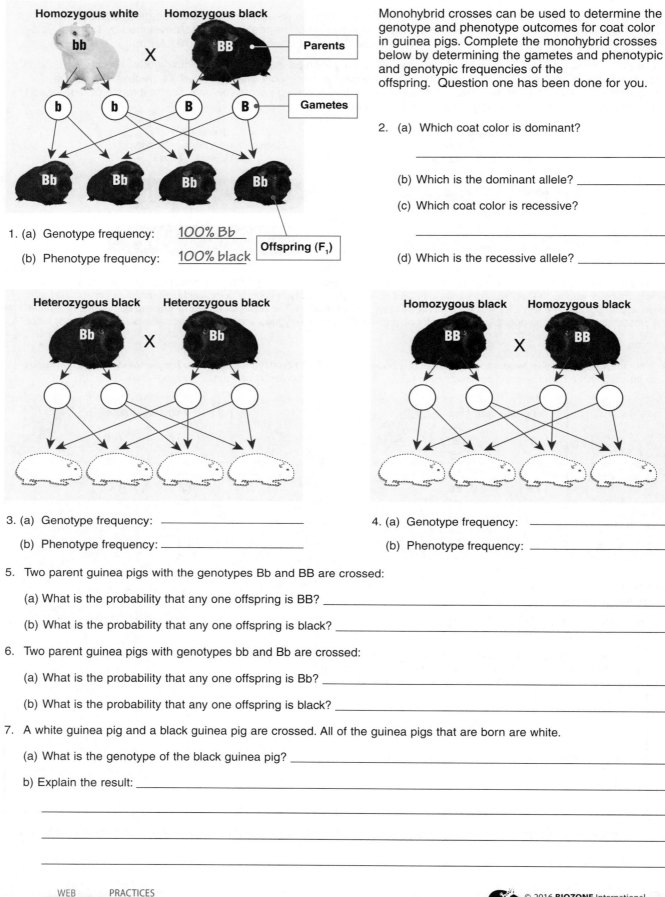

Monohybrid crosses can be used to determine the genotype and phenotype outcomes for coat color in guinea pigs. Complete the monohybrid crosses below by determining the gametes and phenotypic and genotypic frequencies of the offspring. Question one has been done for you.

Homozygous white **Homozygous black**

bb   X   BB — **Parents**

b   b   B   B — **Gametes**

Bb   Bb   Bb   Bb

Offspring (F₁)

1. (a) Genotype frequency: _100% Bb_

   (b) Phenotype frequency: _100% black_

2. (a) Which coat color is dominant?

   _____

   (b) Which is the dominant allele? _____

   (c) Which coat color is recessive?

   _____

   (d) Which is the recessive allele? _____

Heterozygous black **Heterozygous black**

Bb   X   Bb

Homozygous black **Homozygous black**

BB   X   BB

3. (a) Genotype frequency: _____

   (b) Phenotype frequency: _____

4. (a) Genotype frequency: _____

   (b) Phenotype frequency: _____

5. Two parent guinea pigs with the genotypes Bb and BB are crossed:

   (a) What is the probability that any one offspring is BB? _____

   (b) What is the probability that any one offspring is black? _____

6. Two parent guinea pigs with genotypes bb and Bb are crossed:

   (a) What is the probability that any one offspring is Bb? _____

   (b) What is the probability that any one offspring is black? _____

7. A white guinea pig and a black guinea pig are crossed. All of the guinea pigs that are born are white.

   (a) What is the genotype of the black guinea pig? _____

   b) Explain the result: _____

   _____

   _____

WEB   PRACTICES

KNOW   **208**

209 Predicting Traits: Dihybrid Inheritance

Key Idea: A dihybrid cross studies the inheritance pattern of two genes. In crosses involving unlinked autosomal genes, the offspring occur in predictable ratios.

▶ There are four types of gamete produced in a cross involving two genes, where the genes are carried on separate chromosomes and are sorted independently of each other during meiosis.

▶ The two genes in the example below are on separate chromosomes and control two unrelated characteristics, hair color and coat length. Black (B) and short (L) are dominant to white and long.

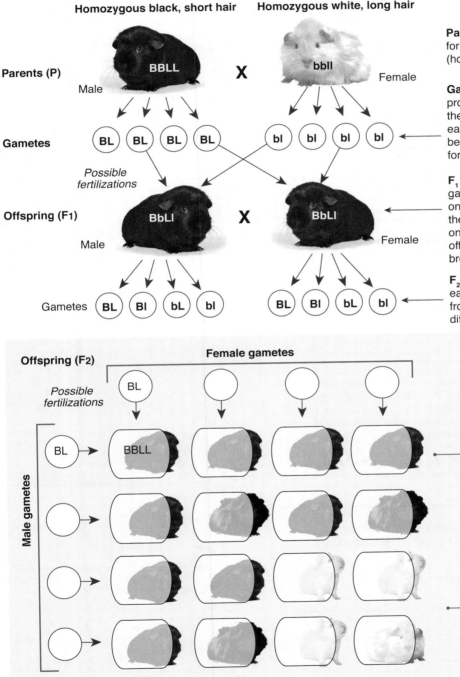

Homozygous black, short hair   **Homozygous white, long hair**

Parents (P)   BBLL   X   bbll
Male   Female

Gametes   BL BL BL BL   bl bl bl bl

Possible fertilizations

Offspring (F1)   BbLl   X   BbLl
Male   Female

Gametes   BL Bl bL bl   BL Bl bL bl

Parents: The notation P is only used for a cross between true breeding (homozygous) parents.

Gametes: Only one type of gamete is produced from each parent (although they will produce four gametes from each oocyte or spermatocyte). This is because each parent is homozygous for both traits.

F_1 offspring: There is only one kind of gamete from each parent, therefore only one kind of offspring produced in the first generation. The notation F_1 is only used to denote the heterozygous offspring of a cross between two true breeding parents.

F_2 offspring: The F_1 were mated with each other (selfed). Each individual from the F_1 is able to produce four different kinds of gamete.

Offspring (F2)

Possible fertilizations

Female gametes: BL

Male gametes: BL → BBLL

Using a grid called a **Punnett square** (left), it is possible to determine the expected genotype and phenotype ratios in the F_2 offspring. The notation F_2 is only used to denote the offspring produced by crossing F_1 heterozygotes.

Each of the 16 animals shown here represents the possible zygotes formed by different combinations of gametes coming together at fertilization.

1. Fill in the gametes and complete the Punnett square above.

2. Use the Punnett square identify the number of each phenotype in the offspring: _____

_____

PRACTICES   CCC   WEB
SPQ   209   **KNOW**

210 Practicing Dihybrid Crosses

Key Idea: Dihybrid crosses produce offspring in predictable ratios. The simplest way of predicting these outcomes is to use a Punnett square.

1. In guinea pigs, rough coat **R** is dominant over smooth coat **r** and black coat **B** is dominant over white **b**. The genes are not linked. A homozygous rough black animal was crossed with a homozygous smooth white:

   (a) State the genotype of the F_1:

   (b) State the phenotype of the F_1:

   (c) Use the Punnett square (top right) to show the outcome of a cross between the F_1 (the F_2):

   (d) Using ratios, state the phenotypes of the F_2 generation:

   _____

   (e) Use the Punnett square (right) to show the outcome of a back cross of the F_1 to the rough, black parent:

   (f) Using ratios, state the phenotype of the offspring of this back cross:

   _____

   (g) A rough black guinea pig was crossed with a rough white guinea pig produced the following offspring: 3 rough black, 2 rough white, and 1 smooth white. What are genotypes of the parents?

   _____

   _____

   _____

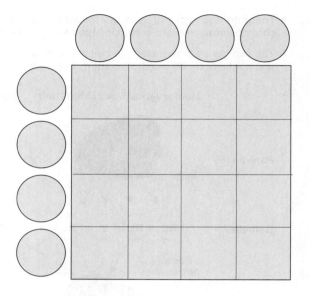

2. In humans, two genes affecting the appearance of the hands are the gene for thumb hyperextension (curving) and the gene for mid-digit hair. The allele for curved thumb, **H**, is dominant to the allele for straight thumb, **h**. The allele for mid digit hair, **M**, is dominant to that for an absence of hair, **m**.

   (a) Give all the genotypes of individuals who are able to curve their thumbs, but have no mid-digit hair:

   _____

   (b) Complete the Punnett square (bottom right) to show the possible genotypes from a cross between two individuals heterozygous for both alleles:

   (c) State the phenotype ratios of the F_1 progeny:

   _____

   _____

   _____

   (d) What is the probability one of the offspring would have mid-digit hair?

   _____

   _____

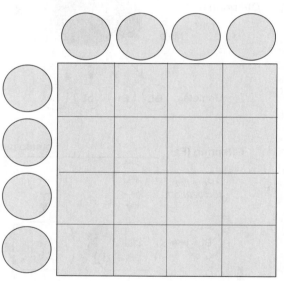

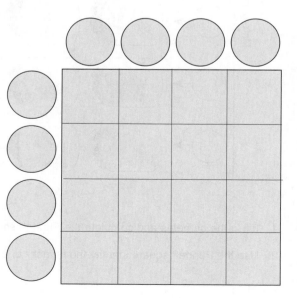

© 2016 **BIOZONE** International
ISBN: 978-1-927309-46-9
Photocopying Prohibited

285

3. In rabbits, spotted coat **S** is dominant to solid color **s**, while for coat color, black **B** is dominant to brown **b**. A brown spotted rabbit is mated with a solid black one and all the offspring are black spotted (the genes are not linked).

(a) State the genotypes:

Parent 1: _____

Parent 2: _____

Offspring: _____

(b) Use the Punnett square to show the outcome of a cross between the F_1 (the F_2):

(c) Using ratios, state the phenotypes of the F_2 generation:

_____

_____

_____

4. The Himalayan color-pointed, long-haired cat is a breed developed by crossing a pedigree (true-breeding), uniform-colored, long-haired Persian with a pedigree color-pointed (darker face, ears, paws, and tail) short-haired Siamese.

Persian   Siamese   Himalayan

The genes controlling hair coloring and length are on separate chromosomes: uniform color **U**, color pointed **u**, short hair **S**, long hair **s**.

(a) State the genotype of the F_1 (Siamese X Persian): _____

(b) State the phenotype of the F_1: _____

(c) Use the Punnett square (right) to show the outcome of a cross between the F_1 (the F_2):

(d) What ratio of the F_2 will be Himalayan? _____

(e) State whether the Himalayan would be true breeding:

_____

(f) What ratio of the F_2 will be color-point, short-haired cats:

_____

5. In cats, the following alleles are present for coat characteristics: black (**B**), brown (**b**), short (**L**), long (**l**). The genes are not linked. Use the information to complete the dihybrid crosses below:

A black short haired (**BBLl**) male is crossed with a black long haired (**Bbll**) female. Determine the genotypic and phenotypic ratios of the offspring:

Genotype ratio: _____

_____

Phenotype ratio: _____

_____

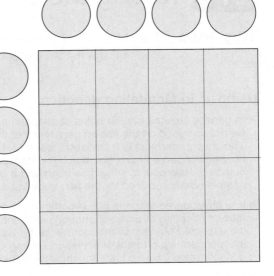
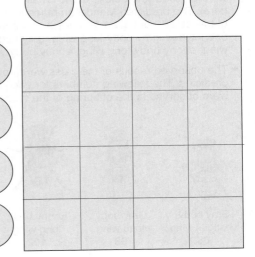
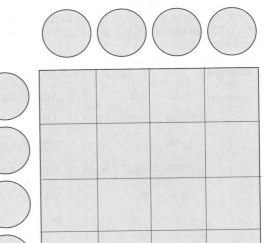

211 Testing the Outcome of Genetic Crosses

Key Idea: The chi-squared test for goodness of fit (χ^2) can be used for testing the outcome of dihybrid crosses against an expected (predicted) Mendelian ratio.

Using χ^2 in Mendelian genetics

▶ In genetic crosses certain ratios of offspring can be predicted based on the known genotypes of the parents The chi-squared test is a statistical test to determine how well observed numbers match (or fit) expected numbers. Raw counts should be used and a large sample size is required for the test to be valid.

▶ In a chi-squared test, the null hypothesis predicts the ratio of offspring of different phenotypes is the same as the expected Mendelian ratio for the cross, assuming independent assortment of alleles (no linkage, i.e. the genes involved are on different chromosomes).

▶ Significant departures from the predicted Mendelian ratio indicate linkage (the genes are on the same chromosome) of the alleles in question.

▶ In a *Drosophila* genetics experiment, two individuals were crossed (the details of the cross are not relevant here). The predicted Mendelian ratios for the offspring of this cross were 1:1:1:1 for each of the four following phenotypes: gray body-long wing, gray body-vestigial wing, ebony body-long wing, ebony body-vestigial wing.

▶ The observed results of the cross were not exactly as predicted. The following numbers for each phenotype were observed in the offspring of the cross:

| Gray body, vestigial wing **88** | Gray body, long wing **98** | Ebony body, long wing **102** | Ebony body, vestigial wing **112** |

Table 1: Critical values of χ^2 at different levels of probability. By convention, the critical probability for rejecting the null hypothesis (H_0) is 5%. If the test statistic is less than the tabulated critical value for $P = 0.05$ we cannot reject Ho and the result is not significant. If the statistic is greater than the tabulated value for $P = 0.05$ we reject (H_0) in favour of the alternative hypothesis.

Degrees of freedom	Level of probability (P)					
	0.50	0.20	0.10	0.05	0.02	0.01
1	0.455	1.64	2.71	3.84	5.41	6.64
2	1.386	3.22	4.61	5.99	7.82	9.21
3	2.366	4.64	6.25	7.82	9.84	11.35
4	3.357	5.99	7.78	9.49	11.67	13.28
5	4.351	7.29	9.24	11.07	13.39	15.09

Do not reject H_0 ⟵   Reject H_0 ⟶

Steps in performing a χ^2 test

1 **Enter the observed value (O).**
Enter the values of the offspring into the table in the appropriate category (column 1).

2 **Calculate the expected value (E).**
In this case the expected ratio is 1:1:1:1. Therefore the number of offspring in each category should be the same (i.e. total offspring/ no. categories). 400 / 4 = 100 (column 2).

3 **Calculate O-E and (O-E)^2**
The difference between the observed and expected values is calculated as a measure of the deviation from a predicted result. Since some deviations are negative, they are all squared to give positive values (column 3 and 4).

4 **Calculate χ^2**
For each category calculate $(O - E)^2 / E$. Then sum these values to produce the χ^2 value (column 5).

$$\chi^2 = \sum \frac{(O - E)^2}{E}$$

5 **Calculate degrees of freedom**
The probability that any particular χ^2 value could be exceeded by chance depends on the number of degrees of freedom. This is simply one less than the total number of categories (this is the number that could vary independently without affecting the last value) In this case 4 - 1 = 3.

6 **Use χ^2 table**
On the χ^2 table with 3 degrees of freedom, the calculated χ^2 value correspond to a probability between 0.2 and 0.5. By chance alone a χ^2 value of **2.96** will happen 20% to 50% of the time. The probability of 0.0 to 0.5 is higher than 0.05 (i.e 5% of the time) and therefore the null hypothesis cannot be rejected. We have no reason to believe the observed values differ significantly from the expected values.

	1	2	3	4	5
Category	O	E	O-E	(O_E)^2	(O_E)^2/E
GB, LW	98	100	-2	4	0.04
GB, VW	88	100	-12	144	1.44
EB, LW	102	100	2	4	0.04
EB, VW	112	100	12	144	1.44
				χ^2 ⟶	2.96

© 2016 **BIOZONE** International
ISBN: 978-1-927309-46-9
Photocopying Prohibited

1. Students carried out a pea plant experiment, where two heterozygous individuals were crossed. The predicted Mendelian ratios for the offspring were **9:3:3:1** for each of the four following phenotypes: round-yellow seed, round-green seed, wrinkled-yellow seed, wrinkled-green seed.

   The observed results of the cross were not exactly as predicted. The numbers of offspring with each phenotype are provided below:

Observed results of the pea plant cross			
Round-yellow seed	441	Wrinkled-yellow seed	143
Round-green seed	159	Wrinkled-green seed	57

   (a) State your null hypothesis for this investigation (H$_0$)

   _____

   _____

   (b) State the alternative hypothesis (H$_A$): _____

   _____

   Use the chi-squared test to determine if the differences observed between the phenotypes are significant. Use the table of critical values for χ^2 at different P values on the previous page.

   (c) Enter the observed and expected values (number of individuals) and complete the table to calculate the χ^2 value.

Category	O	E	O – E	(O – E)²	$\frac{(O-E)^2}{E}$
Round-yellow seed					
Round-green seed					
Wrinkled-yellow seed					
Wrinkled-green seed					
					Σ

   (d) Calculate the χ^2 value using the equation: $\chi^2 = \sum \frac{(O-E)^2}{E}$

   (e) Calculate the degrees of freedom: _____

   (f) Using the χ^2 table, state the P value corresponding to your calculated χ^2 value: _____

   (g) State your decision (circle one):   reject H$_0$ / do not reject H$_0$

2. In another experiment a group of students bred two corn plants together. The first corn plant was known to grown from a kernel that was colorless (c) and did not have a waxy endosperm (w). The second corn plant was grown from a seed that was colored (C) but with a waxy endosperm (W). When the corn ear was mature the students removed it and counted the different phenotypes in the corn kernels.

Observed results of corn kernels			
Colored - waxy	201	Colorless - waxy	86
Colored - not waxy	85	Colorless - not waxy	210

   From the observed results the students argued two points:
   (1) The plant with the dominant phenotype must have been heterozygous for both traits.
   (2) The genes for kernel color and endosperm waxiness must be linked (on the same chromosome).

   (a) Defend the students' first argument: _____

   _____

   (b) On a separate sheet use a chi-squared test to prove or disprove the students second argument:

© 2016 **BIOZONE** International
ISBN: 978-1-927309-46-9
Photocopying Prohibited

212 Pedigree Analysis

Key Idea: Pedigree charts illustrate inheritance patterns over a number of generations. They allow a genetic disorder to be traced back to its origin.

Sample pedigree chart

Pedigree charts are a way of showing inheritance patterns over a number of generations. They are often used to study the inheritance of genetic disorders. The key should be consulted to decode the symbols. Individuals are identified by their generation number and their order number in that generation. For example, **II-6** is the sixth person in the second row. The arrow indicates the person through whom the pedigree was discovered (i.e. who reported the condition).

If the chart on the right were illustrating a human family tree, it would represent three generations: grandparents (I-1 and I-2) with three sons and one daughter. Two of the sons (II-3 and II-4) are identical twins, but did not marry or have any children. The other son (II-1) married and had a daughter and another child (sex unknown). The daughter (II-5) married and had two sons and two daughters (plus a child that died in infancy).

For the particular trait being studied, the grandfather was expressing the phenotype (showing the trait) and the grandmother was a carrier. One of their sons and one of their daughters also show the trait, together with one of their granddaughters.

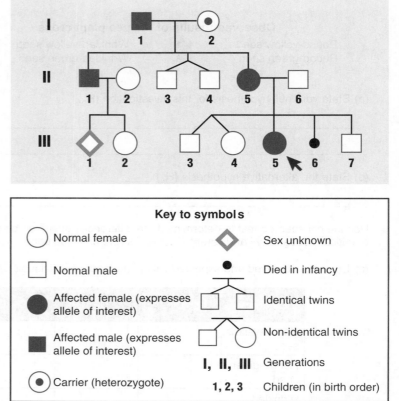

1. Pedigree chart of your family

Using the symbols in the key above and the example illustrated as a guide, construct a pedigree chart of your own family (or one that you know of) starting with the parents of your mother and/or father on the first line. Your parents will appear on the second line (II) and you will appear on the third line (III). There may be a fourth generation line (IV) if one of your brothers or sisters has had a child. Use a ruler to draw up the chart carefully.

© 2016 **BIOZONE** International
ISBN: 978-1-927309-46-9
Photocopying Prohibited

2. The pedigree chart right shows the inheritance of allele A in a flower that can be blue or white. Answer the questions below:

    (a) Which color is produced by the dominant allele?

    _____

    (b) Write on the chart the genotype for each of the generation I individuals.

    (c) III4 is crossed with a white flower. What is the probability that any one offspring also has a white flower?

    _____

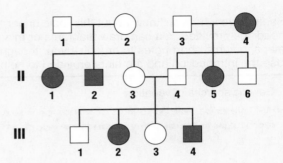

3. The pedigree chart right shows the inheritance of the allele B in a mammal that can have a coat color of black or white.

    (a) Which color is produced by the dominant allele?

    _____

    (b) Explain how you know this: _____

    _____

    _____

    _____

    _____

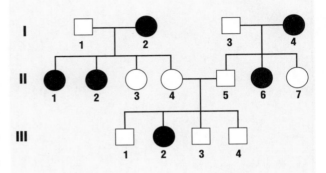

4. **Autosomal recessive traits**
    Albinos lack pigment in the hair, skin and eyes. This trait is inherited as an autosomal recessive allele (i.e. it is not carried on the sex chromosome).

    Albinism in humans

    (a) Write the genotype for each of the individuals on the chart using the following letter codes:
    PP normal skin color; **P-** normal (but unknown if homozygous), **Pp** carrier, **pp** albino.

    (b) Why must the parents (II-3) and (II-4) be **carriers** of a **recessive** allele:

    _____

    _____

    _____

    _____

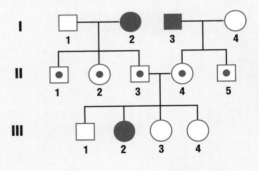

5. **Autosomal dominant traits**
    An unusual trait found in some humans is woolly hair (not to be confused with curly hair). Each affected individual will have at least one affected parent.

    Woolly hair in humans

    (a) Write the genotype for each of the individuals on the chart using the following letter codes:
    WW woolly hair, **Ww** woolly hair (heterozygous), **W-** woolly hair, but unknown if homozygous, **ww** normal hair.

    (b) Describe a feature of this inheritance pattern that suggests the trait is the result of a **dominant** allele:

    _____

    _____

    _____

    _____

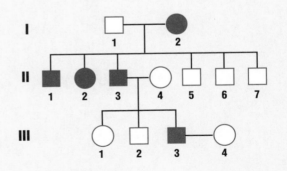

213 Chapter Review

Summarize what you know about this topic under the headings provided. You can draw diagrams or mind maps, or write short notes to organize your thoughts. Use the introduction and the hints provided to help you:

Genetic sources of variation

HINT: Describe genetic sources of variation in sexually reproducing organisms. Why is variation important?

Environmental sources of variation

HINT: How does environment affect expression of genotype.

© 2016 **BIOZONE** International
ISBN: 978-1-927309-46-9
Photocopying Prohibited

214 KEY TERMS AND IDEAS: Did You Get It?

1. Test your vocabulary by matching each term to its definition, as identified by its preceding letter code.

alleles _____

dominant _____

genotype _____

heterozygous _____

homologous chromosomes _____

meiosis _____

mutation _____

phenotype _____

recessive _____

recombination _____

trait _____

A Observable characteristics in an organism.

B Allele that will only express its trait in the absence of the dominant allele.

C Phenotypic characteristic (e.g. red hair).

D Possessing two different alleles of a particular gene, one inherited from each parent.

E Sequences of DNA occupying the same gene locus (position) on different, but homologous, chromosomes.

F The process of double nuclear division (reduction division) to produce four nuclei, each containing half the original number of chromosomes (haploid).

G A change to the DNA sequence of an organism. This may be a deletion, insertion, duplication, inversion or translocation of DNA in a gene or chromosome.

H The exchange of alleles between homologous chromosomes as a result of crossing over.

I Allele that expresses its trait irrespective of the other allele.

J Chromosome pairs, one paternal and one maternal, of the same length, centromere position, and staining pattern with genes for the same characteristics at corresponding loci.

K The allele combination of an organism.

2. The allele for wrinkled seeds in pea plants is considered recessive because:

A It is not expressed in the F_2 generation

B It is not expressed in the heterozygote.

C Individuals who have the allele are less likely to pass genes on to the next generation.

D Round peas are smaller than wrinkled peas.

3. True breeding individuals:

A Only ever breed with others of the same kind.

B Always have the same coat or flower color.

C Are homozygous for a particular allele.

D Always produce heterozygous offspring.

4. Use lines to match the statements in the table below to form complete sentences:

Mutations are the ultimate …

Alleles are variations …

A person carrying two of the same alleles (one on each homologous chromosome) …

If the person carries two different alleles …

Alleles may be …

A dominant allele …

A recessive allele only …

… of a gene.

… dominant or recessive

… for the gene, they are heterozygous.

… is said to be homozygous

… expresses its trait if it is homozygous.

… source of new alleles.

… always expresses its trait whether it is homozygous or heterozygous

5. Complete the following crosses for pea plants, including genotype and phenotype ratios: Round seeds (R) are dominant to wrinkled seeds (r):

(a) RR x Rr:

(b) Rr x rr

TEST

215 Summative Assessment

The pedigree of lactose intolerance

Lactose intolerance is the inability to digest the milk sugar lactose. It occurs because some people do not produce lactase, the enzyme needed to break down lactose. The pedigree chart below was one of the original studies to determine the inheritance pattern of lactose intolerance.

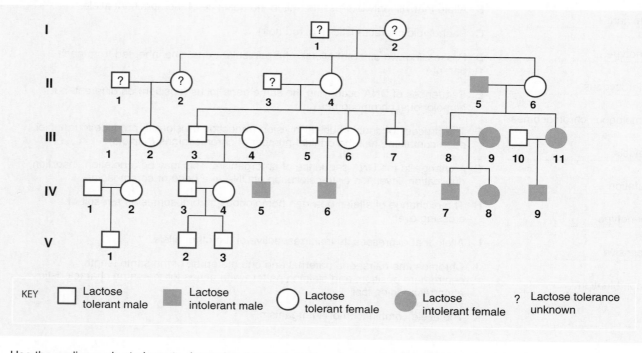

KEY
- ☐ Lactose tolerant male
- ■ Lactose intolerant male
- ○ Lactose tolerant female
- ● Lactose intolerant female
- ? Lactose tolerance unknown

1. Use the pedigree chart above to determine if lactose intolerance is a dominant trait or a recessive trait:

_____

2. Explain your answer: _____

_____

_____

_____

Linked genes

Shortly after the rediscovery of Mendel's work early in the 20th century, it became apparent that his ratios of 9:3:3:1 for heterozygous dihybrid crosses did not always hold true. Experiments on sweet peas by William Bateson and Reginald Punnett, and on *Drosophila* by Thomas Hunt Morgan, showed that there appeared to be some kind of coupling between genes. This coupling, which we now know to be linkage, did not follow any genetic relationship known at the time.

3. The data below is for a cross of sweet peas carried out by Bateson and Punnett. Purple flowers (P) are dominant to red (p), and long pollen grains (L) are dominant to round (l). If these genes were unlinked, the outcome of a cross between two heterozygous sweet peas should be a 9:3:3:1 ratio.

Study the data and use it to test the null hypothesis that the genes for flower color and grain shape are unlinked:

Table 1: Sweet pea cross results

	Observed
Purple long (P_L_)	284
Purple round (P_ll)	21
Red long (ppL_)	21
Red round (ppll)	55
Total	381

© 2016 **BIOZONE** International
ISBN: 978-1-927309-46-9
Photocopying Prohibited

Biological Evolution: Unity and Diversity

Concepts and connections
Use arrows to make your own connections between related concepts in this section of this workbook

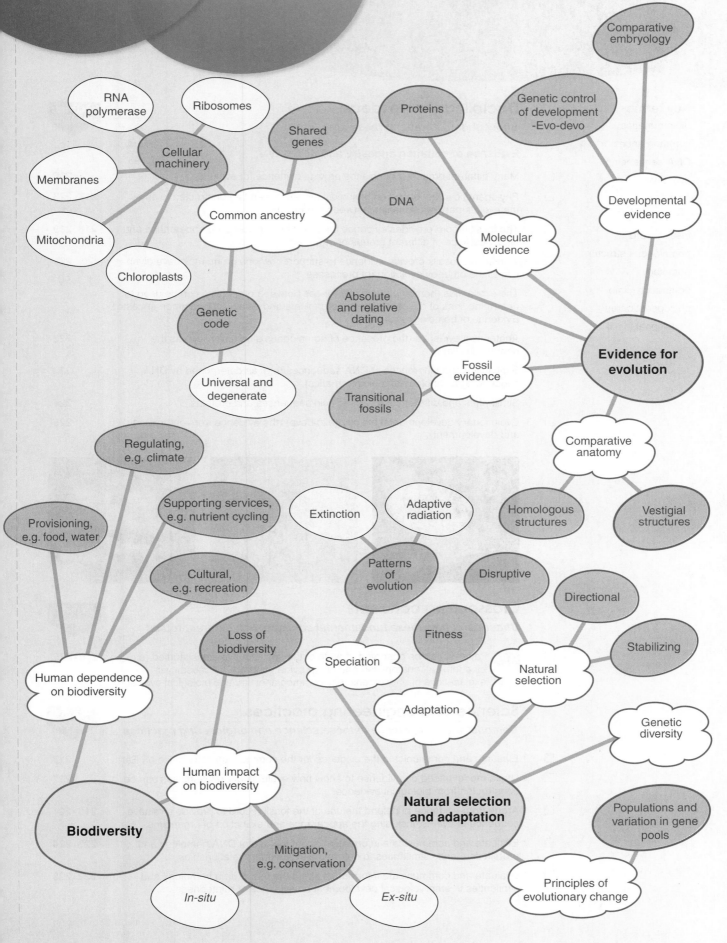

Evidence for Evolution

Key terms

bioinformatics

common ancestor

DNA sequencing

evo-devo

evolution

fossil

fossil record

homologous structure

homology

pentadactyl limb

phylogenetic tree

transitional fossil

Disciplinary core ideas

Show understanding of these core ideas

Evidence of common ancestry and diversity

		Activity number
☐	1 Many different branches of science provide evidence for evolution.	216
☐	2 Phylogenetic trees, based on biological evidence, can be constructed to show the evolutionary relationships between organisms.	217 223
☐	3 The fossil record provides evidence for evolution, including the appearance and disappearance of different groups of organisms.	218 219
☐	4 Transitional fossils provide evidence to support how one group may have given rise to the other by evolutionary processes.	220 221
☐	5 The similarities (homology) and differences between organisms can be used to determine lines of descent between organisms and establish a common ancestor. Evidence of homology includes:	
☐	i Anatomical evidence (the presence of homologous structures such as the pentadactyl limb).	222
☐	ii Similarities and differences of DNA sequences (e.g as determined by DNA sequencing technology and bioinformatics).	223
☐	iii Similarities and differences in amino acid sequences and proteins.	224
☐	iv Evolutionary developmental biology (evo-devo) (the evidence from embryology and development).	225

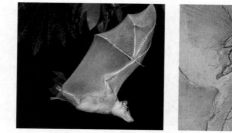 

RA

Crosscutting concepts

Understand how these fundamental concepts link different topics

		Activity number
☐	1 **P** ▶ Patterns can be observed at each of the scales a system is studied (e.g. biochemical, anatomical) and can provide evidence for causality in explanations of phenomena, e.g. common ancestry and biological evolution.	216-225

Science and engineering practices

Demonstrate competence in these science and engineering practices

		Activity number
☐	1 Evaluate and communicate the evidence for the common ancestry of life on Earth.	217
☐	2 Use a model based on evidence to show how evolutionary relationships can be constructed from biological evidence.	217
☐	3 Argue from evidence to defend the use of the fossil record to provide a relative sequence of events, including the appearance and extinction of organisms.	218-221
☐	4 Evaluate and communicate information about the use of DNA homology and protein sequence similarities to determine evolutionary relationships.	223 224
☐	5 Evaluate and communicate information about the use of limb homology and similarities in embryological development as evidence of relatedness.	222 225

216 Evidence for Evolution

Key Idea: Evolution describes the heritable changes in a population's gene pool over time. Evidence for the fact that populations evolve comes from many fields of science.

What is evolution?

Evolution is defined as the heritable genetic changes seen in a population over time. There are two important points to take from this definition. The first is that evolution refers to populations, not individuals. The second is that the changes must be passed on to the next generation (i.e. be inherited). The evidence for evolution comes from many branches of science (below) and includes evidence from living populations as well as from the past.

Comparative anatomy

Comparative anatomy examines the similarities and differences in the anatomy of different species. Similarities in anatomy (e.g. the bones forming the arms in humans and the wings in birds and bats) indicate descent from a common ancestor.

Geology

Geological strata (the layers of rock, soil, and other deposits such as volcanic ash) can be used to determine the relative order of past events and therefore the relative dates of fossils. Fossils in lower strata are older than fossils in higher (newer) strata, unless strata have been disturbed.

DNA comparisons

DNA can be used to determine how closely organisms are related to each other. The greater the similarities between the DNA sequences of species, the more closely related the species are.

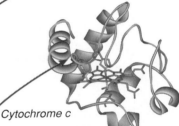

Cytochrome c

Protein evidence

Similarities (and differences) between proteins provides evidence for determining shared ancestry. Fewer differences in amino acid sequences reflects closer genetic relatedness.

EVOLUTION

Fossil record

Fossils, like this shark's tooth (left) are the remains of long-dead organisms. They provide a record of the appearance and extinction of organisms.

Developmental evidence

The study of developmental processes and the genes that control them gives insight into evolutionary processes. This field of study is called evolutionary developmental biology (evo-devo).

Biogeography

The geographical distribution of living and extinct organisms provides evidence of common ancestry and can be explained by speciation, extinction, and continental drift. The biogeography of islands, e.g the Galápagos Islands, provides evidence of how species evolve when separated from their ancestral population on the mainland.

Chronometric dating

Radiometric dating techniques (such as carbon dating) allow scientists to determine an absolute date for a fossil by dating it or the rocks around it. Absolute dating has been used to assign ages to strata, and construct the geological time scale.

© 2016 **BIOZONE** International
ISBN: 978-1-927309-46-9
Photocopying Prohibited

CCC   WEB

 P **216** **REFER**

217 The Common Ancestry of Life

Key Idea: Molecular studies have enabled scientists to clarify the earliest beginnings of the eukaryotes. Such studies provide powerful evidence of the common ancestry of life.

How do we know about the relatedness of organisms?

▶ Traditionally, the phylogeny (evolutionary history) of organisms was established using morphological comparisons. In recent decades, molecular techniques involving the analysis of DNA, RNA, and proteins have provided more information about how all life on Earth is related.

▶ These newer methods have enabled scientists to clarify the origin of the eukaryotes and to recognize two prokaryote domains. The universality of the genetic code and the similarities in the molecular machinery of all cells provide powerful evidence for a common ancestor to all life on Earth.

There is a universal genetic code

DNA encodes the genetic instructions of all life. The form of these genetic instructions, called the **genetic code**, is effectively universal, i.e. the same combination of three DNA bases code for the same amino acid in almost all organisms. The very few exceptions in which there are coding alternatives are restricted to some bacteria and to mitochondrial DNA.

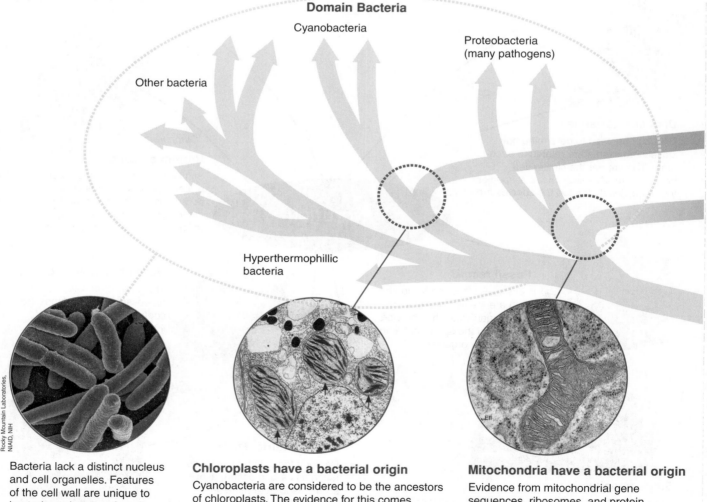

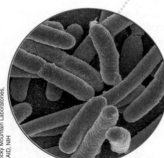

Bacteria lack a distinct nucleus and cell organelles. Features of the cell wall are unique to bacteria and are not found among archaea or eukaryotes. Typically found in less extreme environments than archaea.

Chloroplasts have a bacterial origin

Cyanobacteria are considered to be the ancestors of chloroplasts. The evidence for this comes from similarities in the ribosomes and membrane organization, as well as from genomic studies. Chloroplasts were acquired independently of mitochondria, from a different bacterial lineage, but by a similar process.

Mitochondria have a bacterial origin

Evidence from mitochondrial gene sequences, ribosomes, and protein synthesis indicate that mitochondria have a prokaryotic origin. Mitochondria were probably symbiotic inclusions in an early eukaryotic ancestor.

 © 2016 **BIOZONE** International
ISBN: 978-1-927309-46-9
Photocopying Prohibited

1. Identify three features of the metabolic machinery of cells that support a common ancestry of life:

   (a) _____

   (b) _____

   (c) _____

2. Suggest why scientists believe that mitochondria were acquired before chloroplasts: _____

   _____

   _____

   _____

Eukarya (the eukaryotes) are characterized by complex cells with organelles and a membrane-bound nucleus. This domain contains four of the kingdoms recognized under a traditional scheme.

Domain Eukarya

Animals   Fungi   Plants

Algae

Ciliates

Archaea resemble bacteria but membrane and cell wall composition and aspects of metabolism are very different. They live in extreme environments similar to those on primeval Earth.

Domain Archaea

Bacteria that gave rise to chloroplasts

Bacteria that gave rise to mitochondria

RCN

Eukaryotes have linear chromosomes

Eukaryotic cells all have large linear chromosomes (above) within the cell nucleus. The evolution of linear chromosomes was related to the appearance of mitosis and meiosis.

Xiangyux (PD)

Eukaryotes have an archaean origin

Archaea superficially resemble bacteria but similarities in their molecular machinery (RNA polmerase and ribosome proteins) show that they are more closely related to eukaryotes.

Last Universal
Common Ancestor
(LUCA)

Living systems share the same molecular machinery

In all living systems, the genetic machinery consists of self-replicating DNA molecules. Some DNA is transcribed into RNA, some of which is translated into proteins. The machinery for translation (left) involves proteins and RNA. Ribosomal RNA analysis support a universal common ancestor.

EII

218 The Fossil Record

Key Idea: Fossils provide a record of the appearance and extinction of organisms. The fossil record can be used to establish the relative order of past events.

The importance of the fossil record

► Fossils are the remains of long-dead plants and animals that have become preserved in the Earth's crust.

► Fossils provide a record of the appearance and extinction of organisms, from species to whole taxonomic groups.

► The fossil record can be calibrated against a time scale (using dating techniques), to build up a picture of the evolutionary changes that have taken place.

Fossilized fern frond

Gaps in the fossil record

The fossil record contains gaps and without a complete record, it can sometimes be difficult to determine an evolutionary sequence. Scientists use other information (e.g. associated fossils and changes in morphology) to produce a order of events that best fits all the evidence.

Gaps in the fossil can occur because:

► Fossils are destroyed.

► Some organisms do not fossilize well.

► Fossils have not yet been found.

Profile with sedimentary rocks containing fossils

Rock strata are layered through time

Rock strata are arranged in the order that they were deposited (unless they have been disturbed by geological events). The most recent layers are near the surface and the oldest are at the bottom. Fossils can be used to establish the sequential (relative) order of past events in a rock profile.

New fossil types mark changes in environment

In the strata at the end of one geological period, it is common to find many new fossils that become dominant in the next.
Each geological period had a different environment from the others. Their boundaries coincided with drastic environmental changes and the appearance of new niches. These produced new selection pressures, resulting in new adaptive features in the surviving species as they responded to the changes.

Ground surface

Youngest sediments

Oldest sediments

Recent fossils are found in more recent sediments

The more recent the layer of rock, the more resemblance there is between the fossils found in it and living organisms.

Fossil types differ in each stratum

Fossils found in a given layer of sedimentary rock are generally significantly different to fossils in other layers.

Extinct species

The number of extinct species is far greater than the number of species living today.

More primitive fossils are found in older sediments

Fossils in older layers tend to have quite generalized forms. In contrast, organisms alive today have specialized forms.

1. Discuss the importance of fossils as a record of evolutionary change over time: _____

_____

_____

_____

2. Why can gaps in the fossil record make it difficult to determine an evolutionary history? _____

_____

_____

© 2016 **BIOZONE** International
ISBN: 978-1-927309-46-9
Photocopying Prohibited

219 Interpreting the Fossil Record

Key Idea: Analyzing the fossils within rock strata allows scientists to order past events in a rock profile, from oldest to most recent.

The diagram below shows a hypothetical rock profile from two locations separated by a distance of 67 km. There are differences between the rock layers at the two locations. Apart from layers D and L, which are volcanic ash deposits, all other layers are composed of sedimentary rock. Use the information on the diagram to answer the questions below.

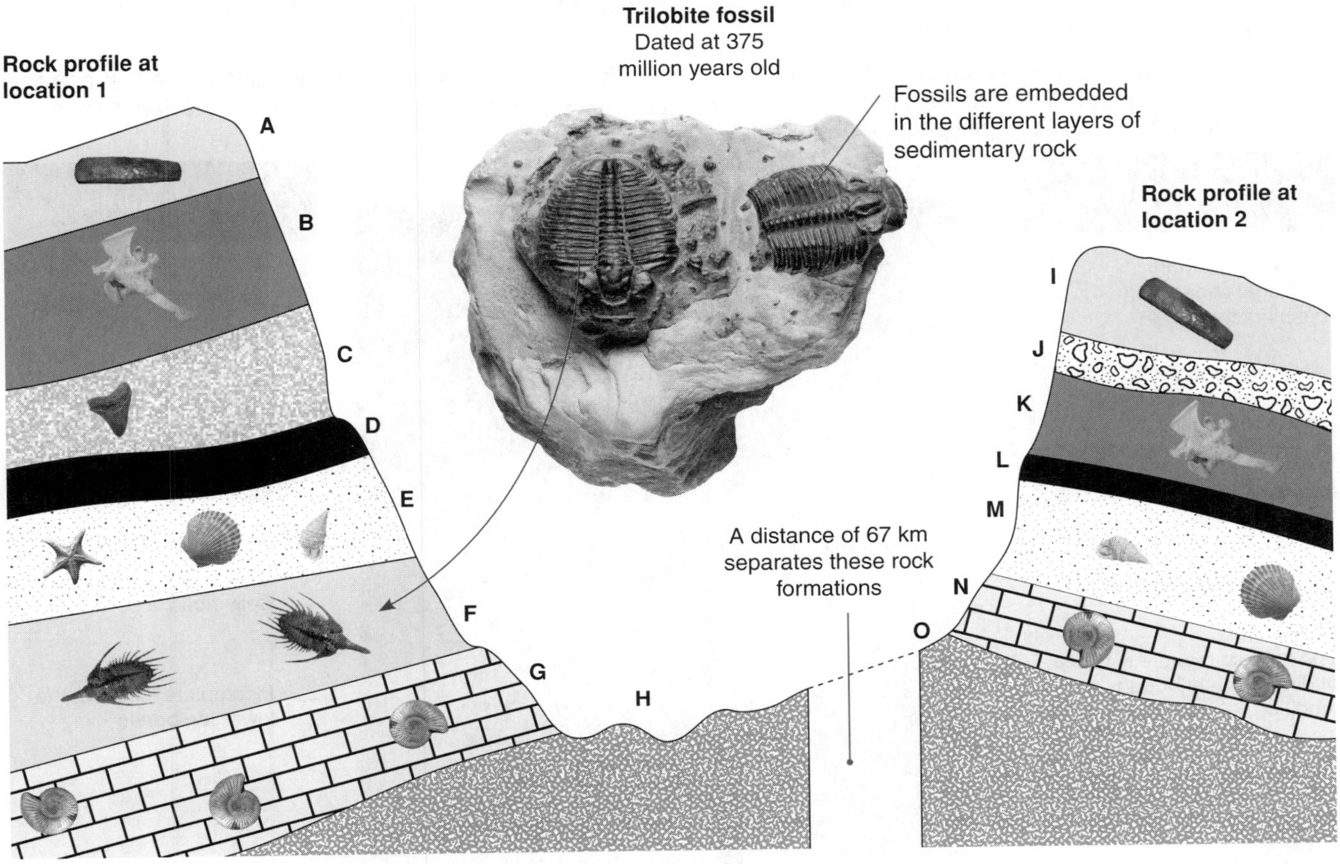

Trilobite fossil
Dated at 375 million years old

Fossils are embedded in the different layers of sedimentary rock

Rock profile at location 1

A
B
C
D
E
F
G
H

Rock profile at location 2

I
J
K
L
M
N
O

A distance of 67 km separates these rock formations

1. Assuming there has been no geological activity to disturb the order of the rock layers, state in which rock layer (A-O) you would find:

    (a) The youngest rocks at Location 1: _____

    (c) The youngest rocks at Location 2: _____

    (b) The oldest rocks at Location 1: _____

    (d) The oldest rocks at Location 2: _____

2. (a) State which layer at location 1 is of the same age as layer M at location 2: _____

    (b) Explain the reason for your answer in 2 (a): _____

    _____

    _____

    _____

3. (a) State which layers present at location 1 are missing at location 2: _____

    (b) State which layers present at location 2 are missing at location 1: _____

4. The rocks in layer H and O are sedimentary rocks. Why are there no visible fossils in these layers? _____

    _____

    _____

© 2016 **BIOZONE** International
ISBN: 978-1-927309-46-9
Photocopying Prohibited

PRACTICES

CCC

WEB

P

219

KNOW

220 Transitional Fossils

Key Idea: Transitional fossils show intermediate states between two different, but related, groups. They provide important links in the fossil record.

Transitional fossils are fossils which have a mixture of features that are found in two different, but related, groups. Transitional fossils provide important links in the fossil record and provide evidence to support how one group may have given rise to the other by evolutionary processes.

Important examples of transitional fossils include horses, whales, and *Archaeopteryx* (below), a transitional form between birds and non-avian dinosaurs.

Archaeopteryx was crow-sized (50 cm length) and lived about 150 million years ago. It is regarded as the first primitive bird and had a number of birdlike (avian) features, including feathers. However, it also had many non-avian features, which it shared with theropod dinosaurs of the time.

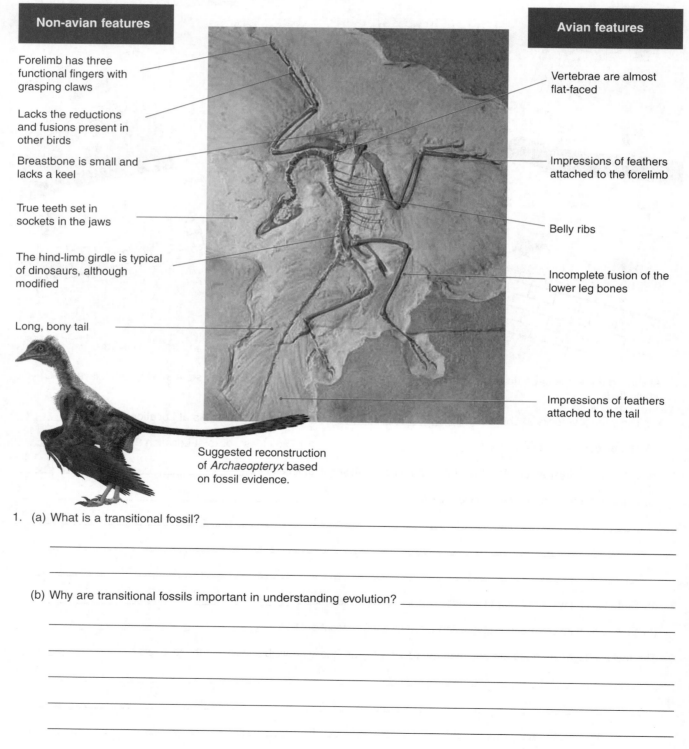

Non-avian features

Forelimb has three functional fingers with grasping claws

Lacks the reductions and fusions present in other birds

Breastbone is small and lacks a keel

True teeth set in sockets in the jaws

The hind-limb girdle is typical of dinosaurs, although modified

Long, bony tail

Avian features

Vertebrae are almost flat-faced

Impressions of feathers attached to the forelimb

Belly ribs

Incomplete fusion of the lower leg bones

Impressions of feathers attached to the tail

Suggested reconstruction of *Archaeopteryx* based on fossil evidence.

1. (a) What is a transitional fossil? _____

_____

_____

(b) Why are transitional fossils important in understanding evolution? _____

_____

_____

_____

_____

_____

WEB   CCC   PRACTICES

KNOW   **220**   **P**

© 2016 **BIOZONE** International
ISBN: 978-1-927309-46-9
Photocopying Prohibited

221 Case Study: Whale Evolution

Key Idea: The evolution of whales is well documented in the fossil record, with many transitional forms recording the shift from a terrestrial to an aquatic life.

Whale evolution

The evolution of modern whales from an ancestral land mammal is well documented in the fossil record. The fossil record of whales includes many transitional forms, which has enabled scientists to develop an excellent model of whale evolution. The evolution of the whales (below) shows a gradual accumulation of adaptive features that have equipped them for life in the open ocean.

Modern whales are categorized into two groups.

▶ Toothed whales have full sets of teeth throughout their lives (e.g. sperm whales and orca).

▶ Baleen whales. These are toothless whales and they use a comb-like structure (called baleen) to filter food (e.g. humpback whale).

Humpback whale

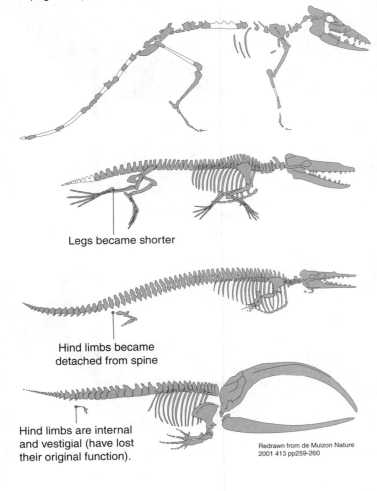

Legs became shorter

Hind limbs became detached from spine

Hind limbs are internal and vestigial (have lost their original function).

Redrawn from de Muizon Nature 2001 413 pp259-260

50 mya *Pakicetus*

Pakicetus was a transitional species between carnivorous land mammals and the earliest true whales. It was mainly terrestrial (land dwelling), but foraged for food in water. It had four, long limbs. Its eyes were near the top of the head and its nostrils were at the end of the snout. It had external ears, but they showed features of both terrestrial mammals and fully aquatic mammals.

45 mya *Rhodocetus*

Rhodocetus was mainly aquatic (water living). It had adaptations for swimming, including shorter legs and a shorter tail. Its eyes had moved to the side of the skull, and the nostrils were located further up the skull. The ear showed specializations for hearing in water.

40 mya *Dorudon*

Dorudon was fully aquatic. Its adaptations for swimming included a long, streamlined body, a broad powerful muscular tail, the development of flippers and webbing. It had very small hind limbs (not attached to the spine) which would no longer bear weight on land.

Balaena (recent whale ancestor)

The hind limbs became fully internal and vestigial. Studies of modern whales show that limb development begins, but is arrested at the limb bud stage. The nostrils became modified as blowholes. This recent ancestor to modern whales diverged into two groups (toothed and baleen) about 36 million years ago. Baleen whales have teeth in their early fetal stage, but lose them before birth.

1. Why does the whale fossil record provide a good example of the evolutionary process? _____

_____

_____

2. Briefly describe the adaptations of whales for swimming that evolved over time: _____

_____

_____

PRACTICES   CCC   WEB

P   221   **KNOW**

222 Anatomical Evidence for Evolution

Key Idea: Homologous structures are anatomical similarities that are the result of common origin. They indicate the evolutionary relationship between groups of organisms.

Homologous structures

Homologous structures are structures found in different organisms that are the result of their inheritance from a common ancestor. Their presence indicates the evolutionary relationship between organisms. Homologous structures have a common origin, but they may have different functions.

For example, the forelimbs of birds and seals are homologous structures. They have the same basic skeletal structure, but have different functions. A bird's wings have been adapted for flight, and a seal's flippers are modified as paddles for swimming.

The pentadactyl limb

A **pentadactyl limb** is a limb with five fingers or toes (e.g. hands and feet), with the bones arranged in a specific pattern (below, left).

Early land vertebrates were amphibians with pentadactyl limbs. All vertebrates that descended from these early amphibians have limbs that have evolved from this same basic pentadactyl pattern. The pentadactyl limb is a good illustration of adaptive radiation (the diversification of an ancestral form into many different forms). The generalized limb plan has been adapted to meet the requirements of organisms in many different niches.

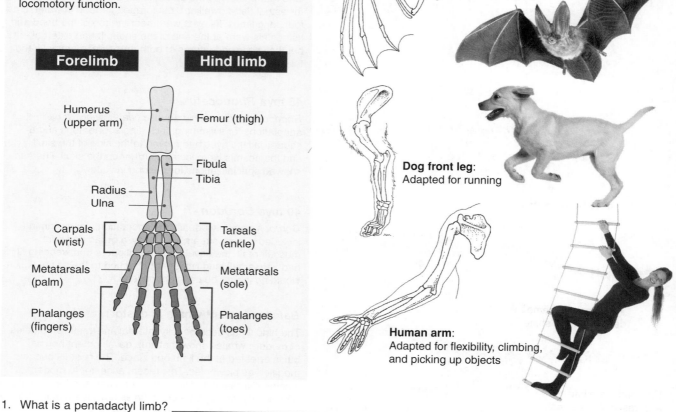

Generalized pentadactyl limb

The forelimbs and hind limbs have the same arrangement of bones. In many cases bones in different parts of the limb have been modified for a specialized locomotory function.

Forelimb	Hind limb

Humerus (upper arm) — Femur (thigh)

Fibula
Tibia

Radius
Ulna

Carpals (wrist) — Tarsals (ankle)

Metatarsals (palm) — Metatarsals (sole)

Phalanges (fingers) — Phalanges (toes)

Specializations of pentadactyl limbs

Bat wing: Adapted for flying

Dog front leg: Adapted for running

Human arm: Adapted for flexibility, climbing, and picking up objects

1. What is a pentadactyl limb? _____

_____

_____

2. Explain how homology in the pentadactyl limb provides evidence for adaptive radiation: _____

_____

_____

_____

_____

_____

PRACTICES

 © 2016 **BIOZONE** International
ISBN: 978-1-927309-46-9
Photocopying Prohibited

223 DNA Evidence for Evolution

Key Idea: DNA sequencing and comparison and the use of computer databases are now frequently used when analyzing the evolutionary histories and relationships of different species.

▶ The advancement of techniques in molecular biology is providing increasingly large amounts of information about the genetic makeup of organisms.

▶ **Bioinformatics** involves the collection, analysis, and storage of biochemical information (e.g. DNA sequences) using computer science and mathematics.

▶ Bioinformatics allows biological information to be stored in databases where it can be easily retrieved, analyzed, and compared. Comparison of DNA or protein sequences between species enables researchers to investigate and better understand their evolutionary relationships.

An overview of the bioinformatics process

▶ A gene of interest is selected for analysis.

...G A G A A C T G T T T A G A T G C A A A A...

▶ High throughput 'Next-Gen' sequencing technologies allow the DNA sequence of the gene to be quickly determined.

Organism 1   ...G A G A A C T G T T T A G A T G C A A A A A...

Organism 2   ...G A G A T C T G T G T A G A T G C A G A A...

Organism 3   ...G A G T T C T G T G T C G A T G C A G A A...

Organism 4   ...G A G T T C T G T T T C G A T G C A G A G...

▶ Powerful computer software can quickly compare the DNA sequences of many organisms. Commonalities and differences in the DNA sequence can help to determine the evolutionary relationships of organisms. The blue boxes indicate differences in the DNA sequences.

▶ Once sequence comparisons have been made, the evolutionary relationships can be displayed as a phylogenetic tree. The example (right) shows the evolutionary relationships of the whales to some other land mammals.

▶ Bioinformatics has played an important role in determining the origin of whales and their transition from a terrestrial (land) form to a fully aquatic form.

▶ This phylogenetic tree was determined by comparing repetitive DNA fragments that are inserted into chromosomes after they have been reverse transcribed from a mRNA molecule. The locations of these repetitive fragments are predictable and stable, so they make reliable markers for determining species relationships. If two species have the same repeats in the same location, they are very likely to share a common ancestor.

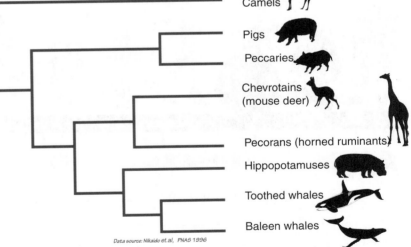

Camels

Pigs

Peccaries

Chevrotains (mouse deer)

Pecorans (horned ruminants)

Hippopotamuses

Toothed whales

Baleen whales

Data source: Nikaido et.al, PNAS 1996

1. The diagram above shows the relatedness of several mammals as determined by DNA sequencing of 10 genes:

   (a) Which land mammal are whales most closely related to? _____

   (b) Mark with an arrow on the diagram above where whales and the organism in (a) last shared a common ancestor.

   (c) Pigs were once considered to be the most closely related land ancestor to the whales. Use the phylogenetic tree above to describe the currently accepted relationship.

   _____

   _____

PRACTICES   PRACTICES   CCC   WEB

P   223   **KNOW**

224 Protein Evidence for Evolution

Key Idea: Protein homology can be used to determine evolutionary patterns. As genetic relatedness increases, the number of amino acid differences decreases.

Protein homology

The amino acid sequence of proteins can be used to establish homologies (similarities) between organisms. Any change in the amino acid sequence reflects changes in the DNA sequence.

Some proteins are common to many different species. These proteins are often highly conserved, meaning they change (mutate) very little over time. This is because they have critical roles (e.g. in cellular respiration) and mutations are likely to be detrimental to their function.

Evidence indicates that these highly conserved proteins are homologous and have been derived from a common ancestor. Because they are highly conserved, changes in the amino acid sequence are likely to represent major divergences between groups during the course of evolution.

The Pax-6 protein provides evidence for evolution

▶ The Pax-6 protein regulates eye formation during embryonic development.

▶ The Pax-6 gene is so highly conserved that the gene from one species can be inserted into another species, and still produce a normally functioning eye.

▶ This suggests the Pax-6 proteins are homologous, and the gene has been inherited from a common ancestor.

An experiment inserted mouse Pax6 gene into fly DNA and turned it on in a fly's legs. The fly developed fly eyes on its legs!

Hemoglobin homology

Hemoglobin is the oxygen-transporting blood protein found in most vertebrates. The beta chain hemoglobin sequences from different organisms can be compared to determine evolutionary relationships.

As genetic relatedness decreases, the number of amino acid differences between the hemoglobin chains of different vertebrates increases (below). For example, there are no amino acid differences between humans and chimpanzees, indicating they recently shared a common ancestor. Humans and frogs have 67 amino acid differences, indicating they had a common ancestor a very long time ago.

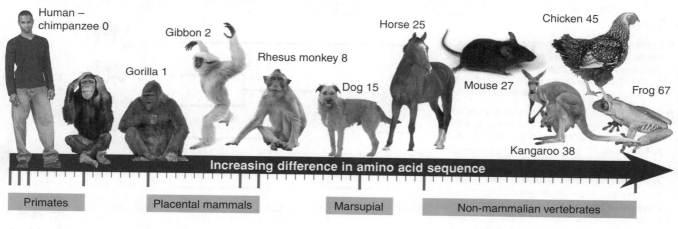

Human – chimpanzee 0 Gorilla 1 Gibbon 2 Rhesus monkey 8 Dog 15 Horse 25 Mouse 27 Kangaroo 38 Chicken 45 Frog 67

Increasing difference in amino acid sequence

Primates   Placental mammals   Marsupial   Non-mammalian vertebrates

1. (a) What is a highly conserved protein? _____

_____

(b) Why are highly conserved proteins good for constructing phylogenies? _____

_____

_____

2. Compare the differences in the hemoglobin sequence of humans, rhesus monkeys, and horses. What do these tell you about the relative relatedness of these organisms?

_____

_____

_____

© 2016 **BIOZONE** International
ISBN: 978-1-927309-46-9
Photocopying Prohibited

225 Developmental Evidence for Evolution

Key Idea: Similarities in the development of embryos, including the genetic control of development, provides strong evidence for evolution.

Developmental biology

Developmental biology studies the process by which organisms grow and develop. In the past, it was restricted to the appearance (morphology) of a growing fetus. Today, developmental biology focuses on the genetic control of development and its role in producing the large differences we see in the adult appearance of different species.

During development, vertebrate embryos pass through the same stages, in the same sequence, regardless of the total time period of development. This similarity is strong evidence of their shared ancestry. The stage of embryonic development is identified using a standardized system based on the development of structures, not by size or the number of days of development. The Carnegie stages (right) cover the first 60 days of development.

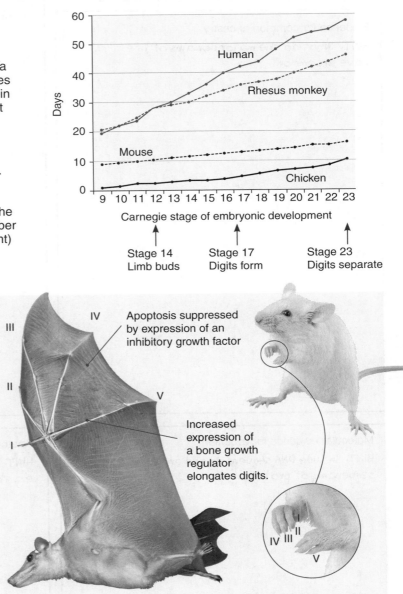

Stage 14 Limb buds · Stage 17 Digits form · Stage 23 Digits separate

Limb homology and the control of development

As we have seen, homology (e.g. in limb structure) is evidence of shared ancestry. How do these homologous structures become so different in appearance? The answer lies in the way the same genes are regulated during development.

All vertebrate limbs form as buds at the same stage of development. At first, the limbs resemble paddles, but apoptosis (programmed cell death) of the tissue between the developing bones separates the digits to form fingers and toes.

Like humans, mice have digits that become fully separated by interdigital apoptosis during development. In bat forelimbs, this controlled destruction of the tissue between the forelimb digits is inhibited. The developmental program is the result of different patterns of expression of the same genes in the two types of embryos.

Apoptosis suppressed by expression of an inhibitory growth factor

Increased expression of a bone growth regulator elongates digits.

Bat wings are highly specialized structures with unique features, such as elongated wrist and fingers (I-V) and membranous wing surfaces. The forelimb structures of bats and mice are homologous, but how the limb looks and works is quite different.

1. Describe a feature of vertebrate embryonic development that supports evolution from a common ancestor:

2. Explain how different specialized limb structures can arise from a basic pentadactyl structure:

 © 2016 **BIOZONE** International ISBN: 978-1-927309-46-9 Photocopying Prohibited

PRACTICES CCC WEB 225 KNOW

226 Chapter Review

Summarize what you know about this topic under the headings provided. You can draw diagrams or mind maps, or write short notes to organize your thoughts. Use the points in the introduction and the hints to help you:

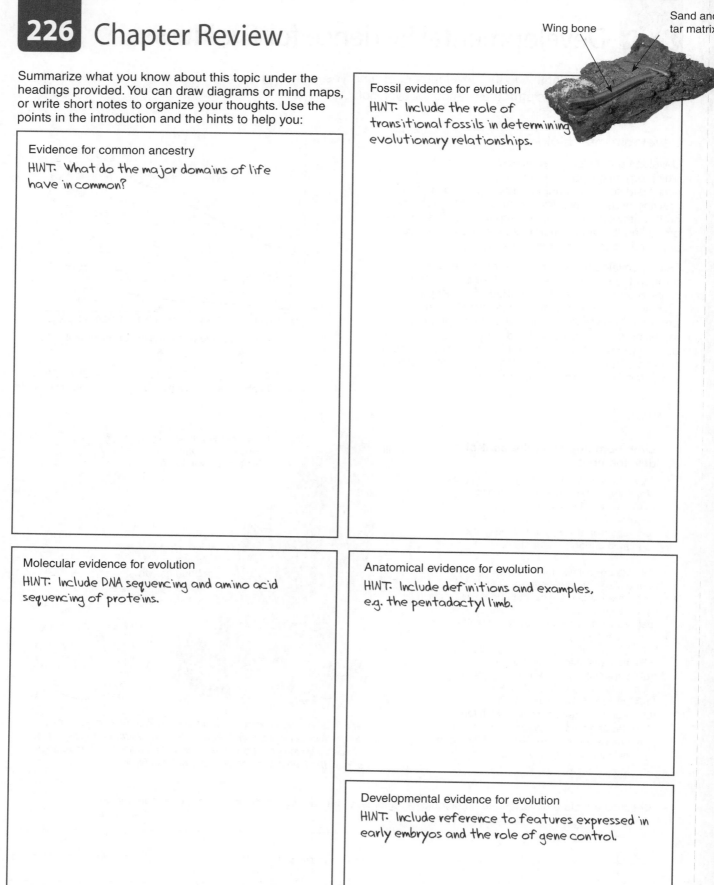

Wing bone

Sand and tar matrix

Evidence for common ancestry
HINT: What do the major domains of life have in common?

Fossil evidence for evolution
HINT: Include the role of transitional fossils in determining evolutionary relationships.

Molecular evidence for evolution
HINT: Include DNA sequencing and amino acid sequencing of proteins.

Anatomical evidence for evolution
HINT: Include definitions and examples, e.g. the pentadactyl limb.

Developmental evidence for evolution
HINT: Include reference to features expressed in early embryos and the role of gene control.

© 2016 **BIOZONE** International
ISBN: 978-1-927309-46-9
Photocopying Prohibited

227 KEY TERMS AND IDEAS: Did You Get It?

1. Test your vocabulary by matching each term to its definition, as identified by its preceding letter code.

common ancestor

evolution

fossil

fossil record

homologous structure

phylogenetic tree

transitional fossil

A The individual from which all organisms in a taxon are directly descended.

B The fossilized remains of organisms that illustrate an evolutionary transition. They possess both primitive and derived characteristics.

C Changes in the allele composition of gene pools over time.

D The preserved remains or traces of a past organism.

E The sum total of current paleontological knowledge. It is all the fossils that have existed throughout life's history, whether they have been found or not.

F Structures in different but related species that are derived from the same ancestral structure but now serve different purposes, e.g. wings and fins.

G The evolutionary history or genealogy of a group of organisms represented as a 'tree' showing descent of new species from the ancestral one.

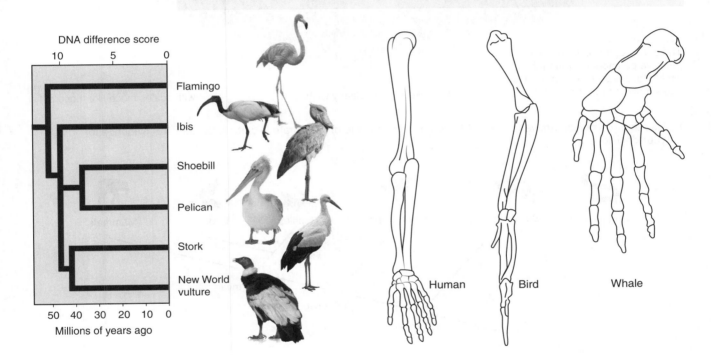

2. The diagram above left shows the evolutionary relationship of a group of birds based on DNA similarities:

(a) Place an X to the last common ancestor of all the birds:

(b) How many years ago did storks diverge from vultures?_____

(c) What are the most closely related birds?_____

(d) What is the difference in DNA (score) between:

i: Storks and vultures: _____ ii: Ibises and shoebills: _____

(e) Which of the birds is the least related to vultures?_____

(f) How long ago did ibises and vultures share a common ancestor: _____

3. (a) The diagrams above right show the forelimbs of a whale, bird, and human. Shade the diagram to indicate which bones are homologous. Use the same color to indicate the equivalent bones in each limb.

(b) What does the homology of these bones indicate? _____

_____

 © 2016 **BIOZONE** International
ISBN: 978-1-927309-46-9
Photocopying Prohibited

TEST

228 Summative Assessment

In 2004, a fossil of an unknown vertebrate was discovered in northern Canada and subsequently called *Tiktaalik roseae*. The *Tiktaalik* fossil was quite well preserved and many interesting features could be identified. These are shown on the photograph of the fossil below.

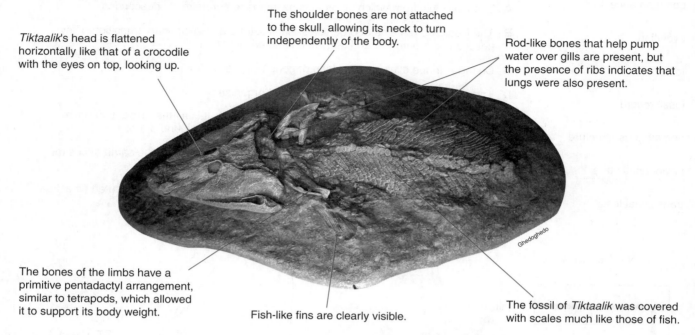

Tiktaalik's head is flattened horizontally like that of a crocodile with the eyes on top, looking up.

The shoulder bones are not attached to the skull, allowing its neck to turn independently of the body.

Rod-like bones that help pump water over gills are present, but the presence of ribs indicates that lungs were also present.

The bones of the limbs have a primitive pentadactyl arrangement, similar to tetrapods, which allowed it to support its body weight.

Fish-like fins are clearly visible.

The fossil of *Tiktaalik* was covered with scales much like those of fish.

1. Use the information above to place *Tiktaalik* on the time line of vertebrate evolution. Discuss the evidence for your decision.

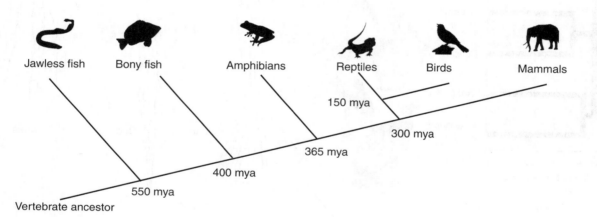

Jawless fish   Bony fish   Amphibians   Reptiles   Birds   Mammals

150 mya

300 mya

365 mya

400 mya

550 mya

Vertebrate ancestor

_____

_____

_____

_____

_____

_____

_____

_____

_____

_____

LS4.B LS4.C Natural Selection and Adaptation

Key terms

adaptation

biodiversity

evolution

extinction

fitness

gene pool

genotype

natural selection

phenotype

trait

variation

Disciplinary core ideas

Show understanding of these core ideas

	Activity number

Natural selection: differential survival of favorable phenotypes

☐ 1 Natural selection is the process by which favorable phenotypes (and therefore genotypes) become relatively more or less common in a population. It is a consequence of variation in the genes and how they are expressed in individuals, which leads to differences in individual survival and reproductive success in the environment (i.e. fitness). 229 232 233 234

☐ 2 Traits that increase fitness are more likely to be passed on to the next generation and become more common over time. 229

Natural selection leads to adaptation

☐ 3 Evolution is a result of four factors: (1) the potential for a species to increase in number (2) the genetic variation between members of the species (3) competition for the limited resources, and (4) survival and reproduction of the individuals that are better able to survive and reproduce. 229 232-234

☐ 4 Natural selection leads to adaptation, i.e. to a population dominated by individuals with the characteristics most suited to survival and reproduction within the prevailing (current) environment. 230-234

☐ 5 Adaptation allows for a shift in the distribution of traits in a population when conditions change. 232-234

☐ 6 Changes in the environment may lead species to expand or contract their range. New species can emerge when populations diverge in different environments. In this way, evolution has produced Earth's past and present biodiversity. 232-234 237-240

☐ 7 Species unable to adapt to a changing environment may become extinct. Extinction is natural process but can be accelerated under certain conditions. 239 241 242

Putneymark cc 2.0

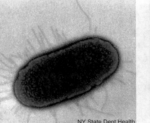
NY State Dept Health

Crosscutting concepts

Understand how these fundamental concepts link different topics

	Activity number

☐ 1 **CE** ▶ Empirical evidence enables us to support claims about how evolution occurs, how new species arise, and how species become extinct. 232-234 240-242

☐ 2 **P** ▶ Observed patterns can provide evidence for causes of evolutionary change. 232-234

Science and engineering practices

Demonstrate competence in these science and engineering practices

	Activity number

☐ 1 Evaluate the evidence for claims that change in environment can lead to an increase in population numbers, emergence of new species, or extinction. 232-234 241

☐ 2 Use concepts of statistics and probability to support explanations for how organisms with advantageous traits become proportionally more numerous. 229 232-234

☐ 3 Construct an evidence-based explanation for how biological evolution occurs and how natural selection leads to adaptation. 230-234

☐ 4 Use models to illustrate natural selection and explain patterns of evolution. 236 239

229 How Evolution Occurs

Key Idea: Evolution by natural selection describes how organisms that are better adapted to their environment survive to produce a greater number of offspring.

Evolution

Evolution is the change in inherited characteristics in a population over generations. Evolution is the consequence of interaction between four factors: **(1) The potential for populations to increase in numbers, (2) Genetic variation as a result of mutation and sexual reproduction, (3) competition for resources, and (4) proliferation of individuals with better survival and reproduction.**

Natural selection is the term for the mechanism by which better adapted organisms survive to produce a greater number of viable offspring. This has the effect of increasing their proportion in the population so that they become more common. This is the basis of Darwin's theory of evolution by natural selection.

We can demonstrate the basic principles of evolution using the analogy of a 'population' of M&M's candy.

In a bag of M&M's, there are many colors, which represents the variation in a population. As you and a friend eat through the bag of candy, you both leave the blue ones, which you both dislike, and return them to bag.

The blue candy becomes more common...

Eventually, you are left with a bag of blue M&M's. Your selective preference for the other colors changed the make-up of the M&M's population. This is the basic principle of selection that drives evolution in natural populations.

Darwin's theory of evolution by natural selection

Darwin's theory of evolution by natural selection is outlined below. It is widely accepted by the scientific community today and is one of founding principles of modern science.

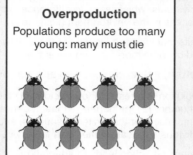

Overproduction
Populations produce too many young: many must die

Populations generally produce more offspring than are needed to replace the parents. Natural populations normally maintain constant numbers. A certain number will die without reproducing.

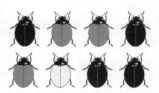

Variation
Individuals show variation: some variations more favorable than others

Individuals in a population have different phenotypes and therefore, genotypes. Some traits are better suited to the environment, and individuals with these have better survival and reproductive success.

Natural selection
Natural selection favors the individuals best suited to the environment at the time

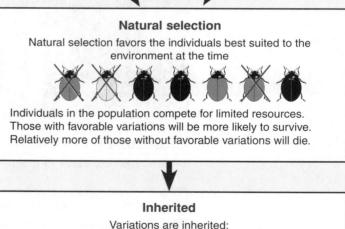

Individuals in the population compete for limited resources. Those with favorable variations will be more likely to survive. Relatively more of those without favorable variations will die.

Inherited
Variations are inherited: the best suited variants leave more offspring

The variations (both favorable and unfavorable) are passed on to offspring. Each generation will contain proportionally more descendants of individuals with favorable characters.

© 2016 **BIOZONE** International ISBN: 978-1-927309-46-9 Photocopying Prohibited

Variation, selection, and population change

1. Variation through mutation and sexual reproduction:
In a population of brown beetles, mutations independently produce red coloration and 2 spot marking on the wings. The individuals in the population compete for limited resources.

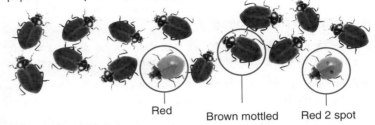

Red   Brown mottled   Red 2 spot

2. Selective predation:
Brown mottled beetles are eaten by birds but red ones are avoided.

3. Change in the genetics of the population:
Red beetles have better survival and fitness and become more numerous with each generation. Brown beetles have poor fitness and become rare.

Natural populations, like the ladybug population above, show genetic variation. This is a result of **mutation** (which creates new alleles) and sexual reproduction (which produces new combinations of alleles). Some variants are more suited to the environment of the time than others. These variants will leave more offspring, as described for the hypothetical population (right).

1. What produces the genetic variation in populations? _____

2. (a) Define evolution: _____

_____

(b) Identify the four factors that interact to bring about evolution in populations: _____

_____

_____

_____

_____

3. Using your answer 2(b) as a basis, explain how the genetic make-up of a population can change over time: _____

_____

_____

_____

_____

_____

_____

230 Adaptation

Key Idea: An adaptation is any heritable trait that suits an organism to its natural functional role in the environment (its niche).

Adaptation and fitness

▶ An adaptation, is any heritable trait that equips an organism for its niche, enhancing its exploitation of the environment and contributing to its survival and successful reproduction.

▶ Adaptations are a product of natural selection and can be structural (morphological), physiological, or behavioral traits. Traits that are not helpful to survival and reproduction will not be favoured and will be lost.

▶ Adaptation is important in an evolutionary sense because adaptive features promote fitness. Fitness is a measure of an organism's ability to maximize the numbers of offspring surviving to reproductive age.

Adaptive features of the North American beaver

North American beavers (*Castor canadensis*) are semi-aquatic and are able to remain submerged for up to 15 minutes. Their adaptations enable them to exploit both aquatic and terrestrial environments.

Beavers are strict herbivores and eat leaves, bark, twigs, roots, and aquatic plants. They do not hibernate. They live in domelike homes called lodges, which they build from mud and branches. Lodges are usually built in the middle of a pond or lake, with an underwater entrance, making it difficult for predators to capture them.

Ears and nostrils

Valves in the ears and nose close when underwater. These keep water out.

Eyes

A clear eyelid (nictitating membrane). This protects the eye and allows the beaver to still see while swimming.

Lips

Their lips can close behind their front teeth. This lets them carry objects and gnaw underwater, but keeps water out and stops them drowning.

Oxygen conservation

During dives, beavers slow their heartbeat and reduce blood flow to the extremities to conserve oxygen and energy. This enables them to stay submerged for 15 minutes even though they are not particularly good at storing oxygen in the tissues.

Front feet

Front paws are good at manipulating objects. The paws are used in dam and lodge construction to pack mud and manipulate branches.

Thick insulating fat

Thick fat layer under the skin. Insulates the beaver from the cold water and helps to keep it warm.

Teeth

Large, strong chisel-shaped front teeth (incisors) grow constantly. These let them fell trees and branches for food and to make lodges with.

Waterproof coat

A double-coat of fur (coarse outer hairs and short, fine inner hairs). An oil is secreted from glands and spread through the fur. The underfur traps air against the skin for insulation and the oil acts as a waterproofing agent and keeps the skin dry in the water.

Large, webbed, hind feet

The webbing between the toes acts like a diver's swimming fins, and helps to propel the beaver through the water.

Large, flat paddle-like tail

Assists swimming and acts like a rudder. Tail is also used to slap the water in communication with other beavers, to store fat for the winter, and as a means of temperature regulation in hot weather because heat can be lost over the large unfurred surface area.

 © 2016 **BIOZONE** International
ISBN: 978-1-927309-46-9
Photocopying Prohibited

Adaptations for diving in air-breathing animals

Air breathing animals that dive must cope with lack of oxygen (which limits the length of the dive) and pressure (which limits the depth of the dive). Many different animal phyla have diving representatives, which have evolved from terrestrial ancestors and become adapted for an aquatic life. Diving air-breathers must maintain a supply of oxygen to the tissues during dives and can only stay underwater for as long as their oxygen supplies last. Their adaptations enable them to conserve oxygen and so prolong their dive time.

Species for which there is a comprehensive fossil record, e.g. whales (right), show that adaptations for a diving lifestyle accumulated slowly during the course of the group's evolution.

Humpback whale

Penguin

Green turtle

Diving mammals

Dolphins, whales, and seals are among the most well adapted divers. They exhale before diving, so that there is no air in the lungs and nitrogen does not enter the blood. This prevents them getting the bends when they surface (a condition in which dissolved gases come out of solution at reduced pressures and form bubbles in the tissues).

During dives, oxygen is conserved by reducing heart rate dramatically, and redistributing blood to supply only critical organs. Diving mammals have high levels of muscle myoglobin, which stores oxygen, but their muscles also function efficiently using anaerobic metabolism.

Diving birds

Penguins show many of the adaptations typical of diving birds. During dives, a bird's heart rate slows, and blood is diverted to the head, heart, and eyes.

Diving reptiles

Sea turtles have low metabolic rates and their tissues are tolerant of low oxygen. These adaptations allow them to remain submerged for long periods and they surface only occasionally.

1. (a) What is an adaptation? _____
   _____
   _____

   (b) How can an adaptation increase an organism's fitness? _____
   _____
   _____

2. The following list identifies some adaptations in a beaver which allow it to survive in its environment. Identify each adaptation as structural, physiological, or behavioral and describe its survival advantage:

   (a) Large front teeth: _____
   _____

   (b) Lodge built in middle of pond: _____
   _____

   (c) Oil secreting glands in skin: _____
   _____

3. (a) What restricts the amount of time diving animals can spend underwater? _____
   _____

   (b) How does reducing heart rate during a dive enable animals to stay underwater longer? _____
   _____
   _____

231 Similar Environments, Similar Adaptations

Key Idea: Unrelated species often evolve similar adaptations to overcome the same environmental challenges.

Sometimes, genetically unrelated organisms evolve similar adaptations in response to the particular environmental challenges they face. The adaptations may result in different species with a very similar appearance. Although the organisms are not closely related, evolution has produced similar solutions in order to solve similar ecological problems. This phenomenon is called **convergent evolution**.

Similar adaptations in unrelated plants

Cactus   *Euphorbia*

▶ The North American cactus and African *Euphorbia* species shown above are both xerophytes. They have evolved similar structural adaptations to conserve water and survive in a hot, dry, desert environment. Although they have a similar appearance, they are not related. They provide an excellent illustration of how unrelated organisms living in the same environment have independently evolved the same adaptations to survive.

▶ Their appearance is so similar at first glance that the *Euphorbia* is often mistaken for a cactus. Both have thick stems to store water and both have lost the presence of obvious leaves. Instead, they have spines or thorns to conserve water (a leafy plant would quickly exhaust its water reserves because of losses via transpiration). In cacti, spines are highly modified leaves. In *Euphorbia*, the thorns are modified stalks. It is not until the two flower that their differences are obvious.

Similar adaptations in unrelated animals

Colugo (related to lemurs)

Flying squirrel (rodent)

The ability for mammals to glide between trees has evolved independently in unrelated animals. The characteristics listed below for the sugar glider are typically found in the gliding mammals shown here.

Sugar glider (marsupial)

Tail acts as a stabilizer and an air brake during flight.

The animals are nocturnal and large eyes help them see at night.

Skin is stretched between the front and back legs to form a wing like flap. This allows them to glide up to 50 m between trees.

By moving their limbs the animal has some control about the direction it flies in.

1. Explain why the North American cactus and African *Euphorbia* species have evolved such similar adaptations:

_____

_____

2. Suggest why gliding between trees (rather than walking) is an advantage to the gliding mammals described above:

_____

_____

3. Tenrecs, echidnas and hedgehogs are examples of unrelated organisms that have all evolved spines. Explain the advantage of this adaptation to these animals and why might it have evolved?

_____

_____

© 2016 **BIOZONE** International
ISBN: 978-1-927309-46-9
Photocopying Prohibited

232 Natural Selection in Finches

Key Idea: The effect of natural selection on a population can be verified by making quantitative measurements of phenotypic traits.

▶ **Natural selection** acts on the phenotypes of a population. Individuals with phenotypes that increase their fitness produce more offspring, increasing the proportion of the genes corresponding to that phenotype in the next generation.

▶ Numerous population studies have shown natural selection can cause phenotypic changes in a population relatively quickly.

▶ The finches on the Galápagos island (Darwin's finches) are famous in that they are commonly used as examples of how evolution produces new species. In this activity you will analyze data from the measurement of beak depths of the medium ground finch (*Geospiza fortis*) on the island of Daphne Major near the center of the Galápagos Islands. The measurements were taken in 1976 before a major drought hit the island and in 1978 after the drought (survivors and survivors' offspring).

Beak depth (mm)	No. 1976 birds	No. 1978 survivors	Beak depth of offspring (mm)	Number of birds
7.30-7.79	1	0	7.30-7.79	2
7.80-8.29	12	1	7.80-8.29	2
8.30-8.79	30	3	8.30-8.79	5
8.80-9.29	47	3	8.80-9.29	21
9.30-9.79	45	6	9.30-9.79	34
9.80-10.29	40	9	9.80-10.29	37
10.30-10.79	25	10	10.30-10.79	19
10.80-11.29	3	1	10-80-11.29	15
11.30+	0	0	11.30+	2

1. Use the data above to draw two separate sets of histograms:

   (a) On the left hand grid draw side-by-side histograms for the number of 1976 birds per beak depth and the number of 1978 survivors per beak depth.

   (b) On the right hand grid draw a histogram of the beak depths of the offspring of the 1978 survivors.

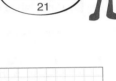

NEED HELP? See Activity 21

2. (a) Mark on the graphs of the 1976 beak depths and the 1978 offspring the approximate mean beak depth.

   (b) How much has the average moved from 1976 to 1978? _____

   (c) Is beak depth heritable? What does this mean for the process of natural selection in the finches? _____

   _____

   _____

   _____

PRACTICES PRACTICES PRACTICES CCC CCC WEB

 CE P 232 DATA

233 Natural Selection in Pocket Mice

Key Idea: The need to blend into their surroundings to avoid predation is an important selection pressure acting on the coat color of rock pocket mice.

Rock pocket mice are found in the deserts of southwestern United States and northern Mexico. They are nocturnal, foraging at night for seeds, while avoiding owls (their main predator). During the day they shelter from the desert heat in their burrows. The coat color of the mice varies from light brown to very dark brown. Throughout the desert environment in which the mice live there are outcrops of dark volcanic rock. The presence of these outcrops and the mice that live on them present an excellent study in natural selection.

▶ The coat color of the Arizona rock pocket mice is controlled by the Mc1r gene (a gene that in mammals is commonly associated with the production of the pigment melanin). Homozygous dominant (DD) and heterozygous mice (Dd) have dark coats, while homozygous recessive mice (dd) have light coats.
Coat color of mice in New Mexico is not related to the Mc1r gene.

▶ 107 rock pocket mice from 14 sites were collected and their coat color and the rock color they were found on were recorded by measuring the percentage of light reflected from their coat (low percentage reflectance equals a dark coat). The data is presented right:

| | | Percent reflectance (%) | |
Site	Rock type (V volcanic)	Mice coat	Rock
KNZ	V	4	10.5
ARM	V	4	9
CAR	V	4	10
MEX	V	5	10.5
TUM	V	5	27
PIN	V	5.5	11
AFT		6	30
AVR		6.5	26
WHT		8	42
BLK	V	8.5	15
FRA		9	39
TIN		9	39
TUL		9.5	25
POR		12	34.5

1. (a) What is the genotype(s) of the dark colored mice? _____

   (b) What is the genotype of the light colored mice? _____

2. Using the data in the table above and the grids below and on the facing page, draw column graphs of the percent reflectance of the mice coats and the rocks at each of the 14 collection sites.

NEED HELP? See Activity 21

DATA 233 P CE

WEB CCC CCC PRACTICES PRACTICES PRACTICES

© 2016 **BIOZONE** International
ISBN: 978-1-927309-46-9
Photocopying Prohibited

3. (a) What do you notice about the reflectance of the rock pocket mice coat color and the reflectance of the rocks they were found on?

_____

_____

(b) Suggest a cause for the pattern in 3(a). How do the phenotypes of the mice affect where the mice live?

_____

_____

_____

_____

(c) What are two exceptions to the pattern you have noticed in 3(a)? _____

_____

_____

_____

_____

(d) How might these exceptions have occurred? _____

_____

_____

_____

4. The rock pocket mice populations in Arizona use a different genetic mechanism to control coat color than the New Mexico populations. What does this tell you about the evolution of the genetic mechanism for coat color?

_____

_____

_____

234 Insecticide Resistance

Key Idea: The application of insecticide provides a strong selection pressure on insects so that resistance to insecticides is constantly evolving in insect populations.

Insecticides are pesticides used to control insects considered harmful to humans, their livelihood, or environment. Insecticide use has increased since the advent of synthetic insecticides in the 1940s.

▶ The widespread but often inefficient use of insecticides can lead to resistance to the insecticide in insects (see right). Mutations may also produce traits that further assist with resistance.

▶ Ineffectual application may include application at the wrong time, e.g. before the majority of the population has established or close to rain, and applying contact sprays that may be avoided by hiding under leaves.

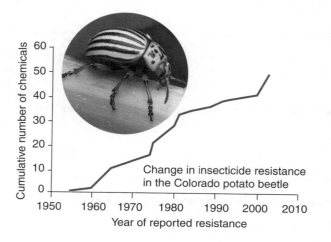

Change in insecticide resistance in the Colorado potato beetle

(graph: y-axis "Cumulative number of chemicals" 0–60; x-axis "Year of reported resistance" 1950–2010)

The Colorado potato beetle (*Leptinotarsa decemlineata*) is a major potato pest that was originally found living on buffalo-bur (*Solanum rostratum*) in the Rocky mountains. It has an extraordinary ability to develop resistance to synthetic pesticides. Since the 1940s, when these pesticides were first developed, it has become resistant to more than 50 different types.

The evolution of resistance

The application of an insecticide can act as a potent selection pressure for resistance in pest insects. Insects with a low natural resistance die from an insecticide application, but those with a higher natural resistance may survive if the insecticide is not effectively applied. These will found a new generation which will on average have a higher resistance to the insecticide.

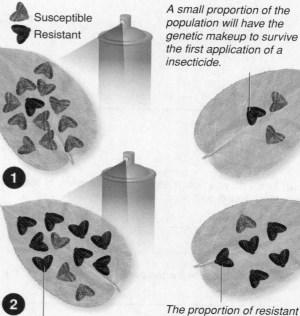

Susceptible
Resistant

A small proportion of the population will have the genetic makeup to survive the first application of a insecticide.

1

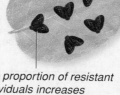

2

The genetic information for insecticide resistance is passed to the next generation.

The proportion of resistant individuals increases following subsequent applications of insecticide. Eventually, almost all of the population is resistant.

Mechanisms of resistance in insect pests

Insecticide resistance in insects can arise through a combination of mechanisms.

▶ Increased sensitivity to an insecticide will cause the pest to avoid a treated area.

▶ Certain genes confer stronger physical barriers, decreasing the rate at which the chemical penetrates the insect's cuticle.

▶ Detoxification by enzymes within the insect's body can render the insecticide harmless.

▶ Structural changes to the target enzymes make the insecticide ineffective.

▶ No single mechanism provides total immunity, but together they transform the effect from potentially lethal to insignificant.

1. Why must farmers be sure that insecticides are applied correctly: _____

_____

_____

2. Describe two mechanisms that increase insecticide resistance in insects: _____

_____

_____

_____

WEB   CCC   CCC   PRACTICES   PRACTICES   PRACTICES

234   P   CE

© 2016 **BIOZONE** International
ISBN: 978-1-927309-46-9
Photocopying Prohibited

235 Gene Pool Exercise

The set of all the versions of all the genes in a population (its genetic make-up) is called the **gene pool**. Cut out the squares below and use them to model the events described in *Modeling Natural Selection*.

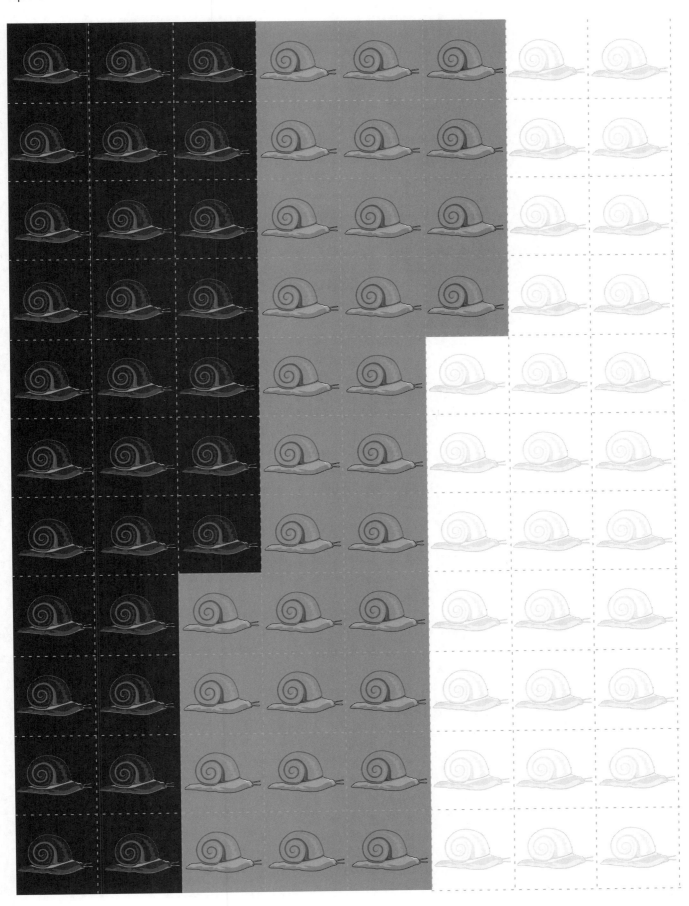

PRAC

236 Modeling Natural Selection

Key Idea: Natural selection acts on phenotypes. Those individuals better suited to an environment will have a greater chance of reproductive success.

Natural selection can be modeled in a simple activity based on predation. You can carry out the following activity by yourself, or work with a partner to increase the size of the population. The black, gray, and white squares on the preceding pages represent phenotypes of a population. Cut them out and follow the instructions below to model natural selection. You will also need a sheet of white paper and a sheet of black paper.

1. Cut out the squares on the preceding pages and record the number of black, gray, and white squares. Work out the proportion of each phenotype in the population (e.g. 0.33 black 0.34 gray, 0.33 white) and place these values in the table below. This represents your starting population (you can combine populations with a partner to increase the population size for more reliable results).

2. For the first half of the activity you will also need a black sheet of paper or material that will act as the environment (A3 is a good size). For the second half of the activity you will need a white sheet of paper.

3. Place 14 each of the black, gray, and white squares in a bag and shake them up to mix them. Keep the others for making up population proportions later.

4. Now take the squares out of the bag and randomly distribute them over the sheet of black paper (this works best if your partner does this while you aren't looking).

5. For 20 seconds, pick up the squares that stand out (are obvious) on the black paper. These squares represent animals in the population that have been preyed upon and killed (you are acting the part of a predator on the snails). Place them to one side and pick up the rest of the squares. These represent the population that survived to reproduce.

6. Count the remaining phenotype colors and calculate the proportions of each phenotype. Record them in the table below in the proportions row of generation 2. Use the formula: Proportion = number of colored squares /total number of squares remaining. For example: for one student doing this activity: proportion of white after predation = 10/30 = 0.33.

7. Before the next round of selection, the population must be rebuilt to its original total number using the newly calculated proportions of colors and the second half of the squares from step 3. Use the following formula to calculate the number of each color: number of colored squares required = proportion x number of squares in original population (42 if you are by yourself, 84 with a partner). For example: for one student doing this activity: 0.33 x 42 = 13.9 = 14 (you can't have half a phenotype). Therefore in generation 2 there should be 14 white squares. Do this for all phenotypes using the spare colors to make up the numbers if needed. Record the numbers in the numbers row of generation 2. Place generation 2 into the bag.

8. Repeat steps 4 to 7 for generation 2, and 3 more generations (5 generations in total or more if you wish).

9. On separate graph paper, draw a line graph of the proportions of each color over the five generations. Which colors have increased, which have decreased?

10. Now repeat the whole activity using a white sheet background instead of the black sheet. What do you notice about the proportions this time?

Generation		Black	Gray	White
1	Number			
	Proportion			
2	Number			
	Proportion			
3	Number			
	Proportion			
4	Number			
	Proportion			
5	Number			
	Proportion			

© 2016 **BIOZONE** International
ISBN: 978-1-927309-46-9
Photocopying Prohibited

PRACTICES

WEB

236 **KNOW**

237 What is a Species?

Key Idea: A species is a very specific unit of taxonomy given to a group of organisms with similar characteristics in order to categorize them.

The **species** is the basic unit of taxonomy. A **biological species** is defined as a group of organisms capable of interbreeding to produce fertile offspring.

▶ There can be difficulties in applying the biological species concept (BSC) in practice as some closely related species are able to interbreed to produce fertile hybrids (e.g. species of *Canis*, which includes wolves, coyotes, domestic dogs, and dingoes).

▶ The BSC is more successfully applied to animals than to plants and organisms that reproduce asexually. Plants hybridize easily and can reproduce vegetatively. For some, e.g. cotton and rice, first generation hybrids are fertile but second generation hybrids are not.

▶ In addition, the BSC cannot be applied to extinct organisms. Some organisms, such as bacteria can also transfer genetic material to unrelated species. Increasingly, biologists are using DNA analyses to clarify relationships between closely related populations.

These five breeds of domestic may look and act very differently, but they can all breed with each other and with the wolf (bottom right), a different species, to produce viable offspring.

Another way of defining a species is by using the **phylogenetic species concept** (PSC). Phylogenetic species are defined on the basis of their evolutionary history. This is determined on the basis of **shared derived characteristics**. These are characteristics that evolved in an ancestor and are present in all its descendants.

▶ The PSC defines a species as the smallest group that all share a derived character state. It is useful in paleontology because biologists can compare both living and extinct organisms.

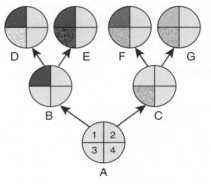

Species B and C are related to species A as they share three of four characteristics with it. However they only share two characteristics with each other.

D and E share characteristics with B, while F and G share characteristics with C.

1. Define a biological species: _____

_____

_____

2. Why is it difficult to fully define a species in practice? _____

_____

_____

_____

3. How is the biological species different to the phylogenetic species? _____

_____

_____

4. There often appear to be greater differences between different breeds of dog than there are between different species of the *Canis* genus. Why then are dogs all considered one species?

_____

_____

_____

 © 2016 **BIOZONE** International
ISBN: 978-1-927309-46-9
Photocopying Prohibited

238 How Species Form

Key Idea: Gene flow is reduced when populations are separated. Continual reduction in gene flow by isolating mechanisms may eventually lead to the formation of new species.

Species formation

▶ Species evolve in response to selection pressures of the environment. These may be naturally occurring or caused by humans. The diagram below represents a possible sequence for the evolution of two hypothetical species of butterfly from an ancestral population. As time progresses (from top to bottom of the diagram below) the amount of genetic difference between the populations increases, with each group becoming increasingly isolated from the other.

▶ The isolation of two gene pools from one another may begin with **geographical barriers**. This may be followed by isolating mechanisms that occur before the production of a zygote (e.g. behavioral changes), and isolating mechanisms that occur after a zygote is formed (e.g. hybrid sterility). As the two gene pools become increasingly isolated and different from each other, they are progressively labelled: population, race, and subspecies. Finally they attain the status of separate species.

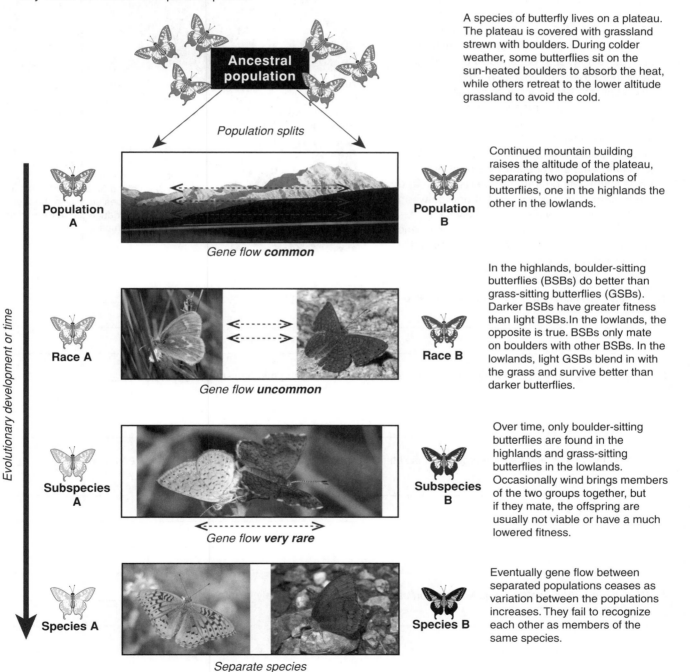

A species of butterfly lives on a plateau. The plateau is covered with grassland strewn with boulders. During colder weather, some butterflies sit on the sun-heated boulders to absorb the heat, while others retreat to the lower altitude grassland to avoid the cold.

Continued mountain building raises the altitude of the plateau, separating two populations of butterflies, one in the highlands the other in the lowlands.

In the highlands, boulder-sitting butterflies (BSBs) do better than grass-sitting butterflies (GSBs). Darker BSBs have greater fitness than light BSBs. In the lowlands, the opposite is true. BSBs only mate on boulders with other BSBs. In the lowlands, light GSBs blend in with the grass and survive better than darker butterflies.

Over time, only boulder-sitting butterflies are found in the highlands and grass-sitting butterflies in the lowlands. Occasionally wind brings members of the two groups together, but if they mate, the offspring are usually not viable or have a much lowered fitness.

Eventually gene flow between separated populations ceases as variation between the populations increases. They fail to recognize each other as members of the same species.

Population splits

Population A — *Gene flow common* — **Population B**

Race A — *Gene flow uncommon* — **Race B**

Subspecies A — *Gene flow very rare* — **Subspecies B**

Species A — *Separate species* — **Species B**

Evolutionary development or time

Geographic isolation

- ▶ Geographical isolation describes the isolation of a species population (gene pool) by some kind of physical barrier, e.g. mountain range, water body, desert, or ice sheet. Geographical isolation is a frequent first step in the subsequent reproductive isolation of a species.

- ▶ An example of geographic isolation leading to speciation is the large variety of cichlid fish in the rift lakes of East Africa (right). Geologic changes to the lake basins have been important in the increase of cichlid fish species.

Reproductive isolating mechanisms

- ▶ Reproductive isolating mechanisms (RIMs) are reproductive barriers that are part of a species' biology (and therefore do not include geographical isolation). They prevent interbreeding (therefore gene flow) between species. Single barriers may not completely stop gene flow, so most species commonly have more than one type of barrier. Most RIMs operate before fertilization (prezygotic RIMs). They include mechanisms such as temporal isolation (e.g. differences in breeding season), behavioral isolation (e.g. differences in mating behaviors), and mechanical isolation (e.g. differences in copulatory structures). Postzyotic RIMs operate after fertilization and are important in preventing offspring between closely related species. They involve a mismatch of chromosomes in the zygote, and include hybrid sterility (right), hybrid inviability, and hybrid breakdown.

The white-tailed antelope squirrel (left) and the Harris' antelope squirrel (right) in the southwestern United States and northern Mexico, are separated by the Grand Canyon (center) and have evolved to occupy different habitats.

Mules are a cross between a male donkey and a female horse. The donkey contributes 31 chromosomes while the horse contributes 32, making 63 chromosomes in the mule. This produces sterility in the mule as meiosis cannot produce gametes with an even number of chromosomes.

1. Identify some geographical barriers that could separate populations: _____

2. Why is a geographical barrier not considered a reproductive isolating mechanism? _____

3. Identify the two categories of reproductive isolating mechanisms and explain the difference between them: _____

4. Why is more than one reproductive isolation barrier needed to completely isolate a species? _____

© 2016 **BIOZONE** International
ISBN: 978-1-927309-46-9
Photocopying Prohibited

239 Patterns of Evolution

Key Idea: Populations moving into a new environment may diverge from their common ancestor and form new species.

▶ The diversification of an ancestral group into two or more species in different habitats is called **divergent evolution**. This is shown right, where two species diverge from a **common ancestor**. Note that another species arose, but became extinct.

▶ When divergent evolution involves the formation of a large number of species to occupy different niches, it is called an **adaptive radiation**.

▶ The evolution of species may not necessarily involve branching. A species may accumulate genetic changes that, over time, result in a new species. This is known as **sequential evolution**.

▶ Evolution of species does not always happen at the same pace. Two models describe the pace of evolution.

Phyletic gradualism proposes that populations diverge slowly by accumulating adaptive features in response to different selective pressures.

Punctuated equilibrium proposes that most of a species' existence is spent in stasis and evolutionary change is rapid. The stimulus for evolution is a change in some important aspect of the environment, e.g. mean temperature.

It is likely that both mechanisms operate at different times for different taxonomic groups.

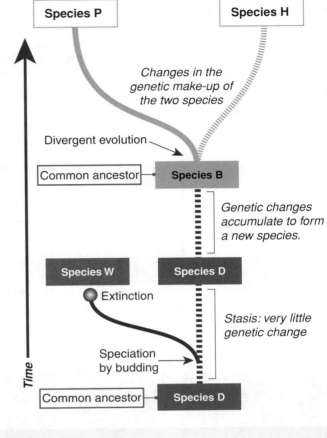

Species P   Species H

Changes in the genetic make-up of the two species

Divergent evolution

Common ancestor → **Species B**

Genetic changes accumulate to form a new species.

Species W   Species D

Extinction

Stasis: very little genetic change

Speciation by budding

Common ancestor → **Species D**

Time

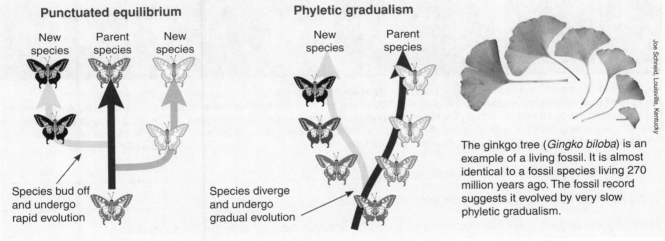

Punctuated equilibrium

New species   Parent species   New species

Species bud off and undergo rapid evolution

Phyletic gradualism

New species   Parent species

Species diverge and undergo gradual evolution

The ginkgo tree (*Gingko biloba*) is an example of a living fossil. It is almost identical to a fossil species living 270 million years ago. The fossil record suggests it evolved by very slow phyletic gradualism.

Joe Schneid, Louisville, Kentucky

1. In the hypothetical example of divergent evolution illustrated at the top of the page:

   (a) Identify the type of evolution that produced species B from species D: _____

   (b) Identify the type of evolution that produced species P and H from species B: _____

   (c) Name all species that evolved from: Common ancestor D: _____ Common ancestor B: _____

2. When do you think punctuated evolution is most likely to occur and why? _____

   _____

   _____

PRACTICES   WEB

 **239**

KNOW

240 Evolution and Biodiversity

Key Idea: Mammals underwent an adaptive radiation around 65 million years ago and became a very diverse group.

▶ The mammals underwent a spectacular adaptive radiation following the sudden extinction of the non-avian dinosaurs, which left many niches vacant. Most of the modern groups of mammals appeared very quickly. Placental mammals have become dominant over other groups of mammals throughout the world, except in Australia where marsupials remained dominant after being isolated following the breakup of Gondwana. Placental mammals (bats) reached Australia 15 million years ago followed by rodents 10 million years ago.

▶ The diagram below shows a simple evolutionary tree for the mammals. The width of the bars indicates the number of species in each of the three mammalian groups.

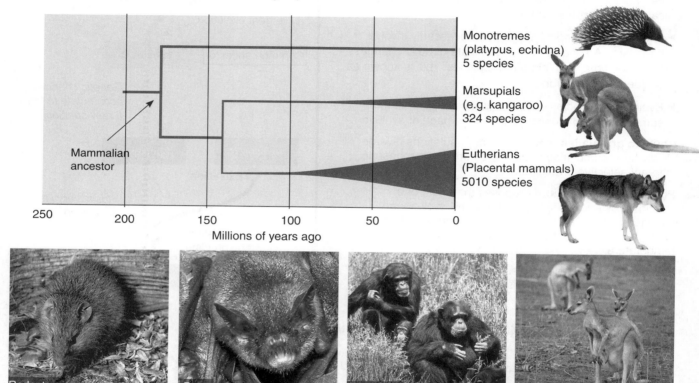

The placental mammals make up the largest group of mammals. The embryos are carried internally and obtain nutrients from the mother via the placenta. The young are born live and well developed. The group consists of 18 orders, although some orders contain many more species than others. Rodents are the largest order, making up nearly 40% of mammalian species, followed by bats. Primates also make up a significant proportion of the mammals.

The five species of monotremes lay eggs whereas marsupials give birth to very undeveloped young that develop in a pouch.

1. (a) When did the mammals first evolve? _____

   (b) When did placental mammals split from the marsupials? _____

   (c) Suggest why the mammals underwent such a major adaptive radiation after the dinosaurs died out? _____

   _____

   _____

   (d) Suggest why marsupials remained the dominant mammalian group in Australia after the split between the marsupial and placental mammals:

   _____

   _____

   _____

   _____

   _____

 © 2016 **BIOZONE** International
ISBN: 978-1-927309-46-9
Photocopying Prohibited

Rodent biodiversity

Rodents make up 40% of mammalian species, making them easily the most successful of the mammalian groups. They have adapted to a huge number of habitats from deserts to forest. All rodents have upper and lower incisor teeth that grow continuously. Fossils with distinctive rodent features first appeared about 66 million years ago.

Squirrel-like rodents

Squirrels are found on many continents. Their lifestyles include tree dwelling, ground dwelling, and gliding forms. Like most rodents they are social, with prairie dogs forming large communities called towns.

Porcupine-like rodents

Capybaras are South American rodents and the largest of all rodents. They occupy habitats from forests to savannahs. Porcupines are found throughout the Old and New Worlds. Their spines make an almost impenetrable defense against predators. The group also includes guinea pigs, which are popular as pets.

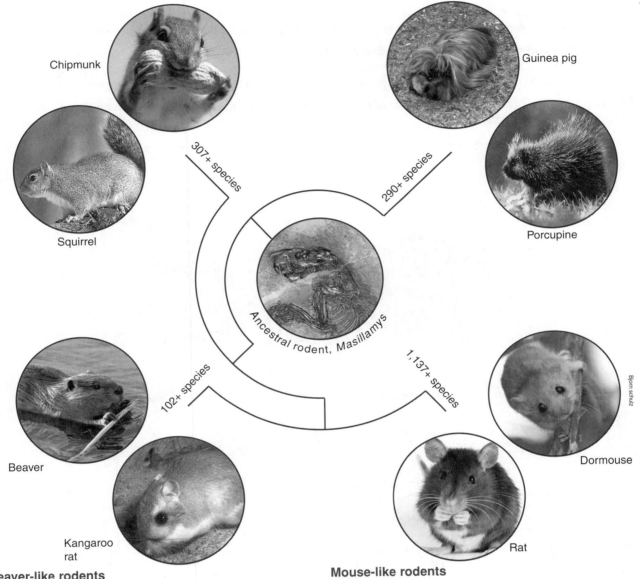

Beaver-like rodents

Beavers are one of the larger types of rodents. They live near rivers, streams and lakes, chewing through small trees to build dams across streams and lodges to live in. Gophers live in burrows, while kangaroo rats are so well adapted to the desert they virtually never need to drink.

Mouse-like rodents

Rats and mice are found in virtually every part of the world thanks to their generalist adaptations and human assisted travel. There are at least 100 species of rats and mice alone. The group also includes voles, lemmings, jerboas and dormice.

2. What anatomical feature do all rodents have?_____

3. (a) Describe some of the habitats rodents have occupied: _____

_____

(b) Describe some adaptations of rodents (you may use extra paper and attach it to this page): _____

_____

_____

 © 2016 **BIOZONE** International
ISBN: 978-1-927309-46-9
Photocopying Prohibited

241 Extinction is a Natural Process

Key Idea: Extinctions are a natural process and may happen gradually or very rapidly.

Extinction is the death of an entire species, no individuals are left alive. Extinction is an important (and natural) process in evolution and describes the loss of a species forever. Extinction provides opportunities, in the form of vacant niches, for the evolution of new species. More than 98% of species that have ever lived are now extinct, most of these before humans were present.

▶ Extinction is the result of a species being unable to adapt to an environmental change. Either it evolves into a new species (and the ancestral species becomes effectively extinct) or it and its lineage becomes extinct.

▶ A **mass extinction** describes the widespread and rapid (in geologic terms) decrease in life on Earth and involves not only the loss of species, but the loss of entire families (which are made up of many genera and species). Such events are linked to major climate shifts or catastrophic events. There have been five previous mass extinctions.

▶ The diagram below shows how the diversity of life has varied over the history of life on Earth and aligns this with major geologic and climatic events.

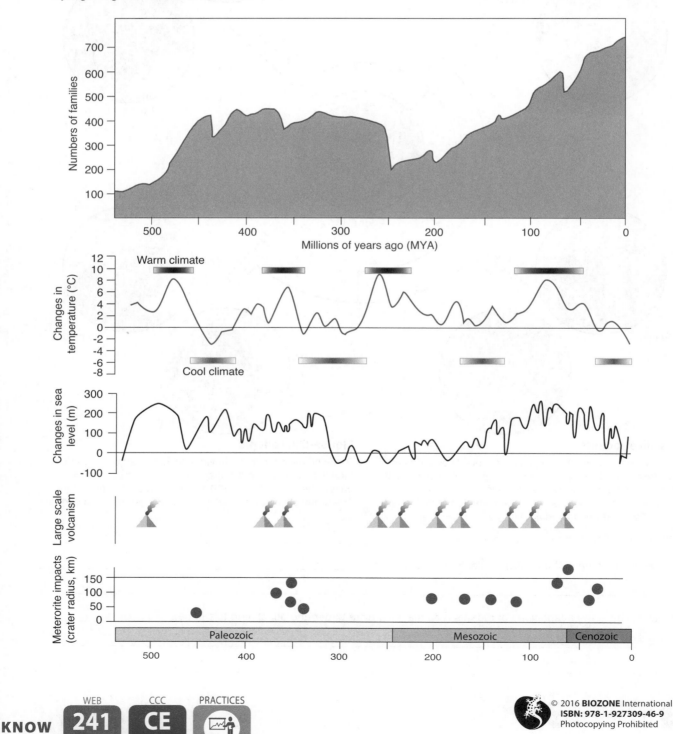

Graptolite

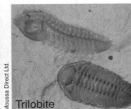
Trilobite

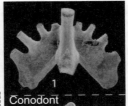
Coral

Conodont

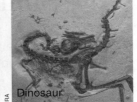
Dinosaur

Ordovician extinction (458-440 MYA). Second largest extinction of marine life: >60% of marine invertebrates died. One of the coldest periods in Earth's history.

Devonian extinction (375-360 MYA). Marine life affected especially brachiopods, trilobites, and reef building organisms.

Permian extinction (252 MYA). Nearly all life on Earth perished. 57% of families and 83% of genera were wiped out. 96% of marine species became extinct.

Triassic extinction (201.3 MYA). At least half of the species present became extinct, vacating niches and ushering in the age of the dinosaurs.

Cretaceous extinction (66 MYA). Marked by the extinction of nearly all dinosaur species (their descendants, the birds, survive).

1. What is the general cause of extinction? _____

_____

_____

2. What is a mass extinction? _____

_____

_____

3. Why is extinction an important natural process? _____

_____

_____

4. Study the data on the opposite page carefully and answer the following questions:

   (a) What is the evidence that extinction is a natural process? _____

   _____

   _____

   (b) Mark the five mass extinctions on the top graph opposite:

   (c) Is there reason to believe that there is any one cause for any of these mass extinctions? Explain your answer:

   _____

   _____

   _____

   _____

   _____

   _____

   (d) What has happened to the diversity of life soon after each mass extinction?

   _____

   _____

   _____

5. Which of the five extinctions appears to be the most severe? _____

6. Which extinctions mark the rise and end of the dinosaurs? _____

_____

242 Humans and Extinction

Key Idea: Human activity is directly responsible for a number of extinctions (e.g. the passenger pigeon) and related to a great many others.

▶ Human activity is the cause of many recent extinctions. Famous extinctions include the extermination of the dodo within 70 years of its first sighting in 1598 and the passenger pigeon, which went from an estimated 3 billion individuals before the arrival of Europeans to North America to being extinct by 1914. Recent extinctions include the Yangtze River dolphin and the Pinta Island tortoise (a giant tortoise).

▶ Human activities such as hunting and destruction of habitat (by logging, pollution, and land clearance) have caused many recent extinctions. The effects of climate change on habitats and life histories (e.g. breeding times) may also drive many other vulnerable species to extinction.

The dodo (skeleton right) was endemic to the island of Mauritius in the Indian Ocean. It was first discovered around 1598 and extinct by 1662, just 64 years later. Its extermination was so rapid that we are not even completely sure what it looked like and know very little about it.

Heinz-Josef Lücking

Species become extinct naturally at an estimated rate of about one species per one million species per year. This is called the **background extinction rate**. Since the 1500s, at least 412 vertebrate or plant species have become extinct. Proportionally, birds have been affected more than any other vertebrate group, although a large number of frog species are in rapid decline.

US Fish and Wildlife Service

The golden toad (left) from Costa Rica was officially declared extinct in 1989.

Organism	Total number of species (approx)*	Known extinctions (since ~1500 AD)*
Mammals	5487	87
Birds	9975	150
Reptiles	10,000	22
Amphibians	6700	39
Plants	300,000	114

* These numbers vastly underestimate the true numbers because so many species are undescribed.

1. (a) What is the total number of known extinctions over the last 500 years?_____

    (b) Why is this number probably an underestimate? _____

    _____

2. (a)   There are approximately 10,000 living or recently extinct types of bird. Assuming a background extinction rate, how many bird species should be becoming extinct per year? Show your working:

    _____

    _____

    (b) Since 1500 AD, 150 birds species are known to have become extinct. How many times greater is this rate of extinction than the background rate for birds (assume 1500 AD to 2015)?

    _____

    _____

3. Identify three reasons for an increase in extinct rates over the last 515 years: _____

    _____

    _____

    _____

 © 2016 **BIOZONE** International
ISBN: 978-1-927309-46-9
Photocopying Prohibited

Threatened and endangered

The International Union for Conservation of Nature (IUCN) began its Red List in 1964. It is the most comprehensive inventory of the conservation status of the Earth's numerous biological species. By continually updating the list it is possible to track the changes in species' populations and therefore their risk of extinction.

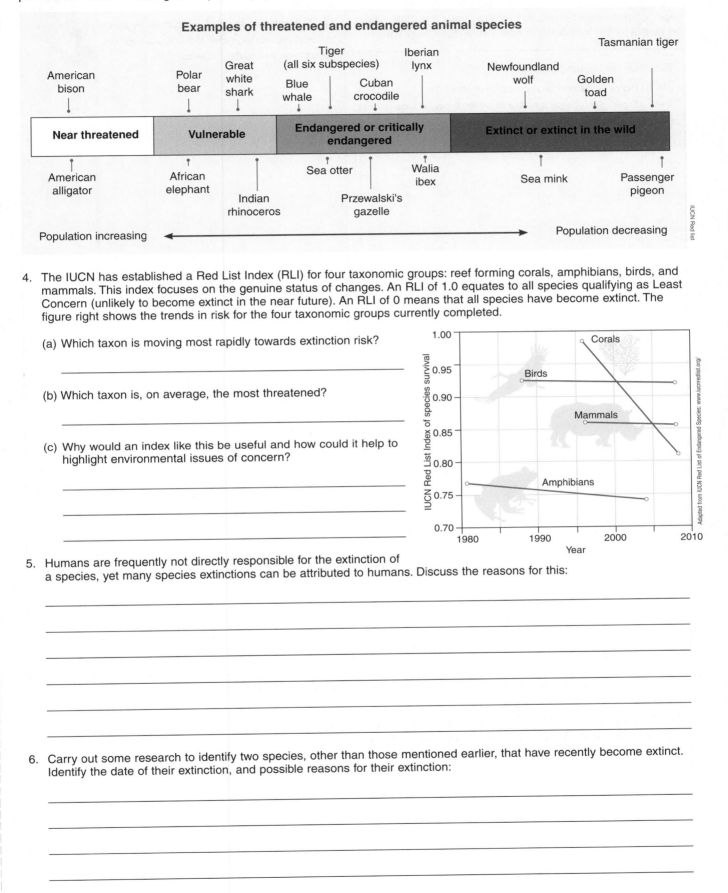

Examples of threatened and endangered animal species

Near threatened | Vulnerable | Endangered or critically endangered | Extinct or extinct in the wild

American bison, Polar bear, Great white shark, Blue whale, Tiger (all six subspecies), Cuban crocodile, Iberian lynx, Newfoundland wolf, Golden toad, Tasmanian tiger

American alligator, African elephant, Indian rhinoceros, Sea otter, Przewalski's gazelle, Walia ibex, Sea mink, Passenger pigeon

Population increasing ← → Population decreasing

IUCN Red list

4. The IUCN has established a Red List Index (RLI) for four taxonomic groups: reef forming corals, amphibians, birds, and mammals. This index focuses on the genuine status of changes. An RLI of 1.0 equates to all species qualifying as Least Concern (unlikely to become extinct in the near future). An RLI of 0 means that all species have become extinct. The figure right shows the trends in risk for the four taxonomic groups currently completed.

(a) Which taxon is moving most rapidly towards extinction risk?

_____

(b) Which taxon is, on average, the most threatened?

_____

(c) Why would an index like this be useful and how could it help to highlight environmental issues of concern?

_____

_____

_____

IUCN Red List Index of species survival — Corals, Birds, Mammals, Amphibians (1980–2010)

Adapted from IUCN Red List of Endangered Species: www.iucnredlist.org

5. Humans are frequently not directly responsible for the extinction of a species, yet many species extinctions can be attributed to humans. Discuss the reasons for this:

_____

_____

_____

_____

_____

_____

6. Carry out some research to identify two species, other than those mentioned earlier, that have recently become extinct. Identify the date of their extinction, and possible reasons for their extinction:

_____

_____

_____

_____

© 2016 **BIOZONE** International
ISBN: 978-1-927309-46-9
Photocopying Prohibited

243 Chapter Review

Summarize what you know about this topic under the headings provided. You can draw diagrams or mind maps, or write short notes to organize your thoughts. Use the points in the introduction and the hints to help you.

Evolution and natural selection

HINT: How does natural selection lead to evolution? What does natural selection act on?

Adaptation

HINT: How is adaptation a consequence of natural selection? How do adaptive features enhance fitness?

Extinction

HINT: Describe natural causes of extinction. How has human activity influenced the current rates of extinction?

REVISE

244 KEY TERMS AND IDEAS: Did You Get It?

1. Test your vocabulary by matching each term to its definition, as identified by its preceding letter code.

adaptation ....

biodiversity ....

evolution ....

extinction ....

fitness ....

genotype ....

natural selection ....

phenotype ....

variation ....

A The complete dying out of a species so that there are no representatives of the species remaining anywhere.

B The observable characteristics in an organism.

C A heritable characteristic of a species that equips it for survival and reproductive success in its environment.

D The allele combination of an organism e.g. AA.

E The differences between individuals in a population as a result of genes and environment.

F The process by which favorable heritable traits become more common in successive generations.

G Change in the genetic makeup of a population over time.

H Biological diversity, e.g. of a region or of the Earth

I A measure of an individual's relative genetic contribution to the next generation as a result of its combination of traits.

2. Complete the sequence below to outline the features of the four factors involved in evolutionary change in a population:

(i) Species population: _____

_____

(ii) Genetic variation: _____

_____

(iii) Competition: _____

_____

(iv) Proliferation: _____

_____

3. Drosophilidae (fruit flies) are a group of small flies found almost everywhere in the world. Two genera, *Drosophila* and *Scaptomyza* are found in the Hawaiian islands and between them there are more than 800 species present on a land area of just 16,500 km². It is one of the densest concentrations of related species found anywhere. The flies range from 1.5 mm to 20 mm in length and display a startling range of wing forms and patterns, body shapes and colors, and head and leg shapes. Genetic analyses show that they are all related to a single species that may have arrived on the islands around 8 million years ago. Older species appear on the older islands and more recent species appear as one moves from the oldest to the newest islands.

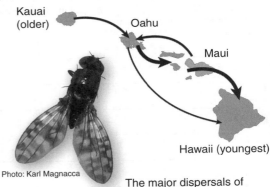

Photo: Karl Magnacca

The major dispersals of *Drosophila* and *Scaptomyza*.

(a) What evolutionary pattern is shown by the Hawaiian fruit flies:

_____

(b) Suggest why so many fruit fly species are present in Hawaii: _____

_____

(c) Describe the relationship between the age of the islands and the age of the fly species: _____

_____

(d) Account for this relationship: _____

_____

TEST

245 Summative Assessment

In this assessment task you will design a living organism and explain how its adaptations help it survive:

You are to think of a new kind of living organism, it can be anything, but not something already living. Your new organism could be a combination of other living things, e.g. a cog, which has some features of a cat and some of a dog. You also need to think of the environment that the organism usually lives in (e.g tropical rainforest, high humidity, etc.)

1. Describe your new organism, including one specific adaptation that helps the organism survive (e.g. thick fur coat helps the organism to …). You will also need to describe the environment that your organism lives in:

_____

_____

_____

_____

_____

_____

_____

_____

_____

_____

_____

_____

_____

_____

_____

_____

2. The adaptations of living organisms vary, usually in a normal distribution (e.g. claws may vary in length through the population). Explain how the specific adaptation for your organism varies through the population, from the most extreme forms to the median form.

_____

_____

_____

_____

_____

3. Describe how the variation in the adaptation affects the ability of the organism to survive in its normal environment, assuming there is no selective pressure for change:

_____

_____

_____

_____

_____

TEST

4. Imagine now that a change in the environment has introduced a slight directional selection pressure on the adaptation of your organism.

(a) Which extreme of your organism's adaptation is negatively affected? _____

_____

(b) Explain why: _____

_____

_____

_____

(c) If the selection pressure remains for many generations describe how your organism will change over time:

_____

_____

_____

5. Imagine now that the selection pressure changes to act upon the median form of your organism's adaptation:

(a) Describe how the population of the organism will be affected over many generations:

_____

_____

_____

(b) Will this affect the ability of the organisms at the extreme ends of the range to breed with one another?

_____

(c) If so, why, and how will this affect the species? _____

_____

_____

_____

_____

6. Now imagine that a mutation in your organism affects one extreme of the phenotypic range so that the organism's fitness increases (i.e. it increases the organism's chance of surviving and reproducing).

(a) What is the effect of the mutation on the adaptation, i.e. what change is there and how does this affect the organism for the better?

_____

_____

_____

(b) Imagine that the mutation only increases chances of survival under certain circumstances (e.g. a mutation from brown to white fur enhances survival in the snow).

i) What is the circumstance? _____

_____

ii) How will this limited enhancement of survival affect the evolution of the species over many generations?

_____

_____

_____

LS4.D
ETS1.B

Biodiversity

Key terms

biodiversity
biodiversity hotspot
climate change
conservation
ecosystem services
extinction
ex-situ conservation
genetic diversity
global warming
in-situ conservation

Disciplinary core ideas

Show understanding of these core ideas

	Activity number

Humans depend on biodiversity

☐ 1 Recall that human-induced changes in the environment may lead species to contract their range and may lead to the decline or extinction of some species. — 242

☐ 2 Recall that the Earth's biodiversity is the result of evolution. There are different methods by which humans evaluate biodiversity but certain regions of the world are more biodiverse than others. These are also often areas with high human population densities. — 240 246 248

☐ 3 Humans depend on the Earth's biodiversity for ecosystem services. These are things that ecosystems provide such as resources, carbon storage, and purification of air and water. Ecosystem services are provided most effectively by well-functioning, biodiverse ecosystems. — 247

☐ 4 The activity of humans may adversely affect the biodiversity through overpopulation, overexploitation, habitat destruction, pollution, introduction of invasive species, and climate change. — 248 249

☐ 5 Sustaining biodiversity to maintain ecosystem functioning and productivity is essential to supporting life on Earth and has positive benefits to humans who depend on it. — 247

Developing possible solutions

☐ 6 Humans must find solutions to sustaining biodiversity. Solutions to problems created by human use of resources must consider practical constraints, such as costs, and social, cultural, and environmental impacts. — 250 251 252 253

☐ 7 Models and computer simulations can be used to test different possible solutions to sustaining biodiversity. — 249

Christian Ziegler cc 2.5 Richard Ling (2004) cc 2.5, 3.0 Heikki Valve cc 3.0

Crosscutting concepts

Understand how these fundamental concepts link different topics

	Activity number

☐ 1 **CE** Empirical evidence enables us to support claims about the ways in which humans are affecting the Earth's biodiversity. — 249

☐ 2 **CE** Empirical evidence enables us to support claims about the best ways to reduce the adverse effects of human activity on biodiversity. — 249 251 252 253

Science and engineering practices

Demonstrate competence in these science and engineering practices

	Activity number

☐ 1 Create or revise a simulation to test a solution to reduce the adverse effects of human activity on biodiversity. — 251 252

☐ 2 Explain based on evidence how a range of conservation methods can be used to restore the species and genetic diversity of endangered populations. — 251 252 253

☐ 3 Devise a solution based on current evidence for the recovery of an endangered species. — 256

246 Biodiversity

Key Idea: Biodiversity describes the biotic variation in an ecosystem. The perception of an ecosystem's diversity may differ depending on how the biodiversity is measured.

What is biodiversity?

Biodiversity is the amount of biotic variation within a given group. It could be the number of species in a particular area or the amount of genetic diversity in a species. Ecosystem biodiversity refers to the number of ecosystems in a given region and is usually correlated with species diversity. To conserve biodiversity, conservation of habitats is important.

Biodiversity tends to be clustered in certain parts of the world, called hotspots, where species diversity is high. Tropical forests and coral reefs (above) are some of the most diverse ecosystems on Earth.

Biodiversity of Earth	
Type of organism	**Estimated number of species**
Protozoa	36,400
Brown algae, diatoms	27,500
Invertebrates	7.6 million
Plants	298,000
Fungi	611,000
Vertebrates	60,000

The latest estimate of the number of eukaryotic species on Earth is 8.7 million, of which only 1.2 million have been formally described. Prokaryotic species are so variable and prolific, the number of species is virtually inestimable and could be hundreds of millions.

Measuring biodiversity

Biodiversity is measured for a variety of reasons, e.g. to assess the success of conservation work or to measure the impact of human activity. One measure of biodiversity is to simply count all the species present (the **species richness**) although this may give an imprecise impression of the ecosystem's biodiversity. **Species evenness** gives a measure of relative abundance of species, i.e. how close the numbers of each species in an environment are. It describes how the biodiversity is apportioned.

Two ecosystems with quite different biodiversity.

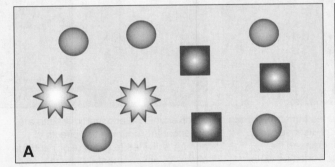

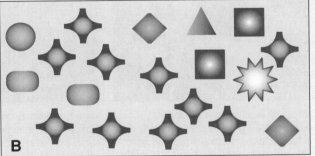

Species richness is a simple method of estimating biodiversity in which the number of species is counted. This does not show if one species is more abundant than others. **Species evenness** is a method in which the proportions of species in an ecosystem are estimated.

Diversity Indices use mathematical formula based on the species abundance and the number of each species to describe the biodiversity of an ecosystem. Many indices produce a number between 0 and 1 to describe biodiversity, 1 being high diversity and 0 being low diversity.

1. Calculate the number and percentage of eukaryotic species that are still to be formally described: _____

_____

2. Describe in words the species richness and species evenness of ecosystem A and B above:

A _____

B _____

247 Humans Depend on Biodiversity

Key Idea: Humans rely on ecosystems and the services they provide for health, well being, and livelihood. The biodiversity of an ecosystem affects its ability to provide these services.

Ecosystems provide services

▶ Humans depend on Earth's ecosystems for the services they provide. These ecosystem services include resources such as food and fuel, as well as processes such as purification of the air and water. These directly affect human health.

▶ Biologically diverse and resilient ecosystems that are managed in a sustainable way are better able to provide the ecosystem services on which we depend.

▶ The UN has identified four categories of ecosystem services: supporting, provisioning, regulating, and cultural.

▶ Regulating and provisioning services are important for human health and security (security of resources and security against natural disasters).

▶ Cultural services are particularly important to the social fabric of human societies and contribute to well being. These are often things we cannot value in monetary terms.

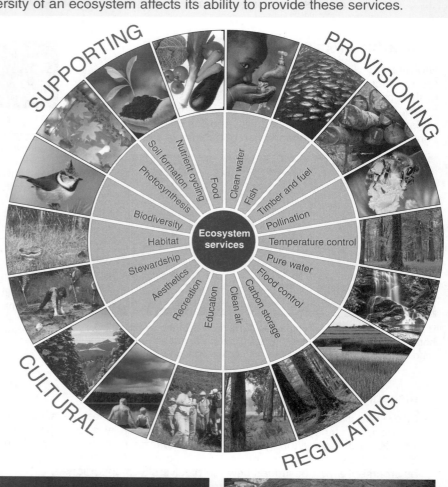

SUPPORTING · PROVISIONING · CULTURAL · REGULATING

Ecosystem services: Soil formation, Nutrient cycling, Photosynthesis, Biodiversity, Habitat, Stewardship, Aesthetics, Recreation, Education, Clean air, Carbon storage, Flood control, Pure water, Temperature control, Pollination, Timber and fuel, Fish, Clean water, Food

Disease resistance in sorghum

Biodiversity is important in crop development, e.g promoting disease resistance. Many medical breakthroughs have come from understanding the biology of wild plants and animals.

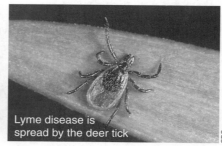

Lyme disease is spread by the deer tick

High biodiversity creates buffers between humans and infectious diseases (e.g. Lyme disease) and increases the efficiency of processes such as water purification.

Landslide

Biodiversity and ecosystem health are essential for reducing the effects of human activities (e.g. pollution) and the effects of environmental disasters (e.g. eruptions and landslides).

1. What are ecosystem services and why are they important to humans? _____

_____

_____

_____

2. What is the relationship between biodiversity and the ability of an ecosystem to provide essential ecosystem services?

_____

_____

_____

© 2016 **BIOZONE** International
ISBN: 978-1-927309-46-9
Photocopying Prohibited

248 Biodiversity Hotspots

Key Idea: Some regions are biodiversity hotspots; they have higher biodiversity than other regions but are under threat from human activity.

▶ Biodiversity is not distributed evenly on Earth. It tends to be clustered in certain parts of the world, called **biodiversity hotspots**. These regions are biologically diverse and ecologically distinct regions under the greatest threat of destruction from human activity. They are identified on the basis of the number of species present, the amount of endemism (species unique to a specific geographic location), and the extent to which the species are threatened.

▶ Biodiversity hotspots make up less than 2% of Earth's land surface but support nearly 60% of the world's plant and vertebrate species. Their conservation is considered central to securing global biodiversity.

▶ Habitat destruction and human-induced climate change are major threats to biodiversity hotspots. The introduction of invasive or predatory species can also place the biodiversity of these regions in danger.

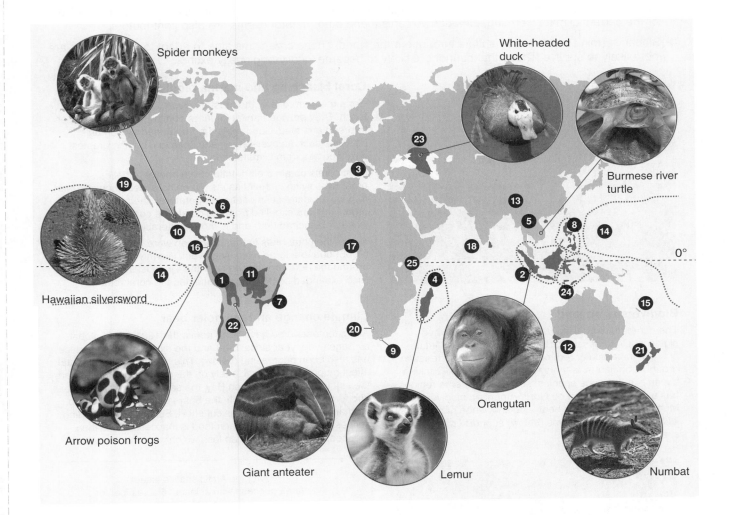

1. Looking at the map, where are most of the hotspots concentrated?_____

_____

2. Many of the biodiversity hotspots coincide with regions of very high human population density. How does high population density create greater risk of biodiversity loss in these regions?

_____

_____

3. Use your research tools (including the Weblinks identified below) to identify each of the 25 biodiversity hotspots illustrated in the diagram above. For one region that interests you, summarize the characteristics that have resulted in it being identified as a biodiversity hotspot. Attach your summary to this page.

© 2016 **BIOZONE** International
ISBN: 978-1-927309-46-9
Photocopying Prohibited

WEB
248 KNOW

249 How Humans Affect Biodiversity

Key Idea: The activities of an expanding human population are contributing to an increase in extinction rates above the natural level and local and global reductions in biodiversity.

▶ As human demand for resources increases, increasing pressure is placed on habitats and their natural populations. The effects of these pressures are often detrimental to ecosystem health and biodiversity.

▶ A decline in biodiversity reduces the ability of ecosystems to resist change and to recover from disturbance. Humans depend both directly and indirectly on healthy ecosystems, so a loss of biodiversity affects us too.

▶ **Global warming**, the continuing rise in the average temperature of the Earth's surface, is a significant contributor to loss of biodiversity.

How does global warming affect biodiversity?

▶ Evidence indicates that human activities, particularly deforestation and the use of fossil fuels, are responsible for the current climate warming, although volcanism and other natural events are also contributors.

▶ Global warming will change habitats throughout the world. Those organisms able to adapt to the changes are more likely to survive. Those that cannot are likely to become locally or globally extinct.

Bleached coral

Healthy coral

Coral bleaching and marine biodiversity

Coral reefs are one of the most biodiverse ecosystems on Earth. They provide habitat for large number of marine species to breed and feed. Coral reefs are threatened by human activities such as over-fishing and pollution, but the greatest threat comes from climate change.

Most corals obtain their nutrition from photosynthetic organisms living in their tissues. When coral becomes stressed (e.g. by an increase in ocean temperature), the photosynthetic organisms are expelled from the coral. The coral becomes white or bleached (left) and is more vulnerable to disease.

Half of the coral reefs in the Caribbean were lost in one year (2005) due to a large bleaching event. Warm waters centered around the northern Antilles near the Virgin Islands and Puerto Rico expanded southward and affected the coral reefs.

Biodiversity on land

The numbers of quaking aspen trees have declined significantly across the US in the last decade. Climate change, in particular reduced rainfall and an increase in drought conditions, is thought to be the cause. Quaking aspen is a keystone species, so its loss in some regions has a significant effect on North American biodiversity. Moose, elk, deer, black bear, and snowshoe hare browse its bark, and aspen groves (below) support up to 34 species of birds.

Climate change and the polar bear

Arctic temperatures have been above the 1981-2010 global average every year since 1988 and the extent of Arctic sea ice has also been decreasing (below). This is having a detrimental effect on polar bears, which rely on the sea ice to hunt their prey (seals). In Canada's Hudson Bay, the sea ice is melting earlier and forming later. As a result, the bears must swim further to hunt and their hunting time is cut short. Survival and breeding success during summer (when food is inaccessible) is then reduced because they put on less weight through the winter.

Quaking aspen grove, Nevada

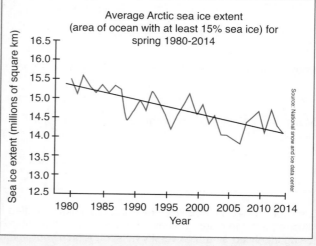

Average Arctic sea ice extent (area of ocean with at least 15% sea ice) for spring 1980-2014

Sea ice extent (millions of square km)

Year

Source: National snow and ice data center

© 2016 **BIOZONE** International
ISBN: 978-1-927309-46-9
Photocopying Prohibited

How land use affects biodiversity

Natural grasslands (top right) are diverse and productive ecosystems. Ancient grasslands may have contained 80-100 plant species, in contrast to currently cultivated grasslands, which may contain as few as three species. Unfortunately, many of the management practices that promote grassland species diversity conflict with modern farming methods. Appropriate management can help to conserve grassland ecosystems while maintaining their viability for agriculture.

Demand for food increases as the population grows. Modern farming techniques favor monocultures (bottom left) to maximize yield and profit. However monocultures, in which a single crop type is grown year after year, are low diversity systems and food supplies are vulnerable if the crop fails. Rice, maize, and wheat alone make up two thirds of human food consumption and are staples for more than 4 billion people. This creates issues of food security for the human population.

▶ Human activities have had major effects on the biodiversity of Earth.

▶ Nearly 40% of the Earth's land surface is devoted to agricultural use. In these areas, the original biodiversity (a polyculture of plants and animals) has been severely reduced. Many of these areas are effectively monocultures, where just one type of plant is grown.

▶ The graph right shows that as the land is more intensively used, the populations and variety of plants and animals fall.

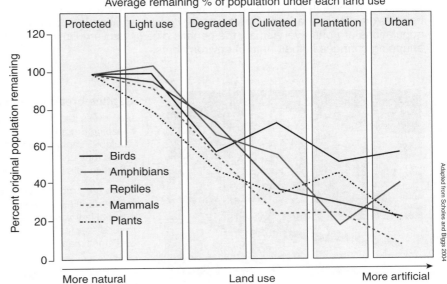

Average remaining % of population under each land use

Adapted from Scholes and Biggs 2004

1. How does coral bleaching affect marine biodiversity? _____

2. (a) Describe the trend in Arctic sea ice extent since the 1980s: _____

   (b) Reduced sea ice means polar bears need to swim longer distances to hunt seals. What effect is this likely to have on the polar bear population numbers?

3. Study the land use graph above and describe the effect of land development on biodiversity: _____

250 *Ex-Situ* Conservation

Key Idea: *Ex-situ* conservation methods operate away from the natural environment and are useful when species are critically endangered.

▶ Conservation aims to maintain the biodiversity of a particular ecosystem. This is achieved by managing species in their habitats (*in-situ*) or by employing conservation programs away from the natural environment (*ex-situ*).

▶ ***Ex-situ*** means off site conservation. Individuals of an endangered species are removed from their natural habitat and placed in a new secure location where their needs for survival are met. Zoos, aquaria, and botanical gardens are common sites for *ex-situ* conservation programs. They house and protect specimens for breeding and, where possible, they reintroduce captive-bred organisms into the wild to restore natural populations. Breeding is carefully recorded and managed to maintain genetic diversity (the diversity of genes and alleles within a species).

▶ *Ex-situ* conservation methods are particularly useful when species are critically endangered (have a very high risk of extinction in the wild). However, *ex-situ* conservation is expensive and labor intensive. Also, because the populations are often very small, the genetic diversity is very limited and the species may not be able to adapt to ongpoing changes in their natural environments.

Above: Captive breeding at the Bronx Zoo in the early 1900s was important for the recovery of American bison in the wild.

Right: A puppet 'mother' shelters a takahe chick. Takahe, a rare species native to New Zealand, were brought back from the brink of extinction through a successful captive breeding program.

Captive breeding and relocation

Individuals are captured and bred under protected conditions. If breeding programs are successful and there is suitable habitat available, captive-bred individuals may be relocated to the wild where they can establish natural populations. Many zoos now have an active role in captive breeding.

There are problems with captive breeding. Individuals are inadvertently selected for fitness in a captive environment and their survival in the wild may be compromised. This is especially so for marine species. However, for some taxa, such as reptiles, birds, and small mammals, captive rearing is very successful.

The important role of zoos and aquaria

As well as their role in captive breeding programs and as custodians of rare species, zoos have a major role in public education. They raise awareness of the threats facing species in their natural environments and gain public support for conservation work. Modern zoos tend to concentrate on particular species and are part of global programs that work together to help retain genetic diversity in captive bred animals.

Right: The okapi is an endangered species of rare forest antelope related to giraffes. Okapi are only found naturally in the Ituri Forest, in the northeastern rainforests of the Democratic Republic of Congo, Africa, an area at the front line of an ongoing civil war. Successful breeding programs have been developed in a number of zoos in the US, including San Diego zoo.

1. What is the purpose of conservation? _____

_____

_____

2. What is *ex-situ* conservation? _____

_____

 © 2016 **BIOZONE** International
ISBN: 978-1-927309-46-9

Ingfbruno CC3.0

The United States Botanic Garden

The role of botanic gardens

Botanic gardens use their expertise and resources to play a critical role in plant conservation. They maintain seed banks, nurture rare species, maintain a living collection of plants, and help to conserve indigenous plant knowledge. They also have an important role in plant research and public education.

In the US, 30% of plant species are threatened. The US Botanic Garden (Washington, D.C) houses more than 500 rare plant species. They are involved in the conservation of endangered species by maintaining live specimen collections, studying wild plants at risk, banking seeds of rare plants, and introducing rare plants to the commercial growers.

Seed banks and gene banks

Seed banks and gene banks help preserve the genetic diversity of species. A seed bank (above) stores seeds as a source for future planting in case seed reserves elsewhere are lost. The seeds may be from rare species whose genetic diversity is at risk, or they may be the seeds of crop plants, in some cases of ancient varieties no longer used in commercial production.

3. Describe the key features of *ex-situ* conservation methods: _____

_____

_____

_____

4. Describe some challenges or disadvantages associated with *ex-situ* conservation: _____

_____

_____

_____

_____

5. Describe three key roles of zoos and aquaria and explain the importance of each:

(a) _____

_____

(b) _____

_____

(c) _____

_____

6. Explain the role of gene and seed banks in the conservation of endangered species: _____

_____

_____

_____

_____

_____

251 *In-Situ* Conservation

Key Idea: *In-situ* (on site) conservation methods manage ecosystems to protect diversity within the natural environment.

What is *in-situ* conservation?

▶ *In-situ* conservation means the conservation of a species in its natural environment.

▶ *In-situ* methods focus on ecological preservation or restoration (cleaning up the ecosystem) and often involves removing predators or invasive species.

▶ *In-situ* conservation protects more species at once, including unknown species.

▶ *In-situ* methods have several disadvantages:

  • Ecological restoration is a long term process. It involves collaboration between local communities and institutions with scientific expertise.

  • Populations may continue to decline during restoration, either because of lag in response or because the population has become critically low.

  • Illegal poaching can be difficult to control.

In the US, the Endangered Species Act (ESA) protects species and the ecosystems on which they depend. The act is administered by the US Fish and Wildlife Service and the National Oceanic and Atmospheric Administration. As a result of this and other efforts, species such as the snowy egret (above right), white tailed deer, and wild turkey (above left), which were all critically endangered, are common again.

Returning bison to the wild

The American bison (also called buffalo) was hunted nearly to extinction in the 1800s. Since the 1900s, both *ex-situ* and *in-situ* conservation methods have been used successfully to boost bison numbers, although their recovery still presents several challenges. Reduced genetic diversity, low numbers, and interbreeding with domestic cattle mean that, even today, the numbers of "wild bison" are still quite low. American bison consists of two subspecies, the plains bison and the wood bison. In this activity, bison refers to both subspecies.

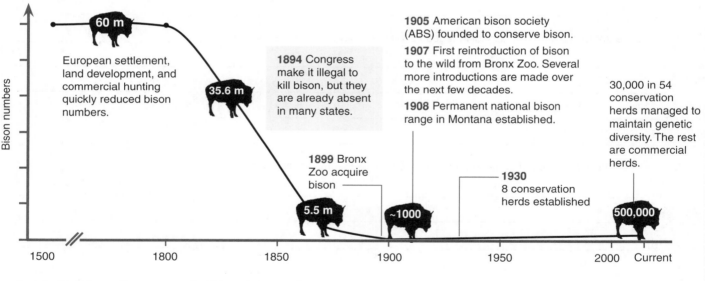

Bison numbers (y-axis)

60 m

European settlement, land development, and commercial hunting quickly reduced bison numbers.

35.6 m

1894 Congress make it illegal to kill bison, but they are already absent in many states.

1905 American bison society (ABS) founded to conserve bison.

1907 First reintroduction of bison to the wild from Bronx Zoo. Several more introductions are made over the next few decades.

1908 Permanent national bison range in Montana established.

30,000 in 54 conservation herds managed to maintain genetic diversity. The rest are commercial herds.

1899 Bronx Zoo acquire bison

5.5 m

~1000

1930 8 conservation herds established

500,000

(x-axis: 1500, 1800, 1850, 1900, 1950, 2000, Current)

1. Identify the *ex-situ* component of bison conservation: _____

_____

2. How have *in-situ* conservation methods and legislation contributed to the success of bison recovery in North America?

_____

_____

3. Identify a possible risk for bison conservation today: _____

_____

4. Use the information on this page and the reports located on the Weblinks page to create or revise a projection for recovery of species and genetic diversity in American bison. Write a brief report and attach it to this page.

WEB   CCC   PRACTICES   PRACTICES

KNOW   251   CE

 © 2016 **BIOZONE** International
ISBN: 978-1-927309-46-9
Photocopying Prohibited

252 Conservation and Genetic Diversity

Key Idea: Maintaining genetic diversity in a population is important in avoiding inbreeding depression, which is a cause of reduced fitness in individuals.

Conservation genetics involves many branches of science and uses genetic methods to restore genetic diversity in a declining species.

One of the biggest problems occurring when a species' population declines is loss of **genetic diversity** (the loss of gene diversity within a species). This increases the relatedness between individuals because there are fewer individuals to breed with. Decreased genetic diversity and increased relatedness can result in inbreeding depression (the reduced fitness of individuals as a result of inbreeding), and can dramatically reduce population viability.

The Florida panther (right) is an example of how conservation genetics has been used to restore genetic diversity in an endangered population.

USFW

▶ In the late 1970s, the Florida panther population had become critically low and occupied just 5% of its historical range. Population models showed it would be extinct within a few decades. Individuals often had several abnormalities including kinked tails, heart defects, and sperm defects. It was determined that these were due to inbreeding depression.

▶ In 1995, eight female panthers were translocated from Texas to increase genetic diversity. Over the last two decades there has been an increase in the population growth rate and an improvement in the survival and health of individuals (graph, right).

Population Florida panther

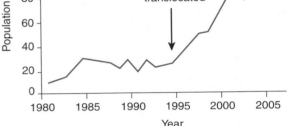

Texan panthers translocated

Conservation genetics is an important concept for conservation and breeding programs. By keeping detailed breeding records of which individuals are bred together and monitoring the genetic relatedness of populations, conservation scientists can make sure endangered species avoid inbreeding depression and maintain their genetic diversity.

Dept of National Resources, Illinois

Conservation genetics were used to rebuild the Illinois prairie chicken population.

DoC

Critically endangered birds, such as the New Zealand kakapo, have lost much of their genetic diversity.

1. What is conservation genetics? _____

_____

2. (a) Why were panthers from Texas translocated to Florida? _____

_____

_____

   (b) Why is it important to maintain genetic diversity in populations? _____

_____

_____

3. Use the reports located on the Weblinks page to create or revise a projection for recovery of the Florida panther. One of these reports provides in-depth information for gifted and talented students. Write a brief report and attach it to this page.

PRACTICES   PRACTICES   CCC   WEB

CE   252   **KNOW**

253 Maasai Mara Case Study

Key Idea: The Maasai Mara case study shows that balancing human and environmental needs can be difficult and requires education, incentive, and compromise.

How farming affected the Maasai Mara region

The Maasai Mara National Reserve is a region in south western Kenya covering 1500 km^2. It is part of the much larger Mara-Serengeti ecosystem, which covers around 25,000 km^2. Considerable change has occurred in the Maasai Mara region since the start of last century. Early in the 20th century, the region was much less populated and the land was used mainly for nomadic agriculture (raising cattle). European settlers forced many of the Maasai off their traditional lands. In 1945, more land was turned into reserves and the Maasai land placed in Trust. As a result of changes in governments and ideals, Trust land was redesignated as group ranches. This encouraged the Maasai to subdivide the land to acquire individual titles in order to secure legal rights to lands, rather than risk losing them outright. Privatization led to an increase in mechanized farming and a reduction in wildlife.

The changes to the land use has had a drastic affect on wildlife (below). Wildebeest numbers dropped from approximately 150,000 in 1977 to 40,000 in 2010. Water buffalo numbers dropped from 40,000 to 5000. Livestock numbers increased (cattle, sheep, and goats).

Wildlife and livestock changes in the Maasai Mara

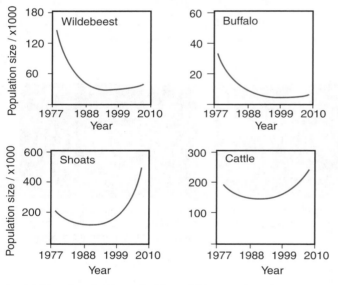

The positive effect of conservancies

Over time, it was realized that modern farming methods limited the range of wildlife and also grazing options for livestock. In 2005, many landowners in the northern Maasai Mara began consolidating their land into conservancies (land set aside for conservation), aiming to generate income through tourism. This included establishing partnerships with tourism operators. The success of this approach has seen a rapid expansion of conservancies.

Although the formation of conservancies has benefited the wildlife and many people, those who do not own any land have been no better off. In addition, livestock are only allowed into the conservancies during certain times (e.g. drought) which has lead to higher stocking rates outside the park and conservancies.

The development of conservancies has had benefits for the community. Payments by tour operators (above) for use of the land are made directly to the land owners, reducing loss of income through bureaucratic handling and corruption. Today there are 8 conservancies representing 92,000 ha. Around $3.6 million is paid to the conservancies each year.

1. (a) How has land use in Maasai Mara region changed since the early 20th century? _____

_____

(b) What effect did farming have on wildebeest and buffalo in the Maasai Mara region? _____

_____

_____

2. In what way did conservancies help reduce conflict between humans are wildlife? _____

_____

_____

_____

CCC   PRACTICES

KNOW   CE

254 Chapter Review

Summarize what you know about this topic under the headings provided. You can draw diagrams or mind maps, or write short notes to organize your thoughts. Use the points in the introduction and the hints to help you:

Biodiversity
HINT: What are the benefits of biodiversity?.

Humans and biodiversity
HINT: How do humans depend on biodiversity and how do they affect biodiversity?

Conservation of biodiversity
HINT: Describe how ex-situ and in-situ conservation methods can be used to maintain biodiversity.

REVISE

255 KEY TERMS: Did You Get It?

1. Test your vocabulary by matching each term to its definition, as identified by its preceding letter code.

biodiversity

biodiversity hotspots

conservation

ecosystem services

extinction

ex-situ conservation

genetic diversity

in-situ conservation

A Conservation efforts that take place on site involving whole ecosystem management.

B The diversity of genes and alleles within a species.

C The act of preserving, protecting, or restoring something (e.g. an organism or habitat).

D The situation in which a species has died out completely.

E Conservation methods that operate away from the natural environment (e.g. zoo breeding programs).

F The economic benefits (goods and services) provided by the ecosystem.

G The number or variety of species living within given ecosystem, biome, or on the entire Earth. Incorporates species richness as well as genetic and habitat diversity.

H Biologically diverse and ecologically distinct regions under the greatest threat of destruction.

2. (a) Identify the two groups of organisms (right) as high diversity or low diversity:

Group A. _____

Group B. _____

(b) Explain your answer:

_____

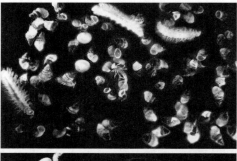

3. Explain how conservation can help sustain biodiversity and how this might affect human well-being:

_____
_____
_____
_____
_____
_____
_____

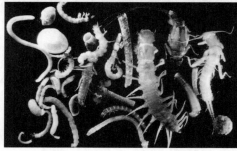

Photos: Stephen Moore

4. Summarize the features of *in-situ* and *ex-situ* conservation methods, identifying advantages and disadvantages of each:

_____
_____
_____
_____
_____
_____
_____
_____

TEST

256 Summative Assessment

About the Indiana bat

The Indiana bat (*Myotis sodalis*) is an insectivorous (insect eating) bat native to North America. They are found in the eastern US from New Hampshire south to northern Florida and west to Iowa, Missouri, and Oklahoma. In winter, Indiana bats hibernate in caves. In spring they migrate north and make their home in tree cavities or under loose tree bark.

Winter hibernation allows the bats to conserve energy when there are too few insects available as food. The bats do not eat during hibernation, but live off stored fat laid down before winter when there is ample insect food available. However, each time a bat is woken during hibernation energy stores are used up more rapidly. If they are disturbed too often, they may not have enough fat reserves to survive the winter.

The Indiana bat has an important ecological role. It is a major predator of night-flying insects and contributes energy to the cave food web through guano (excrement) and decomposition of its body when it dies.

US Fish and Wildlife Service CC2.0

The Indiana bat population is estimated to be 244,000. Very few caves provide suitable hibernation conditions, so the majority of the population hibernate in only a few caves. About 23% hibernate in caves in Indiana. The bats hibernate in tight clusters (above).

Why is the Indiana bat endangered?

Population numbers of the Indiana bat have declined by more than 50% in the last 10 years and they are now listed as endangered. Several factors have contributed to this.

- Although they have a large range, they have very few suitable winter hibernation sites. They must roost in cool, humid caves with temperatures above freezing but below 10°C.

- Changes in cave structure (e.g. blocking an entrance) can change the temperature in a cave making it unsuitable for hibernating bats.

- Human disturbance can rouse the bats from hibernation. Some caves are fitted with gates to stop people getting in, but a poor gate design also stops the bats entering the cave.

- White nose syndrome (WNS) is a fungal disease and grows around the mouth and wings of hibernating bats. The disease causes abnormal behaviors in the bats such as daytime flights during winter. Nearly 6 million bats (of different species) have been killed by WNS since 2007.

- WNS is spread by direct contact with other infected bats or with infected material in the cave (e.g. guano). The fungus can be infectious for a very long time in the cave and can be spread to new caves by people and bats.

- Increased pesticide use kills the insects the bats eat. Indiana bats only eat insects and have no other food sources.

Caves Closed

Help stop the spread of **white-nose syndrome** a condition that is fatal to bats. A cave closure advisory has been issued by the U.S. Fish and Wildlife Service for the eastern U.S. in an effort to minimize the spread of white-nose syndrome.

The caves on this property are closed to public access until further notice. Please contact the property owner if you must enter. Violators may be prosecuted under penalty of the law.

For more information on white-nose syndrome visit: www.fws.gov/northeast/white_nose.html

Cave gate

1. A number of factors have contributed to the severe reduction in the Indiana bat population in the last decade. Using the information above, your task is to design a solution to halt their population decline and ultimately restore their numbers so that they are no longer listed as endangered. You should consider the effect of biological factors and human activities on bat conservation and use both scientific and engineering solutions in your conservation plan.

   Once you have established your plan, you should present the information as an educational resource (e.g. a pamphlet, poster, or slide presentation) designed to teach the general public about the ecological importance of the Indiana bat and how they can help protect it.

ETS   PRACTICES

TEST

Appendix

Questioning terms in biology

The following terms are often used when asking questions in examinations and assessments.

Analyze:	Interpret data to reach stated conclusions.
Annotate:	Add brief notes to a diagram, drawing or graph.
Apply:	Use an idea, equation, principle, theory, or law in a new situation.
Calculate:	Find an answer using mathematical methods. Show the working unless instructed not to.
Compare	Show similarities between two or more items, referring to both (or all) of them throughout.
Construct:	Represent or develop in graphical form.
Contrast:	Show differences. Set in opposition.
Define:	Give the precise meaning of a word or phrase as concisely as possible.
Derive:	Manipulate a mathematical equation to give a new equation or result.
Describe:	Define, name, draw annotated diagrams, give characteristics of, or an account of.
Design:	Produce a plan, object, simulation or model.
Determine:	Find the only possible answer.
Discuss:	Show understanding by linking ideas. Where necessary, justify, relate, evaluate, compare and contrast, or analyze.
Distinguish:	Give the difference(s) between two or more items.
Draw:	Represent by means of pencil lines. Add labels unless told not to do so.
Estimate:	Find an approximate value for an unknown quantity, based on the information provided and application of scientific knowledge.
Evaluate:	Assess the implications and limitations.
Explain:	Provide a reason as to how or why something occurs.
Identify:	Find an answer from a number of possibilities.
Illustrate:	Give concrete examples. Explain clearly by using comparisons or examples.
Interpret:	Comment upon, give examples, describe relationships. Describe, then evaluate.
List:	Give a sequence of answers with no elaboration.
Measure:	Find a value for a quantity.
Outline:	Give a brief account or summary. Include essential information only.
Predict:	Give an expected result.
Solve:	Obtain an answer using numerical methods.
State:	Give a specific name, value, or other answer. No supporting argument or calculation is necessary.
Suggest:	Propose a hypothesis or other possible explanation.
Summarize:	Give a brief, condensed account. Include conclusions and avoid unnecessary details.

Credits

We acknowledge the generosity of those who have provided photographs for this edition: • Louisa Howard and Katherine Connolly, Dartmouth College Electronic Microscope Facility • PASCO for photographs of probeware • D. Dibenski for the photo of the flocking auklets • Stephen Moore for his photos of aquatic invertebrates.

We also acknowledge the photographers that have made their images available through Wikimedia Commons under Creative Commons Licences 2.0, 2.5. or 3.0: • James Hedberg • Brocken Inaglory • Matthias Zepper • Temsabulta • Jerald E. Dewey USDA • Mike Baird • Paul Whippey • Wendy Kaveney • Gina Mikel • Alex Wild Public Domain • Jeffmock • Alastair Rae • Rocky Mountain Laboratories, NIAID, NIH • Xiangyux (PD) • Putney Mark • Allan and Elaine Wilson • UtahCamera • Dario • Jpbarrass • Kristian Peters • Al Aumuller. NY World telegram and the Sun, Public Domain • Joe Schneid, Louisville, Kentucky • Pengo • Lusb • Suseno • Kaldari PD • dsworth Center: New York State Department of Health • Dr. Graham Beards • Luc Viatour www.Lucniz.be • BirdPhotos.com • Haplochromis • Wallombi • CSIRO • Wipeter • U.S. Bureau of Reclamation • BS Thurner Hof • Matt Reinbold • Scott Edhardt, Public Domain • USDA/ Scott Bauer • Derek Quinn • Nicholls H • US Fish and Wildlife Service • ATamari • Takahashi • Adrian A. Smith • Yasunori Koide • it:Utente:Cits • Ghedoghedo • NY State Dept of Health • Lip Kee Yap • NASA Earth Observatory • Bjorn schulz • Wilson44691 • Moussa Direct Ltd. • Heinz-Josef Lücking • Karl Magnacca • Famartin • Reverend Edward Brain, D.D. • D. Eason (DOC) • Raul654 • Ingfbruno • Dept of National Resources, Illinois.

Contributors identified by coded credits are: BF: Brian Finerran (University of Canterbury), BH: Brendan Hicks (Uni. of Waikato), CDC: Centers for Disease Control and Prevention, Atlanta, USA, EII: Education Interactive Imaging, MPI: Max Planck Institute, NASA: National Aeronautics and Space Administration, NIH: National Institutes of Health, NOAA: National Oceanic and Atmospheric Administration www.photolib.noaa.gov, NYSDEC: New York State Dept of Environmental Conservation, RCN: Ralph Cocklin, RA: Richard Allan, USGS: United States Geological Survey, WBS: Warwick Silvester (Uni. of Waikato), WMU: Waikato Microscope Unit, USDA: United States Department of Agriculture, USFW: United States Fish and Wildlife Service, NPS: National Park Service, DoC: New Zealand Department of Conservation.

Royalty free images, purchased by Biozone International Ltd, are used throughout this workbook and have been obtained from the following sources: Corel Corporation from their Professional Photos CD-ROM collection; IMSI (Intl Microcomputer Software Inc.) images from IMSI's MasterClips® and MasterPhotos™ Collection, 1895 Francisco Blvd. East, San Rafael, CA 94901-5506, USA; ©1996 Digital Stock, Medicine and Health Care collection; © 2005 JupiterImages Corporation www.clipart.com; ©Hemera Technologies Inc, 1997-2001; ©Click Art, ©T/Maker Company; ©1994., ©Digital Vision; Gazelle Technologies Inc.; PhotoDisc®, Inc. USA, www.photodisc.com. • TechPool Studios, for their clipart collection of human anatomy: Copyright ©1994, TechPool Studios Corp. USA (some of these images were modified by Biozone) • Totem Graphics, for their clipart collection • Corel Corporation, for use of their clipart from the Corel MEGAGALLERY collection • 3D images created using Bryce, Vue 6, Poser, and Pymol • iStock images • Art Today.

Index

A

95% confidence intervals 101
Abiotic factors 156
Accuracy, of measurement 6
Acetylation of histones 256
Active site of enzyme 69
Active transport 52-53
Adaptations 312-314
 - plant 105
Adaptive radiation 325-327
Adenosine triphosphate 132
Aerobic system 187
Albinism, mutation 271
Alien species 226
Allele 263
Alleles, new 264-265, 270
Altitude, effect on phenotype 277
Altruism 241-242
Amino acid 63, 65
Anabolic reactions 68
Anaerobic respiration 142
Anaerobic system 187
Anatomy, evidence for evolution 302
Animal cell 43
Animal cell, specialization 55
Annotated diagram 30
Antibiotic resistance 272
Ants, cooperative attack 243
Apolipoprotein mutation 273
Apparatus, use of 11
Aquaria, conservation role of 342
Archaea domain 297
Archaean origin of eukaryotes 297
Archaeopteryx 300
Asexual reproduction
 - and variation 264
 - by mitosis 117
Assumptions 5
ATP 131-132, 140-141, 143, 186
 - and active transport 52-53
Autotroph 189
Avery, experiments 253

B

Baboon home range 176
Bacteria growth, investigating 180
Bacteria, domain 296
Bacterial cell 38
Bacteria
 - beneficial mutations in 271-272
 -growth in 179-180
Balsam fir, fluctuations in 216
Bar graph, rules for 23
Base pairing rule 113
Bases in nucleotides 57-59
Beak size, selection for 315
Beneficial mutations 271-273
Bias in samples 28
Biodiversity 326-327, 337
 - and ecosystem services 338
 - and ecosystem stability 215
 - hotspots 339
 - human impact on 340
Biogeochemical cycles 200
Biogeography 295
Bioinformatics 303
Biological drawing, rules 29
Biological species, definition 322
Biotic factors 156
Bison, recovery program 344
Blood glucose regulation 96-97
Blood vessels 74
Body heat, sources 90
Body shape, and heat loss 95
Body systems, homeostasis 86-87
Botanic gardens, role of 343
Breathing rate and exercise 100-101
Bumblebees, competition 170

C

Calvin cycle 134, 137
Canada lynx 181
Capping, control of transcription 257
Captive breeding 342
Carbon cycle 202

Carbon cycle, modeling 203
Carbon cycling 188
Carbon isotopes, use of 139
Cardiovascular system 74-76
Carnivore 190-192
Carrier mediated diffusion 47
Carrying capacity 173
 - and logistic growth 178
 - Coronation Island 174
Catabolic reactions 68
Catalase activity 14, 71
Catalytic proteins 65, 67, 69
Cell cycle 118
Cell differentiation 123, 255
Cell organelles 41, 43
Cell specialization 54-55
Cell types 38
Cell wall 41
Cellular respiration 131,140-145, 186
 - and carbon cycling 205
 - modeling 146
Centrioles 41
Channel mediated diffusion 47
Chargaff's rules 59
Chemical effects on phenotype 276
Chi squared test 286-287
Chimpanzee wars 243
Chlorophyll, role 133
Chloroplast 41, 133, 136-137
 - bacterial origin 296
Chromatin 252
Chromosome structure 252
Chromosomes, during mitosis 120
Circulatory system 74-76, 87
Climate change 221-222
 - effects on biodiversity 340-341
Codon 63
Color-pointing 276
Colorado River, effects of dams 225
Common ancestor 296, 325
Comparative anatomy 295
Competition 162, 164-170
Competition, reducing 169-170
Competitive exclusion 168
Computer model, predator prey
 fluctuations 171
Concentration gradient 47, 52
Conservancies, conservation role 346
Conservation 342-346
 - and genetic diversity 345
Consumers 190-191, 195-198
Contest competition 165
Continuous data 12
Control center in homeostasis 85
Control, experimental 13
Controlled variable 13
Convergent evolution 314
Conversion factors 7
Cooperative behavior 241-245
Coral bleaching 340
Correlation 22
Cotransport 53
Crossing over, chromosomal 267-268
Cycles, of matter 188, 200
Cystic fibrosis mutation 274
Cytokinesis 118-121
Cytoplasm 41, 43

D

Damming, effects of 224-225
Darwin, contributions to science 3
Darwin's finches 315
Darwin's theory 310
Data loggers 15
Data transformation 8
Data, distribution of 25
Data, modelling 4
Data, types 12
Dating, chronometric 295
Deafness, genetic mutation for 270
Decimal form 7
Decomposer 190
Deforestation 230-231
Denaturation of proteins 65
Density, of populations 160

Dependent variable 13, 17
Descriptive statistics 25-27
Detritivore 190-192
Development, human 111
Developmental biology 305
Diabetes mellitus 97-98
Differentiation, of cells 123
Diffusion 47
 - Diffusion, effect of cell size 49-51
Digestive system 76-77, 87
Dihybrid cross 283-285
Dingo habitat 159
Dingo home range 175
Discontinuous data 12
Discrete data 12
Distribution, of populations 161
Disturbance frequency, effect on
 ecosystems 215
Divergent evolution 325
Diversity indices 337
Diving, adaptations for 313
DNA 56-64
DNA as heritable material 253
DNA homology 295-296, 303
DNA methylation 256
DNA model 59
DNA replication 112-113
 - modeling 114-115
DNA sequencing,
 evidence for evolution 303
DNA, experiments 253
DNA, packaging 252, 256
Dolly Sods ecosystem 219
Domains of life 296-297
Double helix structure of DNA 58-59

E

Ecological niche 158
Ecological pyramids 195
Ecosystem services 338
Ecosystems 156
 - and global warming 221-222
 - changes in 213-216, 219
 - dynamics of 213-214
 - human impact 223
 - resilience 214-216
Ectotherm 90-91
Effector 85
Electron transport chain 143
Embryological evidence, for
 evolution 295, 305
Endangered species 331, 342-345
Endoplasmic reticulum 41, 43
Endotherm 90-91
Energy flow, ecosystem 196-197
Energy for muscle contraction 144
Energy in ecosystems 186
Energy transfer 140
Energy, in cells 131
Environment
 - and variation 275-277
 - influence on phenotype 265
Environmental impact of dams 224
Enzyme reaction rates 70
Enzymes 65, 67, 69
 - role in DNA replication 113
Epigenetics 275
 - effects on phenotype 278
Error, sources of 11
Eruption and ecosystem change 219
Estimation 7
Eukarya domain 297
Eukaryotic cell 38, 41-44
Eukaryotic organelles 41, 43
Eusocial animals 240
Everglades and global warming 222
Evo-Devo 295, 305
Evolution 310-311
 - and biodiversity 326-327
 - convergent 314
 - evidence for 295-305
 - patterns of 325
Evolutionary fitness 312
Evolutionary thought, timeline 3
Ex-situ conservation 342-343

Exercise, and homeostasis 99-100
Exercise, effect on body systems 75
Exon 252, 254
Exon splicing 257
Experimental control 13
Exponential functions 10
Exponential growth 177, 179
Extinction 328-331

F

Facilitated diffusion 47
Fermentation 142
Fibrous proteins 65
Finch evolution 315
Fish stocks, human impact on 227
Fitness, evolutionary 312
Flocking, birds 236-238
Florida panther 345
Fluid mosaic model 46
Food chain 191
Food gathering, cooperative 244
Food webs 192
Foraging, cooperative behavior 244
Fossil record 295, 298-299
Fractions 9

G

G_1 (first gap phase) 118
G_2 (Second gap phase) 118
Galapagos finches, evolution in 315
Gas exchange system 74
Gas exchange, plants 79
Gene 63, 252
Gene expression 63, 252, 254-256
Gene flow, role in speciation 323
Gene regulation 254-255
Genes and environment 278
Genetic code, determination of 64
Genetic code, universality 296
Genetic crosses 280-285
Genetic diversity, conservation 345
Genetic variation, sources 264-266
Genotype 265, 275
Geographical isolation 323-324
Ghost fishing 227-228
Global warming
 - and ecosystem change 221-222
 - effects on biodiversity 340-341
Globular proteins 65
Glucagon 96-97
Glucose breakdown 187
Glucose, metabolism of 141-145
Glucose, uses 138
Glycolysis 143
Golgi apparatus 41, 43
Gradients on line graphs 20
Grana of chloroplasts 136-137
Graph, types 18
Gregor Mendel 262
Griffith, experiments 253
Gross primary production 195
Growth and development 111
Growth curves, population 177-178
Growth, role of mitosis 117

H

Habitat 157, 159
Harmful mutations 270-271, 274
Harvesting, effect on ecosystems 215
Heart rate and exercise 100-101
Heat loss, and body shape 95
Hemoglobin homology 304
Herbivore 190-192
Herbivory 162
Herding, mammals 236, 238
Heterozygous 263
Hierarchy of life 37
Histogram, rules for 24
Histone modification 256
Histone proteins 252
Home range, and resources 175-176
Homeostasis 85-87 90-100
 - in plants 102
Homeotherm 91-94
Homologous structures 302

Index

Homology, DNA 296, 303
Homozygous 263
Honeybees, cooperative defense 243
Human activity and extinction 330
Human impact 221-225, 227-231
Huntington's disease mutation 274
Hydrogen bonds in DNA 58
Hydrologic cycle 201
Hypothalamus, in thermoregulation 93
Hypothermia 95
Hypothesis 5

IJK
Ice-albedo effect 221
In-situ conservation 344-345
Inactivation of genes 255
Independent assortment 268
Independent variable 13, 17
Induced fit model 69
Insecticide resistance 318
Insulin 96-97
Insulin resistance 98
Intercepts on line graphs 20
Interphase 118
Interspecific competition 167-170
Intraspecific competition 165-166
Intron 252, 254
Invasive species 226
Ion pump 53
Isolation, and species formation 323
Isotopes, tracing glucose 139
J shaped population growth 177
K (carrying capacity) 173, 178
Keystone species 214, 217-218
Kin selection 241-242
Krebs cycle 143
Kudzu, invasive plant 226

L
Labor, positive feedback 89
Lactose tolerance, mutation 273
Lake food web, constructing 193
Land use and biodiversity 341
Light in/dependent phase 137
Light microscope 39
Limiting factors, carrying capacity 173
Line graphs 19-20
Line of best fit 22
Linear magnification, calculating 40
Lions, cooperative attack 243
Liver, role 76-77
Log books 15
Log transformations 10
Logistic growth 178
Lysosome 43

M
Maasai Mara region 346
Magnification 39
Malaria resistance 273
Mammalian evolution 326
Marine biodiversity, decline in 340
Mass extinction 328-329
Mathematical symbols 7
Matter, cycling in ecosystems 188
Mayr, contributions to science 3
Mean, of data 25
Median, of data 25
Meiosis 267-268
Membrane permeability 51
Meristems 119
Meselson and Stahl, experiment 114
Methanogenesis 187
Microscopy 39-40
Microvilli 43
Migration, benefits of 239
Mitochondria 41, 43, 131
  - bacterial origin 296
Mitosis 111, 117-120
  - modeling 122
Mobbing behavior 242
Mode, of data 25
Models 4
  - energy flow in an ecosystem 197
  - energy and matter 190

- population growth 178
- mitosis 122
- natural selection 321
Modeling
  - solutions to human impact 232
  - the carbon cycle 203
  - respiration & photosynthesis 146
Monohybrid cross 280-282
mRNA 254
Mt St Helens eruption 219
Multipotent stem cells 124
Muscle contraction, energy for 144
Muscular system 73
Musk oxen, group defense 243
Mutations 264-265, 270-274
Mutualism 162

N
Natural selection 310-311
  - and adaptation 315-316, 318
  - modeling 321
Negative feedback 88, 93, 96-97
Net primary production 195
Neutral mutation 271
Newborns, thermoregulation in 94
Niche 158
Niche differentiation, warblers 169
Nitrogen cycle 207
Nitrogen cycling 188
Northern spotted owl 231
Nuclear membrane 41, 43
Nuclear pore 41, 43
Nucleosome 252
Nucleotides 57
Nucleus 41, 43
Nutrient cycle 200-202, 204

O
Observation 5
Old growth forest 231
Omnivore 190-191
Optical microscope 39
Optimal conditions, enzymes 70
Organ systems,interacting
Organ systems 37
  - interactions of 73-77, 79, 126
Organelles, eukaryote 41, 43
Osmosis, in cells 48
Osmotic potential 48
Overfishing 227-229
Oxygen cycling 188, 204

P
Pacific Northwest deforestation 231
Pancreas, endocrine role 96-97
Parasitism 162
Pea plant experiments 262
Pedigree analysis 288
Pentadactyl limb 302
Peptide bond 65
Per capita growth rate 177
Percentage error, calculating 11
Percentages, calculating 8
Phenotype 265, 275
Phospholipid 46
Photosynthesis 133-134, 140
  - and carbon cycling 205
  - in ecosystems 186, 189
  - modeling 146
Photosynthetic rate, investigating 135
Phyletic gradualism 325
Phylogeny 296
Plan diagram 30
Plant cell 41
Plant cell, specialization 54
Plasma membrane 41, 43, 46
Pluripotent stem cells 124
Pocket mice 316
Poikilotherm 91-92
Poly-A tails, role in transcription 257
Polyphenism 275
Population regulation 181
  - and predator-prey cycles 171, 181
  - and competition 165-166
Population size, and K 173

Populations 160-161
Positive feedback 89
  - in climate systems 221
Post-transcriptional modifications to
    messenger RNA 257
Potency of stem cells 123-124
Potometer 103
Precision, of measurement 6
Predation 162
Predator prey cycles 171-172, 181
Presocial animals 240
Primary forest, defined 230
Primary production 195
Producers 189, 191, 195-198
Prokaryotic cell 38
Protein 63, 65
Protein homology 295
Protein homology 304
Protein, membrane 46
Proteins 66-67
  - and gene expression 63, 254
Pulse, measuring 100
Punctuated equilibrium 325
Purine bases 57
Pyramids, ecological 198-199
Pyrimidine bases 57

Q
Qualitative data 12
Qualitative traits 266
Quantitative data 12
Quantitative traits 266

R
Ranked data 12
Rates, calculating 8
Ratios 9
Raw data 8
Reaction rates, factors affecting 70
Reactions in cells 68
Reactions of photosynthesis 137
Receptor 85
Recombination of alleles 268
Red fire ant 226
Repair, role of mitosis 117
Replication fork 113
Reproductive isolation 324
Resilience of ecosystems 214-216
Resistance, evolution of 272, 318
Resolution 39
Resources, and populations 175-176
Resources, competition 162, 164-170
Respiration, measuring 145
Respiratory system 74
Respirometer 145
Results, recording of 15
Ribosome 41, 43
Risk factors, diabetes 98
RNA 58
RNA processing 254
Rock pocket mice 316
Rock strata 295
Rodents, adaptive radiation in 327
Root system, of plants 78
rRNA 254

S
S phase 118
Sampling bias 28
Scale, in microscopy 40
Scatter graph, rules for 21
Schooling, fish 236-237
Science, the process of 2
Scramble competition 165
Sea level rise 221-222
Sea otters, keystone species 218
Secondary production 195
Seed banks, role of 343
Semi-conservative replication 112-114
Sequential evolution 325
Sexual reproduction 264-265
Shoot system, of plants 78
Significant figures 6
Silent mutations 271
Sixth extinction 330-331

Skeletal system 73
Snowshoe hare 181
Social behavior, benefits 240-245
Social groups 236
Social organization 236, 240
Sodium-potassium pump 53
Specialization, of cells 37, 54-55
Speciation 323
Species evenness 337
Species interactions 162-163
Species richness 337
Species, formation of 322
Spruce budworm, fluctuations in 216
Squirrels, competition 168
Stability of ecosystems 214
Standard deviation 27
Standard form 7
Statistical analysis
  - heart rate data 101
  - genetic outcomes 286-287
Stem cell 123
Stomach emptying 88
Stomata, role of 102
Structural proteins 65, 67
Surface area to volume ratio 49-50
Survival, and social behavior 236, 240
Systems 4

T
Tables, constructing 17
Tallies 8
Temperature, and phenotype 276
Ten percent rule 196
Territories,
    role population regulation 166
Test cross 281
Thermoregulation 91-94
Threatened species 331
Thylakoids of chloroplasts 136-137
Tissues, interaction of 125-126
Tolerance range 157
Totipotent stem cells 124
Traits 262, 265
Traits, predicting 280-285
Transcription 63, 254
  - regulation of 256
Transitional fossils 300-301
Translation 63, 254
Transpiration 79, 102-105
Transport proteins 67
Transport, active 52-53
Triose phosphate, 137
Triplet 63
tRNA 254
Trophic efficiency 196
Trophic level 191, 196
Tropical deforestation 230
Type 1 diabetes 96
Type 2 diabetes 98

U
Universal code 303
Urinary system 87

V
Vacuole 41, 43
Variables, types 13
Variation 264, 310-311
  - and environment 275-277
  - and meiosis 267-268
  - sources 264-265, 275
Vascular issue, plants 79

W
Warblers, niche partitions 169
Water balance, plants 102
Watson and Crick, DNA model 58
Whales, evolution of 301

XYZ
Yangtze River, effects of dams 225
Zoos, conservation role of 342
Zygote 111